ECOLOGY OF MANGROVES

Patricia Hutchings is senior research scientist at the Australian Museum in Sydney. She was born and educated in England, arriving in Australia in 1970 to take up her present position at the Museum. Dr Hutchings has published widely in many scientific and semi-popular journals on a variety of topics including taxonomy and ecology of polychaetes, the role of polychaete worms in coral reefs, fauna of mangroves and seagrass beds and estuarine management. She has also been involved in the practical means of conserving wetlands, and has commented on environmental impact statements. Her main interests are polychaete worms, coral reefs and ecology. Her current major research involves the taxonomy of particular polychaete worms and the determination of the rates and mechanisms by which worms bore into coral substrates.

Peter Saenger is senior lecturer in coastal management at the Northern Rivers College of Advanced Education in Lismore. He received his Ph.D. in Marine Botany at the University of Melbourne and went on to lecture at Rhodes University, South Africa. After completing his post-doctoral fellowship at the University of Uppsala, Sweden, he became an environmental consultant and travelled widely; studying mangrove ecology in China, Mozambique, Papua New Guinea, Indonesia, Malaysia, Hawaii, Singapore and Sri Lanka. He has published extensively in international and Australian scientific journals. Dr Saenger's current research interests include a ten year study of the effects of Queensland's Gladstone power station on the mangrove and estuarine communities of the area, marine plants of the Great Barrier Reef; and the response of mangroves to gradients of climatic and hydrological characteristics.

Australian Ecology Series

General Editor: Harold Heatwole

Other titles in this series

Reptile Ecology, by Harold Heatwole
Insect Ecology, by E.G. Matthews and R.L. Kitching
Ecology of Marine Parasites, by Klaus Rohde
Seashore Ecology, by Thomas Carefoot and Rodney Simpson
Ecology of Reptiles, by Harold Heatwole and Janet Taylor

ECOLOGY OF MANGROVES

Patricia Hutchings

Peter Saenger

University of Queensland Press
ST LUCIA • LONDON • NEW YORK

First published 1987 by University of Queensland Press
Box 42, St Lucia, Queensland, Australia

© P. Hutchings and P. Saenger 1987

This book is copyright. Apart from any fair dealing for the
purposes of private study, research, criticism or review, as
permitted under the Copyright Act, no part may be reproduced
by any process without written permission. Enquiries should
be made to the publisher.

Typeset by University of Queensland Press
Printed in Australia by The Book Printer, Melbourne

Distributed in the UK and Europe by University of Queensland Press
Dunhams Lane, Letchworth, Herts. SG6 1LF England

Distributed in the USA and Canada by University of Queensland Press
250 Commercial Street, Manchester, NH 03101 USA

Cataloguing in Publication Data

National Library of Australia

Hutchings, Patricia, 1946– .
  Ecology of mangroves.

  Bibliography.
  Includes index.

  1. Mangrove swamp ecology. 2. Mangrove swamps –
Australia. I. Saenger, Peter, 1943– . II. Title.
(Series: Australian ecology series).

574.5'26325

British Library (data available)

Library of Congress

Hutchings, P. A.
  Ecology of mangroves.

  (Australian ecology series)
  Bibliography: p.
  Includes index.
  1. Mangrove swamp ecology – Australia. 2. Mangrove
swamp ecology. I. Saenger, P. II. Title. III. Series.

QH197.H87   1987   574.5'26325   86-1723

ISBN 0 7022 2015 9

Contents

List of Figures *ix*
List of Plates *xiii*
List of Tables *xv*
Foreword *xix*
Preface *xxi*

Introduction *1*
1 Mangrove Biogeography *6*
2 Adaptations of Mangroves *14*
   Coping with High Salt Concentrations *14*
     Salt Secretion *15*
     Salt Exclusion *17*
     Salt Accumulation *18*
   Conserving Desalinated Water *20*
     Xeromorphic Features *20*
     Transpiration *24*
   Root Specializations *25*
   Responses to Light *31*
     Light and Form *31*
     Photosynthesis *32*
     Light and Other Physical Factors *33*
   Living with Wind, Waves and Frosts *34*
   Reproductive Adaptations *36*
     Flowering and Pollination *36*
     Propagule Production *38*
     Vivipary and Cryptovivipary *39*
     Propagule Dispersal and Establishment *41*
3 Mangroves and Their Environment *45*
   Physico-Chemical Environment–Plant Interactions *45*
     Temperature *45*
     Insolation *53*
     Wind and Evaporation *55*
     Drainage/Aeration *60*
     Salinity of the Soil Water *66*
     Height of the Watertable *73*
     Nature of the Soil *77*
     Proximity of Freshwater Source *84*

Plant–Plant Interactions 92
　Parasitism 92
　Antagonism (Ammensalism) 94
　Mutualism 95
　Competition 97
Plant–Animal Interactions 104
　Sediment Turnover 105
　Grazing and Trampling 106
Interactions Expressed as Structure 108
　Parallel Shoreline Zonation 108
　Longitudinal Upriver Zonation 113
　Unifying Both Zonation Types 115
Classification of Mangrove Communities 119
　Classification Using Structural Attributes 119
　Classification Using Physiographic and Structural Attributes 122
　Classification Using Geomorphological Settings 124

4　Associated Flora 129
Bacteria 129
Fungi 130
Algae 136
Lichens 140
Mangrove Epiphytes 144
Mistletoes 146
Salt Marshes 150
Fringing Species 154

5　The Fauna of Mangroves 155
Composition of the Fauna 163
Terrestrial Fauna 165
　Vertebrates 165
　Invertebrates 171
Freshwater Fauna 178
　Vertebrates 178
Marine Fauna 179
　Vertebrates 179
　Aquatic Invertebrates 182
　Distribution of Marine Fauna 203

6　Adaptations of the Mangrove Fauna 205
Terrestrially Derived Fauna 206
　Mammals 206
　Birds 207
　Amphibians 207
　Reptiles 207
　Insects 210
Marine-Derived Fauna 211
　Morphological Adaptations 211

    Behavioural Adaptations   *221*
    Reproductive Adaptations   *230*
    Physiological Adaptations   *234*
7 Productivity of Mangrove Ecosystems   *245*
    Definition of Primary Production   *245*
    Methods of Measurement and Results   *246*
        Biomass   *246*
        Litter Production   *249*
        Gas Exchange of Leaves   *249*
        Chlorophyll; Light Attenuation   *252*
        Utilization of All Components   *253*
    Factors Influencing Primary Production   *253*
    Seasonal Variation in Primary Production   *256*
    Tidal Control of Primary Production   *259*
    Role of Nutrient Supply of Primary Production   *261*
    Leaf Production   *261*
    Wood Production   *262*
    Mathematical Models   *262*
    Primary Production Budgets   *263*
    Biomass Available for Export or Reuse   *266*
8 The Role of Mangroves and Other Wetlands in Estuarine Ecosystems   *268*
    Primary Productivity   *268*
    Types of Food Chains   *271*
    Fate of Primary Production   *273*
        Mangroves   *276*
        Salt Marshes   *279*
        Seagrasses   *279*
    Detrital Export   *285*
    Feeding Strategies in an Estuarine Community   *286*
    Estuarine Fish Communities   *287*
    Exit Links   *289*
    Generalized Detrital Cycle   *292*
    Exploitation at Higher Trophic levels   *293*
9 Conservation and Management of Mangrove Communities   *296*
    Are Mangroves Endangered?   *297*
    Are Mangroves Worth Managing?   *298*
    Management — Whose Responsibility?   *299*
    Management — On What Basis?   *302*
    Some Specific Management Problems   *303*
        Discharges of Wastes   *304*
        Foreshore Development   *305*
        Flood Mitigation and Swamp Draining Works   *306*
        Reclamation and Dredging   *306*
        Bund Walls   *307*

The Future of Mangroves and Salt Marshes in
Australia *308*

Glossary *311*
Bibliography *315*
Index *371*

Figures

1. World distribution of mangroves in relation to the 24°C isotherm *4*
2. Australian mangrove biogeographic regions *6*
3. Geographical distribution of continents in the early Cretaceous period, showing the probable migration routes of the mangrove flora *9*
4. Australian distribution of major mangrove species *10, 11*
5. Relationship along the Queensland coastline between rich mangrove vegetation and the 1,250 mm annual isohyet *12*
6. Salt glands in mangroves *15*
7. Leaf hairs and scales in mangroves *22*
8. Transverse sections of leaves of Australian mangroves with isobilateral leaves *23*
9. Major morphological root types found in mangroves *27*
10. Flowering times of mangroves and mangrove associates in Port Curtis *37*
11. Fruiting times of mangroves in Port Curtis *38*
12. Interrelationships between major physico-chemical factors and the extent and nature of the mangrove plant cover *46*
13. Rates of leaf formation in nine species of mangroves *48*
14. Stomatal opening in *Rhizophora mangle* in response to various air temperatures *49*
15. Light-saturation curve for *Avicennia marina* at Westernport Bay *53*
16. Number of months per year during which mean radiation level between sunrise and sunset falls below 350 watts m^{-2} *55*
17. Annual rainfall and evaporation distribution in Australia *57*
18. Seasonal mean sea-level changes at Gladstone *59*
19. Soil water content from three study areas at Gladstone *62*
20. Growth of *Avicennia marina* at various seawater concentrations *67*
21. Growth of *Aegiceras corniculatum* and *Rhizophora stylosa* at various seawater concentrations *68*
22. Relationship of soil chlorinity to tidal levels at four study areas at Gladstone *69*

23. Characteristics of the soils from the vegetation-free salt flats and adjacent mangrove communities at Gladstone   *74*
24. Results of model simulation of Florida mangrove ecosystem   *82*
25. Classification of Queensland drainage basins on the basis of their run-off coefficients and mean rainfall   *89*
26. Queensland drainage basin groups showing those that can be considered as having reliable rainfall   *90*
27. Depletion curves for mangroves at Gladstone   *93*
28. Open shoreline zonation at Princess Charlotte Bay   *99*
29. Strategic ordination of mangrove species at Repulse Bay   *101*
30. Superimposition of various characteristics on the ordination shown in figure 29   *103*
31. Upriver distribution patterns of mangroves in three areas   *111*
32. Integration of vegetational boundaries with gradient-related and tidally induced boundary conditions   *116*
33. Stylized zonational sequences along open shorelines and into adjacent river mouths   *120*
34. Classification of mangrove environments using physiographic characteristics   *123*
35. Classification of mangrove environments using geomorphological characteristics   *126*
36. Successional stages in the fungal breakdown of seedlings of *Rhizophora mangle* in Florida   *134*
37. Algal standing crop on mudflats at Gladstone   *137*
38. Percentage of lichens growing on mangroves in three zones   *141*
39. Replacement of species with several lichen genera with change in latitude   *143*
40. Distribution in Australia of mistletoes confined to or commonly growing on mangrove hosts   *149*
41. Relationship between the number of species of mangroves and saltmarsh plants and latitude   *151*
42. Coastal distribution in Australia of selected saltmarsh plants   *152*
43. (a) Schematic diagram of partitioning of the mangrove habitat, as it affects the fauna   *155*
    (b) Vertical zonation of the more abundant animals at the seaward edge of the Pandan mangrove forest, Malaysia   *156*
44. Distribution of mangroves and tropical rainforests and major gaps in mangrove vegetation   *158*
45. Data on molluscs and crustaceans to illustrate higher diversity in tropics   *164*
46. Species of birds occurring regularly in mangroves   *167*

47. Pattern of distribution of *Uca* crab species in relationship to the zoning of mangroves in north Queensland   *185*
48. Densities of mangrove mollusc species in various mangrove zones in the Kimberley region of Western Australia   *189*
49. Density and diversity of molluscs in various mangrove zones in the Kimberley region   *190*
50. Data on species numbers for mangrove stands at Patonga on the Hawkesbury River   *191*
51. Distribution of crabs at Exmouth Gulf in relation to
    (a) percentage of sand and clay   *194*
    (b) texture of sediment   *195*
52. (a) Abundance and penetration of crabs along the Brisbane River during normal weather conditions   *200*
    (b) Numbers of species of crabs at selected localities along the Brisbane River   *200*
53. General distribution patterns of some Australian molluscan wood borers   *202*
54. Branching pattern for a single *Rhizophora mangle* root from Clam Key, Florida   *203*
55. Schematic drawing of *Periophthalmus* showing adaptations of the eyes and pelvic fins   *211*
56. Cycle of fin movements during locomotion on land by crutching of *Periophthalmus*   *213*
57. Respiratory ventilation of *Metaplax crenulatus*   *216*
58. Figures of Spionid, Eunicid and Nereidid (polychaetes) showing the development of gills along the body   *217*
59. Size distribution histograms for three species of crab   *218*
60. Maxillipeds of three species of fiddler crabs   *220*
61. Types of setae found on the mouthparts of hermit crabs   *221*
62. Third maxilliped of *Clibanarius taeniatus*   *222*
63. Third maxilliped of *Clibanarius virescens* and *Paguristes squamosa*   *223*
64. Third maxilliped of *Dardanus setifer*   *224*
65. Crab mating displays analyses by cine film   *226*
66. Schematic distribution of mudskippers along Three Mile Creek   *228*
67. Diagrammatic elevation of the mangroves in Pallarenda swamp showing distribution of mudskippers   *229*
68. Nuptial dance of *Periophthalmus*   *231*
69. Survival of Sydney rock oysters exposed to high water temperatures   *234*
70. Tissue temperature of Sydney rock oysters exposed directly to sunlight in air   *235*
71. Permeability of crab shells to water in air   *238*

72. Comparative osmoregulatory response of marine, estuarine, freshwater and terrestrial animals   240
73. Blood osmo-concentration as a function of salinity in crabs
    (a) *Macrophthalmus setosus* and *Paracleistostoma mcneilli*   241
    (b) *Australoplax tridentata* and *Macrophthalmus crassipes*   241
74. Blood osmo-concentration of *Mictyris longicarpus* as a function of salinity   242
75. Approximate distribution of crabs along the Brisbane River   243
76. Rates of photosynthesis, respiration and export in the Puerto Rican mangrove forest components   251
77. Vertical profile structure of the biomass of *Rhizophora apiculata*   254
78. Vertical distribution of leaf biomass, leaf area, chlorophyll and light intensity in the red mangrove forest of Puerto Rico   255
79. Leaf production and leaf drop in nine species of Australian mangroves   257
80. Diurnal sequence of light intensity in the top of the forest and under the forest canopy   258
81. Potential pathways of energy flow in mangrove ecosystems   269
82. Decomposition of *Avicennia marina* leaves at Roseville, Sydney   277
83. *Avicennia marina* litter fall at Roseville   278
84. Litter fall beneath tall *Avicennia* and low *Avicennia* in Tuff Crater, Auckland   280–81
85. Decomposition of *Avicennia* leaf litter in Tuff Crater   282
86. Decrease in nitrogen percentages of *Avicennia* and *Bruguiera* leaves during decomposition   283
87. Decrease in mass of nitrogen per litter bag during decomposition   283
88. Schematic diagram of the life cycle of the sea mullet   288
89. Schematic diagram of the life cycle of the banana prawn   290
90. Schematic diagram of the life cycle of the barramundi   291
91. Schematic diagram of the life cycle of the mud crab   292
92. Production of organic matter per year by the land vegetation of the world   294

Plates

Following page 170

1. Mangrove communities of Repulse Inlet near Proserpine, Queensland
2. *Nypa fruticans*, Harmer Creek, Cape York
3. Mangrove and freshwater communities on the west coast of Cape York
4. *Avicennia marina* with a low understorey of *Aegiceras corniculatum*, Shoalhaven River
5. *Avicennia marina*, Leschenault Inlet, Western Australia
6. Mangrove zonation on the foreshore of Admiralty Gulf, Western Australia
7. Salt-flat development in the south-eastern Gulf of Carpentaria
8. Zonation on the openshore around the mouth of the Wildman River
9. *Sporobolous virginicus*, Hayes Inlet, north of Brisbane
10. Algal growth on the mudflats, Raglan Creek, Central Queensland
11. *Dischidia nummularia* hanging from the branches and trunk of *Xylocarpus granatum*
12. *Myrmecodia antoinii*, the ant plant
13. *Derris trifoliata*, of the pea-family
14. *Tecticornia cinerea*
15. *Phytophthora* at Port Curtis, Queensland
16. *Avicennia marina* having been blown over has developed a series of new trunks
17. *Thalassina anomala* plays a significant role in the turnover of mangrove muds, Gladstone, Queensland
18. *Cassidula angulifera*
19. Sugar cane crops on coastal wetlands near Cairns
20. Wasteland development, Raby Bay, near Brisbane
21. Cut-and-fill canal estates in coastal wetlands
22. Bank stabilization of the new floodway at the Brisbane International Airport

LIST OF PLATES

23. *Periophthalmus vulgaris* (mudskipper) with orobranchial chamber expanded
24. *Periophthalmus vulgaris* showing well-positioned turret eyes
25. The crab *Heloecius cordiformes* at entrance to its burrow, feeding
26. Close up of *Heloecius cordiformes*
27. The gastropod *Austrocochlea* sp. on the trunk of *Avicennia marina*
28. Crab hole at landward of mangroves, Fullerton Cove
29. The grazing gastropod *Telescopium* on mudflats
30. Dense aggregations of *Telescopium* on mudflats
31. Variety of gastropods grazing among the pneumatophores on mudflats
32. Encrusted pneumatophores, mainly with oysters, on mudflats
33. Base of *Avicennia marina* on seaward margin encrusted with oysters
34. Undersurface of submerged logs, covered in encrusting organisms, mainly oysters
35. (a) and (b) Encrusting organisms on *Rhizophora*
36. Grazing molluscs, showing trails across the mudflats
37. Wood-boring fauna
38. Close up of *Teredo* burrows
39. The spider *Nephila* sp. in web spun between two leaves of *Avicennia marina*
40. The green tree ant
41. Insect damage caused to leaves of *Rhizophora*
42. Unknown insect which has laid its eggs on the undersurface of *Rhizophora* leaves
43. The gastropods *Littoraria* complex grazing on the surface of *Rhizophora* leaves

Tables

1. Geological time scale of fossil mangroves in relation to other evolutionary events   *8*
2. Occurrence of different root types in Australian mangroves   *28*
3. Early development of *Avicennia marina* seedlings grown under different salinities   *33*
4. Reproductive units of Australian mangroves and associated genera   *40*
5. Production, establishment and mortality rates for propagules of Queensland mangroves   *44*
6. Relationship between physico-chemical factors and the essential life processes of mangroves   *47*
7. Classification of Australian mangroves into thermal groups   *51*
8. Shade tolerance of Australian mangroves   *54*
9. Storm surge heights recorded during cyclones in northern Australia   *58*
10. Comparison between sand and clay of various soil characteristics   *61*
11. Soil infiltration rates of mangrove soils in Sydney   *63*
12. Comparison of soils supporting stands of *Rhizophora*, *Avicennia* and *Bruguiera*   *64*
13. Soil water content at which various species occurred at Proserpine   *65*
14. Salinity data for Australian mangroves   *71*
15. Microbial reactions involved in the availability of nitrogen for plant growth   *79*
16. Species groups and their characteristics as derived from floristic data from northeastern Australian coastal systems   *84, 85*
17. Site groups and their group characteristics   *86, 87*
18. Hydrological classification of the Queensland east coast drainage basins   *91*
19. Data used to derive growth and dominance indices for the strategic analysis of Proserpine mangroves   *101*
20. References dealing with zonation of mangrove communities in Australasia   *109*

21. Boundary conditions for the various intertidal plant communities at Port Curtis and Repulse Bay *117*
22. Comparative boundary conditions for the lower limit of salt marshes *118*
23. Tolerance of mangrove, saltmarsh and fringing plants to soil salinity and waterlogging *118*
24. Structural formations of Australian mangrove communities *121*
25. Bacterial numbers in water and sediments of mangrove and associated communities *130*
26. Bacteria recorded from the Lake Macquarie estuarine system *131*
27. Basidiomycetes from Australian mangroves *132*
28. Numbers of algal species recorded from various substrates in Australian mangrove communities *139*
29. Epiphytes recorded from Australian mangroves *145*
30. Mangrove mistletoes with mangrove and non-mangrove host species *147*
31. Distribution of birds adapted to mangroves and their origins *160, 161*
32. Area of specialization of birds to mangrove habitat *168*
33. Density, biomass and number of molluscs and crustaceans in the Bay of Rest *186*
34. Major crab habitats in Mangrove Bay *196*
35. Environment, food and type of respiration of crabs in Mangrove Bay *197*
36. Feeding types of molluscs collected in four intertidal habitats in the Bay of Rest *198*
37. Crabs, with their average gill areas per gram, arranged by habitat *215*
38. Primary productivity estimates for plant communities in Botany Bay *246*
39. Estimates of biomass for non-Australian mangroves *247*
40. Estimates of biomass for areas of mangrove forest of known age in Malaysia *247*
41. Estimates of biomass for mangroves in temperate Australia *248*
42. Mangrove litter production at various localities in the world *250*
43. Variations in *Rhizophora mangle* leaf sizes *252*
44. Net primary production of *Rhizophora apiculata* in Thailand *254*
45. Preliminary annual production budget for *Avicennia* in Westernport Bay *264*
46. Experimentally determined rates of primary production of selected terrestrial and marine plant communities *265*

47. Productivity of *Posidonia australis* in Port Hacking   284
48. Inventory of selected New South Wales estuaries   295
49. Population density and economic status of countries with the world's major mangrove areas   309

Foreword

Ecology is the science involved with the interactions of organisms and their physical and biotic environments. This field always has been a source of fascination to professional biologists, naturalists and conservationists. In recent years, as human population has progressively increased, environmental problems have become of vital interest and importance to the public as well. It has become imperative now that ecological principles, and the ecology of specific regions, be understood by a wide variety of people. The present series is designed to help fill this need.

It is felt that the volumes of this series will serve as a source of information for university students, teachers and the interested public who require a basic factual knowledge to broaden their understanding of ecology, and for those conservationists, agriculturists, foresters, wildlife officers, politicians, planners, engineers, etc. who may need to apply ecological principles in solving specific environmental problems. In addition, it is hoped that the series will be a valuable reference work and source of stimulation for professional ecologists, botanists and zoologists.

The study of ecology can be approached on various levels. For example, one can emphasize the biotic community and analyze the kinds and numbers of organisms living together in a particular habitat, the way they are organized in space and time and the interactions they have with each other. This type of ecology is known as synecology.

Another way of studying ecology is by systems analysis. In this method the biotic community and the physical environment, which together make up what is known as an ecosystem, are looked upon as a functioning unit. In such an approach the main emphasis is on the cycling of energy, minerals or organic materials within the ecosystem and the factors influencing these processes, rather than specifically upon the organisms themselves. Often mathematical or theoretical models are constructed and tested, frequently with the aid of computers.

Both of the above approaches are synthetic; they take an overview of entire communities or systems and do not emphasize individual species. By contrast the following two approaches, collectively known as autecology, are concerned mainly with particular species.

The population approach, often called demography, is concerned with: (1) fluctuation in the abundance and distribution of individuals of a given species in an area, (2) the contributing phenomena such as birth and death rates, immigration, emigration, longevity and survival, and (3) the influence of the physical environment and of other species on these characteristics. Of major interest are mechanisms regulating population density and factors influencing population stability.

The final approach to ecology is one primarily concerned with the effect of the environment on the individuals of a species, that is, how they are affected by temperature, moisture, light or other external factors. This approach is known variously as environmental physiology or physiological ecology. The keynote is adaptation to specific environments.

All of the above approaches are employed with varying emphasis in the volumes of this series.

Certain topics, such as ecology of grasslands, ecology of forests and woodlands, or ecology of deserts lend themselves to a community approach; grassland, forest and desert are types of communities and if studied as an entity must be approached on the community or ecosystem level. On the other hand, where specific taxa such as reptiles, birds or mammals are treated, the autecological approach is used more often. The particular aspect emphasized varies from group to group, depending on the information available.

Regardless of emphasis, in each book of this series the available information in a particular field is reviewed critically and summarized, so that the reader might be brought abreast of current knowledge and developments. Recent trends are indicated and the foundations for future developments are prepared by highlighting conspicuous gaps in knowledge and pointing out what appear to be fruitful avenues for research.

HAROLD HEATWOLE

Preface

Any book dealing with communities or ecosystems must draw information from a wide range of sources. This is especially true of one treating mangroves, for not only are both plants and animals important as in any community, but marine, terrestrial and freshwater habitats and their biotas are all involved to some extent. Detailed treatment of such a variety of conditions and organisms lies beyond the expertise of any one individual and multiple authorship of this book was essential for maintaining even-handed treatment of all aspects of the topic. Inevitably the relative contributions of the two authors varied from chapter to chapter. The first author, a zoologist, had the greatest input into those chapters treating the faunal component and the functioning of the community (Chapters 5-8) whereas the second author, a botanist with experience in management, was primarily responsible for those dealing mostly with plants (Chapters 1-4) and management (Chapter 9). However, each read, revised and re-wrote sections of the other's chapters a number of times and made suggestions for change. We are grateful to the series editor, Harold Heatwole, for assisting in the melding of ideas, styles and approaches and for writing several small sections. His input into the integration of the material was greater than that normally contributed by an editor.

We acknowledge Colin Field, Harry Recher and William Dunson who read and commented on parts of the book; they provided many helpful suggestions and criticisms.

Various people provided advice, literature, special expertise or unpublished data. They are Hal Cogger, Harry Recher, Doug Hoese, Alan Greer, Winston Ponder, Ian Loch, Bill Rudman, Paul Adam, Colin Field, James Elsol, John Tierney, Barry Clough, Roger Springthorpe, Di Jones, Fred Wells, Mike Gray, Courtney Smithers, Mabel Griffiths, Elizabeth Marks, Robert Taylor, Ron Straughan, David McAlpine, Ivor Thomas, Peter Davie, Eric Reye, Jim Davie, David Rentz, Tony Watson, Leigh Miller, Roger Kitching, Ian Common, Elwood Zimmerman, Margaret Cook, Ralph Nursall, David Reid, Norm Milward, Ron West, Rob Williams and Helen Tranter.

We are indebted to Viola Watt and Sandra Pont (Department of Zoology, University of New England) and June Adam (Australian

Museum) who patiently typed the manuscript in its various drafts and eventually put it on a word processor. They never complained about the many changes and reorganization occasioned by numerous versions of the book bouncing back and forth between authors and series editor. Some of the figures are the handi-craft of Robyne Jones and Stephen Perry. Grace Hart, Lyn Albertson and Lexie Walker checked parts of the bibliographies.

Introduction

While the term "mangrove" is generally well understood, it is difficult to define precisely what constitutes a mangrove. The word "mangrove" is used in at least two different ways. It can refer either to an individual species of plant or to a stand, or forest, of plants that contains many species. These two meanings are traditionally used interchangeably; that tradition, although perhaps initially confusing, is maintained in this book. Mangrove communities comprise plants belonging to many different genera and families, many of which are not closely related to one another phylogenetically. What they do have in common is a variety of morphological, physiological and reproductive adaptations that enable them to grow in a particular kind of rather unstable, difficult environment. On the basis of the common possession of these various adaptations, approximately eighty species of plants belonging to about thirty genera in over twenty families are recognized throughout the world as being mangroves. Different species vary in their dependence on the littoral habitat. Of the total number of species accepted worldwide as mangroves plants, fifty-nine are exclusive to the mangrove ecosystem and twenty-two are important but non-exclusive (Saenger et al. 1983).

Mangroves are the characteristic littoral plant formations of sheltered tropical and subtropical coastlines. They have been variously described as "coastal woodland", "mangal", "tidal forest" and "mangrove forest". Where conditions are suitable, they form extensive and productive forests.

Given suitable conditions for growth, propagules of mangrove species colonize and establishment begins. Species interact among themselves and respond to environmental conditions, with the result that a characteristic grouping of species, called a community, is formed. Such a community, in combination with the physical environment with which it interacts, makes up an ecosystem. It is the mangrove ecosystem which is the subject of this book.

The mangrove ecosystem occurs at the interface of land and sea. Loren Eiseley (1971) captured this essential feature in a passage of the book *The Night Country*:

> The beaches on the coast I had come to visit are treacherous and the tides are always shifting things about among the mangrove roots. . . . A

world like that is not really natural. . . . Parts of it are neither land nor sea and so everything is moving from one element to another, wearing uneasily the queer transitional bodies that life adopts in such places. Fish, some of them come out and breathe air and sit about watching you. Plants take to eating insects, mammals go back to the water and grow elongate like fish, crabs climb trees. Nothing stays put where it began because everything is constantly climbing in, or climbing out, of its unstable environment.

This quotation illustrates graphically the mangrove ecosystem in which tides and coastal currents bring unremitting variation to the forest, and where plants and animals adapt continuously to the changing chemical, physical and biological characteristics of their environment. Many species use the environment dominated by mangroves for food and shelter during part or all of their life cycle. There is constant movement of living and non-living matter into and out of the mangrove ecosystem (Walsh 1974).

A major difficulty in delimiting the mangrove ecosystem is that, because it lies at the land-sea interface, many of the processes that regulate it have their origin elsewhere. These external processes, governing water availability, the pool of available nutrients and the stability of the habitat, often are not seen as part of the ecosystem — and if they are, then the physical boundaries of the ecosystem become virtually impossible to define. In view of the above, it seems preferable to leave the delimitation of the mangrove ecosystem rather loose. This can be justified in that there is general agreement on the suite of species which invariably characterize it.

The existence of extensive mangrove communities appears to depend on a number of basic requirements, although there is some disagreement as to the exact number. Jennings and Bird (1967) described the six most important geomorphological characteristics which affect estuaries, and in so doing provided the first summary of the main factors relating to mangrove establishment. The characteristics were: (1) aridity, (2) wave energy, (3) tidal conditions, (4) sedimentation, (5) mineralogy and (6) neotectonic effects.

Walsh (1974) identified five characteristics as essential mangrove prerequisites on a global scale, and Chapman (1975, 1977) added two others. These seven, apart from their biological slant, are very similar to the six derived from geomorphological considerations by Jennings and Bird (1967). They are: (1) air temperature within a certain range, (2) mud substrate, (3) protection, (4) salt water, (5) tidal range, (6) ocean currents and (7) shallow shores. These will be reviewed in turn.

1. Temperature: Walsh (1974) and Chapman (1975, 1977) maintained that extensive mangrove development occurs only when the average air temperature of the coldest month is higher than 20°C and where the seasonal range does not exceed 10 degrees. Also, the world distribution of mangroves (figure 1), particularly at the northern and southern limits, appears to correlate reasonably well with the 16°C isotherm for the air temperature of the coldest month (Chapman 1977). However, Barth (1981) has shown that equally good correlations can be obtained using water temperatures; the presence of mangroves seems to correlate with those areas where the water temperature of the warmest month exceeds 24°C, and the limits occur in those waters that never exceed 24°C throughout the year. The occurrences of mangroves in southwestern Western Australia and Victoria and in the North Island of New Zealand appear to be exceptions regardless of whether air or sea temperatures are used; these mangroves are discussed in more detail below.

2. Mud substrate: Although mangroves are able to grow on sand, peat and coral, the most extensive mangroves are invariably associated with mud and muddy soils. Such soils are usually found along deltaic coasts, in lagoons, and along estuarine shorelines. The mangroves themselves may influence the sediment composition, even accelerating mud accretion on coral islands (Steers 1977).

3. Protection: Walsh (1974) and Chapman (1975, 1977) argued that protected coastlines are essential as mangrove communities cannot develop on exposed coasts where wave action prevents establishment of the seedlings. Bays, lagoons, estuaries and shores behind barrier islands and spits are suitable localities.

4. Salt water: While there is increasing evidence that most mangroves are not obligate halophytes, there is evidence that a number of them have their optimal growth in the presence of some additional sodium chloride (Stern and Voigt 1959; Connor 1969; Sidhu 1975a). Chapman (1977) suggested that *Rhizophora* is probably an obligate halophyte, with growth being poor or reduced in the absence of salt, and Vu-van-Cuong (1964) reported that *Ceriops tagal* and *Avicennia officinalis* would not grow in the absence of salt. However, Walsh (1974) and Chapman (1975, 1977) maintained that the real importance of salt lies in the fact that mangroves are slow-growing and that they cannot compete with faster-growing species unless these species are eliminated or reduced by salt. In this sense, they argued, salt is an essential requirement for mangrove development.

5. Tidal range: Tidal range, coupled with local topography, influences primarily the lateral extent of mangrove development. The greater the tidal range, the greater the vertical range available for

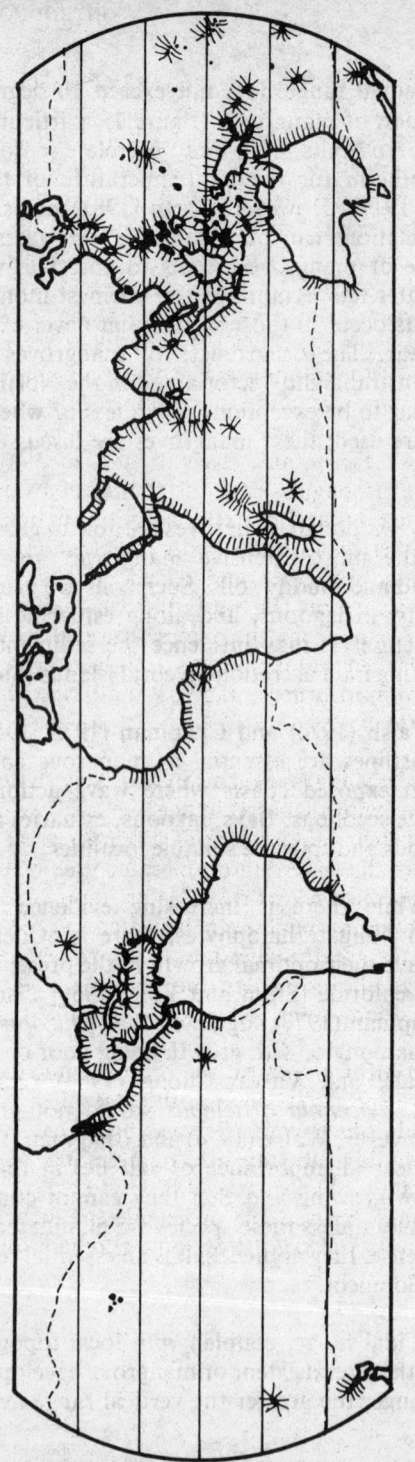

Figure 1 World distribution of mangroves in relation to the 24° isotherm.

mangrove communities. For a given tidal range, steep shores tend to have narrower mangrove zones than do gently sloping ones. Although Walsh (1974) and Chapman (1975, 1977) considered tidal range to be important, there are exceptions in Australia. For example, considerable mangrove development occurs on the microtidal coasts (mean spring range of less than 2 metres) of Cape York Peninsula and in the Gulf of Carpentaria, and Galloway (1982) has shown that similar patterns of mangrove development can exist under a wide range of tidal environments. Mangroves have been reported from tideless areas as well (Beard 1967; Stoddart, Bryan and Gibbs 1973). Although not a direct physiological requirement, tides play an important role in the functioning of the ecosystem.

6. *Ocean currents*: Favourable currents are essential since they disperse mangrove propagules and distribute them along coasts. Chapman (1975) noted that the southern limit of mangroves on the western coast of Africa coincides with the boundary between a southern cold-water upwelling and warm currents, and that a similar situation occurs on the western coasts of Australia and South America. Apart from the temperature of cold currents, Chapman (1975) argued that in all cases in the southern hemisphere such currents flow northwards, thereby inhibiting the southerly drift of floating propagules.

7. *Shallow shores*: Mangroves grow in relatively shallow water as seedlings cannot become anchored in deep water. The physical size of mangroves and their requirement of having a great proportion of their body above the water but at the same time being anchored in the soil makes occupancy of deep water impossible. Chapman (1975) maintained that the shallower the water and the more extensive the shallows, the greater the extent of mangrove development; on steeply shelving shores, where the zone of shallow water is narrow, only fringe communities develop.

Although detailed information on the prerequisites for all the individual mangrove species is lacking, the statement can be made that if certain conditions prevail, such as a protected shoreline with suitable climate, muddy substrate and suitable tidal regime, then a mangrove community is likely to develop, provided, of course, that there is a proximal source of propagules. Furthermore, this mangrove community will consist of some combination of characteristic plant species.

1. Mangrove Biogeography

Based primarily on floristic data, Saenger et al. (1977) divided the mangrove coastlines of Australia into twelve biogeographic zones. More recently, Semeniuk, Kenneally and Wilson (1978) subdivided one of the Western Australian zones into two, resulting in a total of thirteen (figure 2). While it can be expected that a further refinement of boundaries will occur, these thirteen zones correlate closely with certain environmental (particularly meteorological and tidal) as well as physiognomic features for both the mangrove and saltmarsh vegetation. This correlation suggests that meteorological

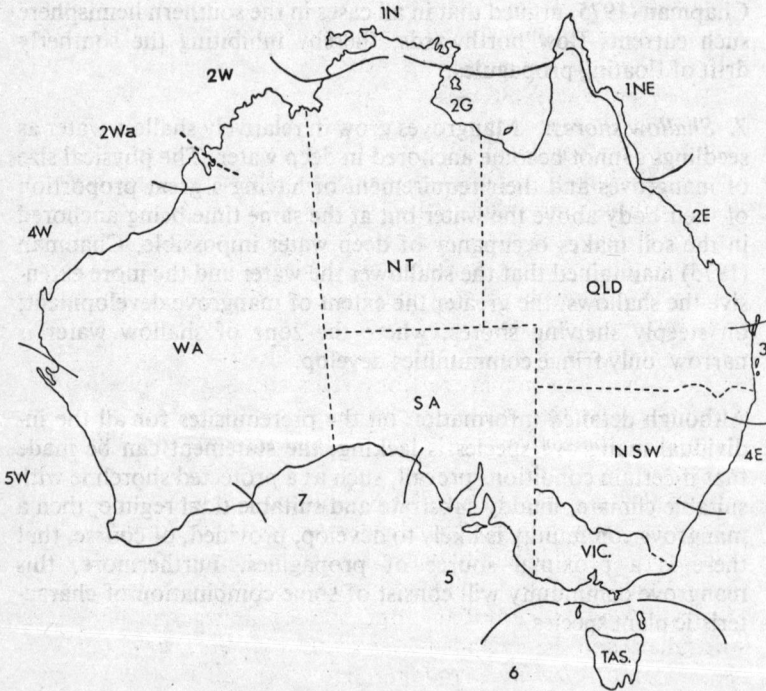

Figure 2 Australian mangrove biogeographic regions based on Saenger et al. (1977) and Semeniuk, Kenneally and Wilson (1978).

and tidal features of the coastline are involved in the distribution and the physiognomy of the mangrove vegetation and its constituent species.

The present-day distribution of mangroves suggests that the region between Malaysia and Northern Australia was the centre of evolution of the mangrove flora (Ding Hou 1958, 1972; Chapman 1976, 1977; Specht 1981b; Mepham 1983). However, Muller (1964) demonstrated an unbroken succession of tropical mangrove vegetation from the Lower Tertiary to the Recent in northwestern Borneo, and Churchill (1973) recorded late Eocene fossils of *Nypa*, *Sonneratia*, *Avicennia* and species of the Rhizophoraceae in southwestern Australia. The geological ages of these fossils are indicated in table 1. On the basis of this fossil evidence, Specht (1981) postulated that the centre of the origin of mangroves is more likely to be the region of southwestern and northern Australia to Papua New Guinea rather than the Malayan Archipelago, and that the present-day distribution could be satisfactorily explained only if the early ancestors evolved in the Early Cretaceous (or even earlier) and were dispersed as shown in figure 3. With the later closure of the Mediterranean Sea as a dispersal route, two isolated groups of mangroves would have been formed, and this accords with the present-day situation; whereas three genera (*Avicennia*, *Hibiscus* and *Rhizophora*) are shared between the Indo-Pacific region and the New World–West African region, only four species are common to both.

Fossilized mangrove pollen and wood from southwestern Australia (Churchill 1973) indicates that tropical coastal waters extended along these shores during the Middle to Late Eocene. Several of the species recorded as fossils in southwestern Australia do not occur there today, and it appears that there has been a loss of these elements from southern Australia since the Eocene. In view of the more restricted distribution of these species today, past changes in climate and coastal conditions appear to have had a sifting effect on the Australian mangrove flora. This sifting had undoubtedly contributed to the existing species gradients, not only of the mangroves themselves but also of their associated plants and animals.

In Australia and Papua New Guinea, approximately thirty species of trees and shrubs, belonging to fourteen families of angiosperms, are generally considered to be part of the mangrove flora. None of these species is endemic to the Australian region; Macnae (1966) stated that three species appear to be purely Australian and he included *Aegialitis annulata*, *Bruguiera exaristata* and *Osbornia octodonta*. However, all of these species occur throughout the Indo–West Pacific region and *Osbornia* occurs as far north as the northern Philippines (Van Steenis 1979).

Table 1 Geological time scale of fossil mangroves in relation to other evolutionary events

Era	Period	Millions of years before present	Fossil Records
Cainozoic	Holocene	0.01	Modern man
	Pleistocene	1.8	Giant mammals
	Pliocene	5	Earliest man-like apes; earliest *Acacias* in Australia
	Miocene	24	Earliest *Eucalyptus* in Australia; earliest *Avicennia* and *Sonneratia* pollen in Borneo
	Oligocene	36	Earliest *Rhizophora* pollen in Asia, New Guinea and South America
	Eocene	54	Earliest *Nypa* pollen in Europe, Asia and Australia; earliest *Rhizophora*, *Avicennia* and *Sonneratia* pollen in Australia. Earliest fossils of *Nypa* fruits and hypocotyls of *Ceriops* and *Palaeobruguiera* in London Clay.
	Palaeocene	65	Earliest *Nypa* pollen in Brazil
Mesozoic	Cretaceous	140	Earliest flowering plants; Extinction of dinosours
	Jurassic	210	Earliest birds
	Triassic	245	Earliest dinosaurs
Palaeozoic	Permian	285	Diverse reptiles and amphibians
	Carboniferous	365	Earliest major coal forests; earliest reptiles and winged insects
	Devonian	415	Earliest trees and amphibians
	Silurian	440	Earliest land plants
	Ordovician	505	Earliest coral reefs and fishes
	Cambrian	570	Earliest invertebrates
Precambrian	Proterozoic	1000	Earliest algae, protozoa and sponges
	Archaeozoic	4000	Earliest bacteria

In Australia, an additional ten species from eight families have been noted as associated lianas, epiphytes, or understorey species (Saenger et al. 1977), and a further ten to fifteen species, although occasionally occurring in the mangrove community, find their greatest development away from it. A large number of other plants such as algae and seagrasses, fungi and lichens also have been recorded from mangrove communities (Saenger et al. 1977; Stevens and Rogers 1979; Cribb 1979; Stevens 1979), but most of these species are not restricted to mangrove environments (see chapter 4).

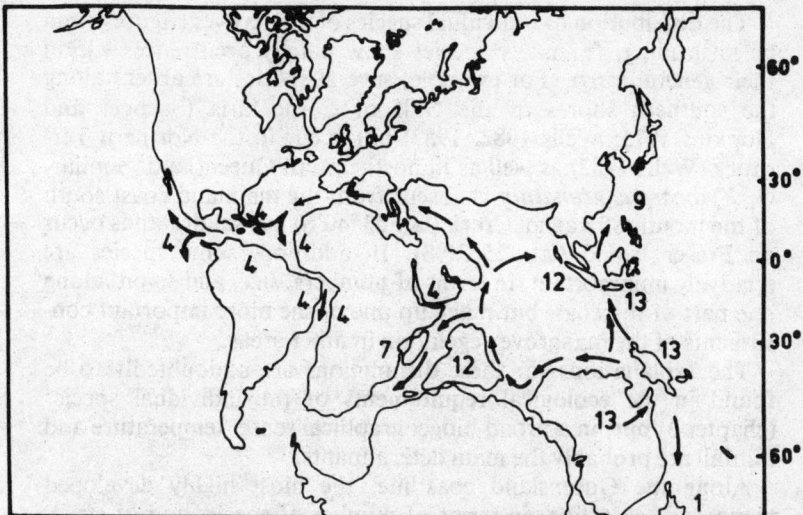

Figure 3 Geographical distribution of continents in the Early Cretaceous period, showing the probable migration routes of the mangrove flora. Number of genera recorded today in various parts of the world are also shown (redrawn from Specht 1981).

The distributions of the abundant species of mangroves around the Australian coastline are shown in figure 4. It is apparent that the largest number of species occurs on the northern and northeastern coastlines. This concentration of mangrove species and associated plants in the northeastern area of Australia can be attributed to three main factors:
1. This region was the centre of origin of mangroves and the point of their secondary dispersal into and out of Australia by virtue of its land connections with southeastern Asia (Walker 1972) during the various changes in palaeo-sealevels. This interpretation accords with other floristic elements (Burbidge 1960).
2. The climatic regime of this area is similar to that under which mangrove vegetation first developed, consequently little or no sifting of species has occurred there. In fact, Mepham (1983) argued that the northeastern coastline provides refuges for the once widespread and diverse Australian mangrove flora as it withdrew northwards with the onset of arid conditions in the Oligocene. Consequently, these northeastern mangrove forests are best regarded as relicts.
3. Coastline configuration in this region, with its numerous estuaries generally sheltered by the offshore Great Barrier Reef, provides large areas of low-energy coastline suitable for mangrove colonization and development.

The distribution of individual species of mangroves (figure 4) can be misleading, for many species show patchy occurrences within their general range. For example, several species are absent along the southern shores of the Gulf of Carpentaria (Saenger and Hopkins 1975; Wells 1982, 1983) but occur in the Northern Territory (Wells 1982) as well as in northeastern Queensland. Similarly, *Xylocarpus granatum* is absent from the mainland coast south of the mouth of Raglan Creek (lat. 23°40'S) but small stands occur on Fraser Island (lat. 25°20'S). In addition, some species are relatively unimportant (in terms of numbers, size, and so on) along one part of the coast but make up one of the more important constituents of the mangrove vegetation in other areas.

The explanations for these disjunctions are undoubtedly to be found in the ecological requirements of the individual species (chapter 3) but, in a broad biogeographical sense, temperature and rainfall are probably the main determinants.

Along the Queensland coastline, the most highly developed mangrove vegetation, in terms of number of species and of structural complexity, is found in those areas where the annual rainfall exceeds 1,250 mm (figure 5); these areas are generally where elevations greater than 700 metres occur in proximity to the coast. With increasing latitude, both on the eastern and western coastlines, the number of species declines rapidly. Lower water and air

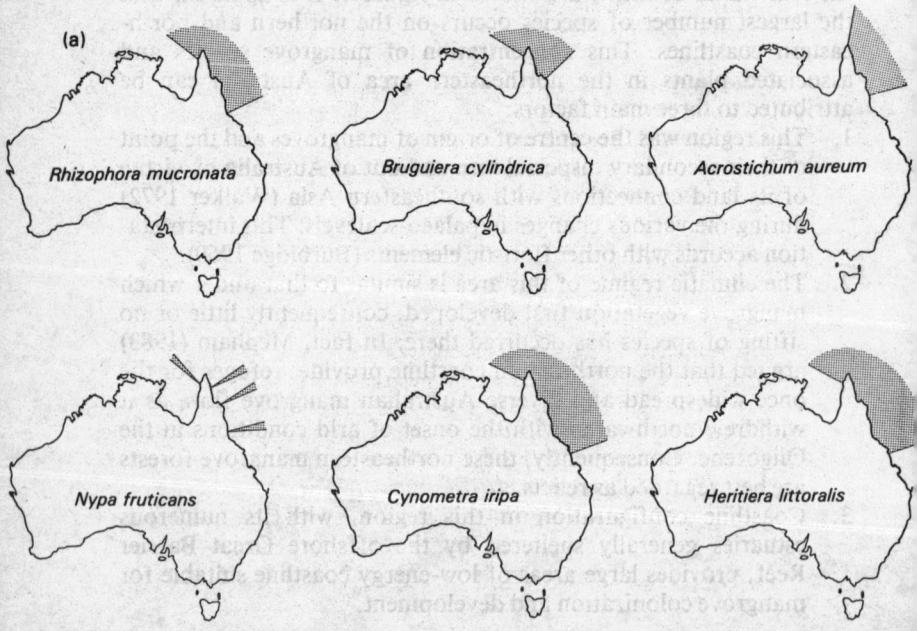

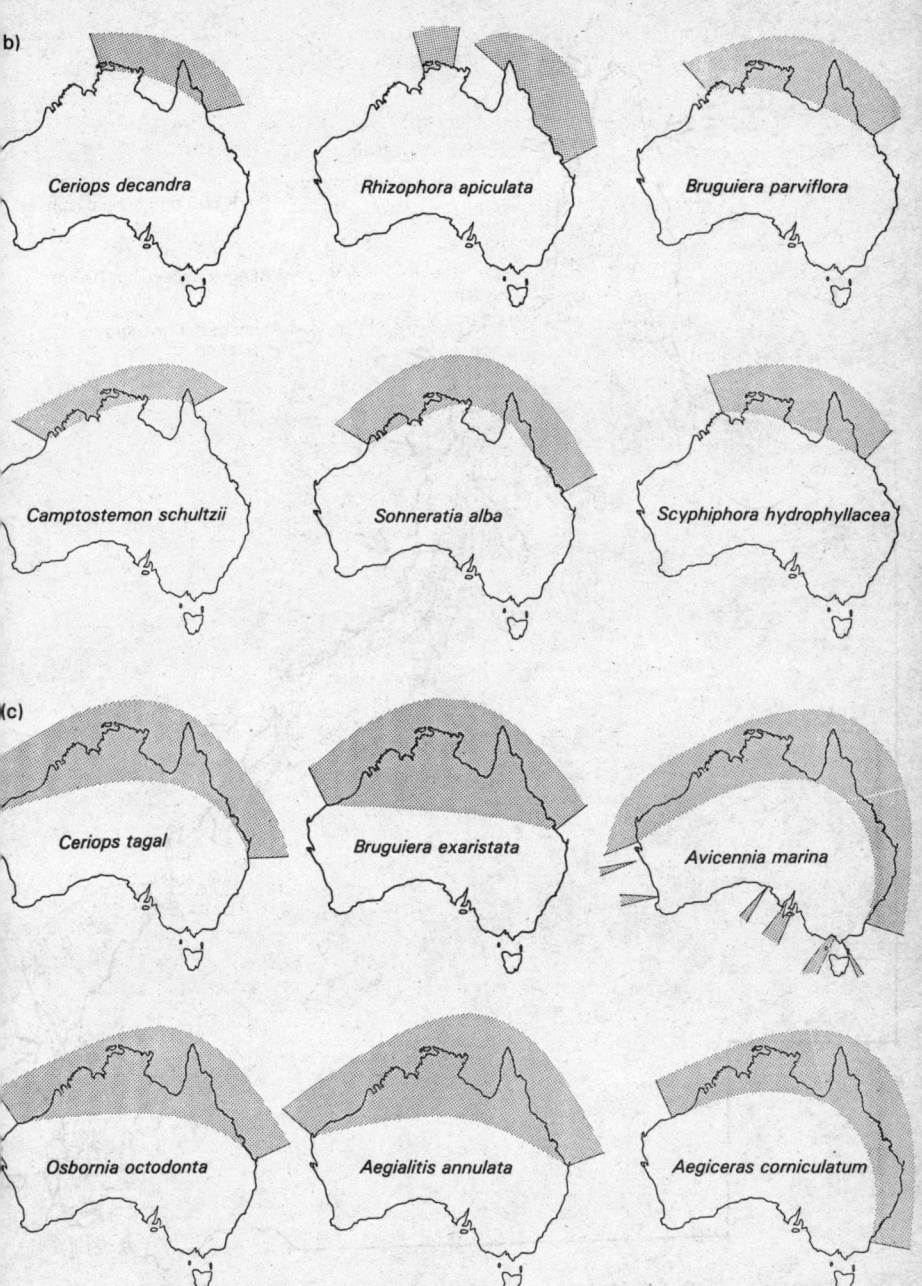

Figure 4 Australian distribution of major mangrove species: (a) species confined to northeastern Australia; (b) species occurring in northern Australia but absent from parts of the Gulf of Carpentaria; and (c) species widespread around northern and eastern coastlines.

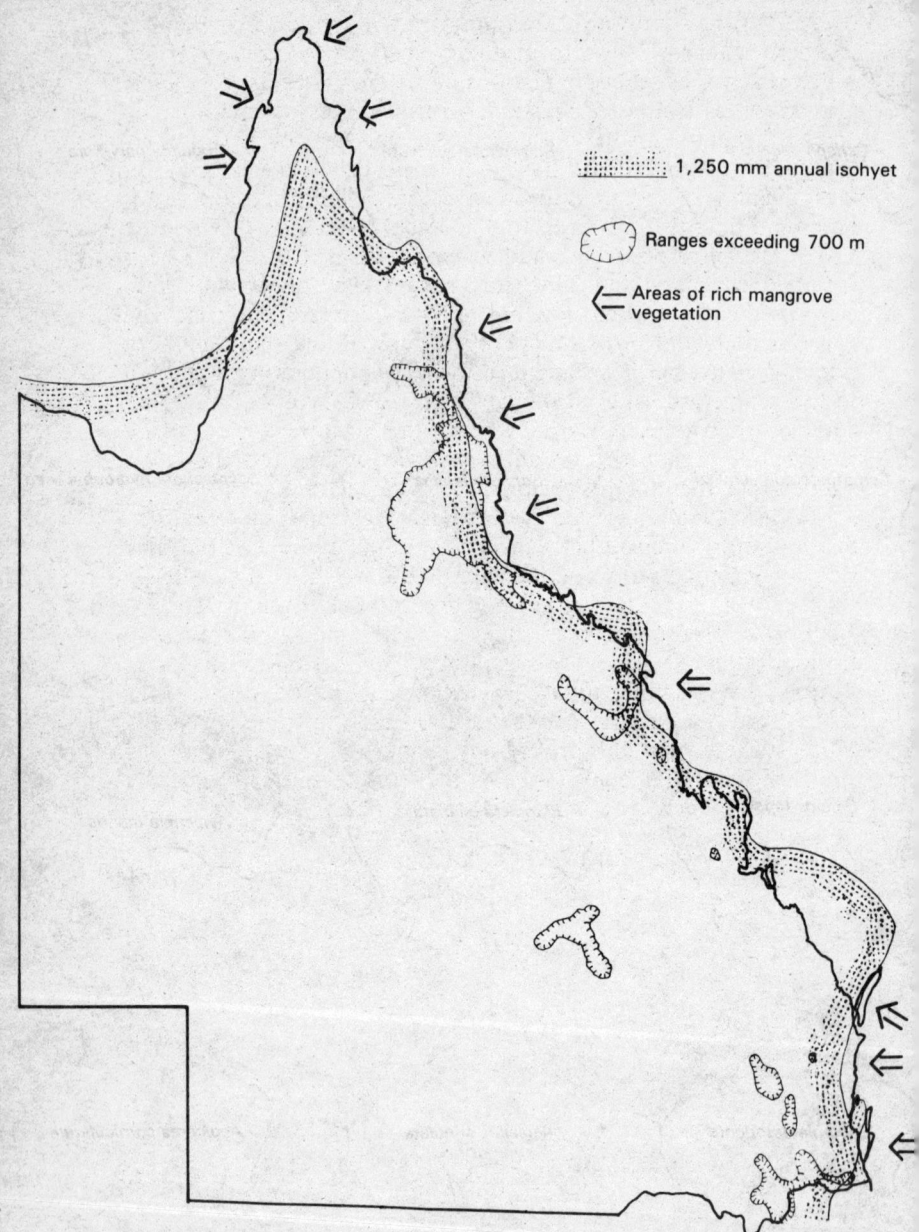

Figure 5 Relationship along the Queensland coastline between areas of rich mangrove vegetation and the 1,250 mm annual isohyet and mountain ranges exceeding 700 m in height.

temperatures as well as the predominance of winter rainfall effectively limit the southward extension of many species, reducing the southern mangrove flora to a solitary species (*Avicennia marina*) which survives as relict populations in disjunct pockets (for example, Abrolhos Islands in Western Australia, Ceduna and Spencer and St Vincent Gulfs in South Australia, and Barwon Heads, Port Phillip Bay, Westernport Bay and Corner Inlet in Victoria). In these localities, *Avicennia* grows in areas where the daily mean minimum temperatures drop to 4°C and 7°C in July (Melbourne and Adelaide respectively), and where minima of 0°C have been recorded (Macnae 1966). These data suggest that, once established, *Avicennia* can withstand low but not sub-zero temperatures. The experimental work of McMillan (1975) showed that these southern populations are hardier than more northerly populations in relation to low temperatures, even though their growth may be somewhat stunted. In southern Africa, where other factors appear to be similar, *Avicennia* occurs only in areas where the mean air temperature does not drop below 13°C (Macnae 1963).

Macnae (1966) suggested two explanations for the present-day distribution of the southerly mangrove populations: (1) transmission by ocean currents and (2) persistence of relicts of previously warmer seas. He preferred the latter explanation. From the work of Ludbrook (1963) and others, it is clear that during the Tertiary (including the Pliocene) the seas around southern Australia were warmer than they are today. The occurrence of other mangrove fossils from the late Eocene in southwestern Australia (Churchill 1973) suggests that the present-day mangrove vegetation on the southern Australian coastline is a relict from these earlier, warmer conditions which has managed to maintain itself in a few favourable localities.

2. Adaptations of Mangroves

The mangrove environment is a variable one owing to a combination of periodic fluctuations and extremes in physico-chemical parameters. Despite such variability, however, the mangrove flora has successfully colonized this environment, apparently aided by the development of numerous morphological, reproductive and physiological adaptations (Macnae 1968; Saenger 1982; Clough, Andrews and Cowan 1982). Many of these adaptations are inferred; that is, adaptations of mangrove species have generally been identified simply by comparing the characteristics of mangroves with those of species from non-mangrove environments. Experimental investigation of the efficiency of many of these adaptations remains to be carried out.

Coping with High Salt Concentrations

The abundance of salt is the single most important characteristic of the mangrove environment, and most mangroves absorb some sodium and chloride ions. Sea water, containing about 35 grams of dissolved salts per litre, has an osmotic potential of approximately -2.5 MPa, and the soil water may have an even lower (more negative) one. The fact that mangroves are able to grow in such highly saline substrates and even grow better in the presence of some salt (Connor 1969; Downton 1982) suggests that they are able to control the intake of salt and maintain a water balance which is physiologically acceptable. Although these processes are understood in general terms, reliable data are lacking on many details.

Jennings (1968) reviewed the mechanisms whereby mangroves deal with excess environmental salt. It appears that three are operative: (1) they take up highly saline water and then secrete the salt (extrusion); (2) they take up water but prevent the entry of salt (exclusion); or (3) they develop tolerance to high salt loads and allow salt to accumulate in the tissues (accumulation). Scholander et al. (1962) classified mangroves functionally into "salt-secretors" and "salt-excluders", although the various mechanisms of dealing with salt are not mutually exclusive. Some species emphasize one, others emphasize another.

Salt Secretion

Salt secretion occurs by means of salt glands (figure 6) in the leaves of *Avicennia* (Baylis 1940), *Sonneratia* (Walter and Steiner 1936), *Aegiceras* (Cardale and Field 1971), *Aegialitis* (Atkinson et al. 1967), *Acanthus* (Mullan 1931) and *Laguncularia* (Biebl and Kinzel 1965), and possibly via cork warts in the leaves of *Rhizophora* (Baijnath and Charles 1980).

Ultrastructural studies of the glands of *Aegiceras* (Cardale and Field 1971; Bostrom and Field 1973) have shown that they consist

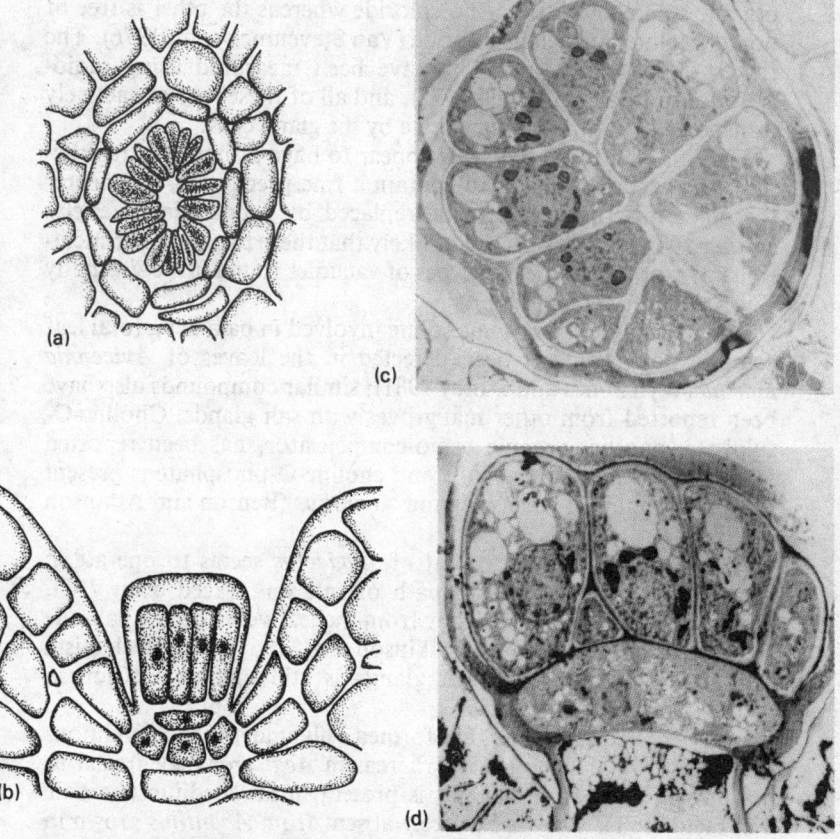

Figure 6 Salt glands in mangroves: (a) salt glands from leaves of *Aegiceras corniculatum* seen in dorsal view (mag. x 1,000); (b) salt glands of *Avicennia marina* on the upper leaf surface seen in transverse section (mag. x 1,000); (c) salt glands of *Acanthus ilicifolius* seen in transverse section (mag. x 5,000); (d) salt glands of *Acanthus ilicifolius* in transverse view (mag. x 5,000). (Figures (c) and (d) courtesy of J.E. Ong and C.H. Wong)

of 24-40 secretory cells situated over a single large, basal cell. The secretory cells are densely packed with mitochondria and other organelles, suggesting some metabolically active function. The living contents of the basal cell and the secretory cells are linked by fine cytoplasmic threads (plasmodesmata) that pass through the cell walls. On the other hand, the junction between the basal cell and the sub-basal cells, which form a layer above the palisade mesophyll, seems to be partially cutinized. Field, Hinwood and Stevenson (1984) showed, however, that there is a small slit-like opening between the cuticle of the gland and that of the leaf; it is through this slit that salt secretion occurs. The mesophyll cells contain two types of vacuoles: one type contains large amounts of an organic solute and little or no chloride whereas the other is free of organic solute but rich in chloride (Van Steveninck et al. 1976). The fluxes of Na^+, K^+ and Cl^- have been measured using radio-isotopes (Cardale and Field 1975), and all of these ions are actively transported out of the parenchyma by the gland cells.

The salt glands of *Acanthus* appear to have a similar ultrastructure; the vacuoles appear to contain a fine precipitate, but nearer the epidermis this seems to be replaced by round, dark vesicles (Wong and Ong 1984). It seems likely that these two vacuolar inclusions correspond to the two types of vacuoles found in *Aegiceras* by Van Steveninck et al. (1976).

Glycinebetaine, an organic solute involved in balancing total leaf osmotic potential, has been detected in the leaves of *Avicennia marina* (Wyn Jones and Storey 1981); similar compounds also have been reported from other mangroves with salt glands. Choline-O-sulphate, another organic osmo-compensator, has been reported from *Avicennia* and *Aegialitis* and choline-O-phosphate is present in large amounts in *Aegiceras* and *Acanthus* (Benson and Atkinson 1967).

A mechanism similar to that of *Aegiceras* seems to operate in *Aegialitis annulata*; the flow-path of salt was traced using ^{36}Cl, and it was found to pass directly from the leaf veins via the palisade mesophyll to the salt glands (Atkinson et al. 1967). The mechanism of the salt pump in the salt glands is still unknown (Clough, Andrews and Cowan 1982).

In *Avicennia*, salt glands are formed only under saline conditions (Mullan 1931; Macnae 1968), whereas in *Aegiceras* they appear to be formed whether or not salt is present in the medium (Cardale and Field 1971). They are entirely absent from *Acanthus* grown in fresh water (Mullan 1931). Joshi et al. (1975) concluded that among salt-secreting species *Avicennia* is the most efficient, and consequently is able to grow in highly saline conditions, whereas the less efficient *Acanthus* and *Aegiceras* are restricted to less salty habitats.

Loetschert and Liemann (1967) found that changes in the contents of Cl, Na, K, Ca and N in *Rhizophora mangle* seedlings indicated that there is a barrier between the cotyledonary body and the peripheral tissues. The outer layer of the cotyledonary body consists of small, nearly spherical cells which according to Pannier (1962) are characterized by an increased phosphatase activity, a condition generally indicative of secretory tissues. Loetschert and Liemann (1967) concluded that the reduced salt uptake by seedlings of *R. mangle* is accomplished by the activity of this glandular tissue. Similar glandular tissue is present on the outside of the cotyledonary body in the Australian *Rhizophora stylosa* (Saenger 1982), and may be identical to the papillose layer described from *R. stylosa* and *Ceriops tagal* by Carey (1934). Highly vacuolated, metabolically active cells also have been described from the outer cotyledons of the propagules of *Avicennia marina* (Butler and Steinke 1976); these cells may have a similar regulatory role.

Salt Exclusion

Salt-excluders possess an effective mechanism, presumably an ultra-filter in the roots (Rains and Epstein 1967; Scholander 1968), whereby water is taken up and salt is largely excluded. Species found to be able to exclude salt are *Rhizophora*, *Ceriops*, *Sonneratia*, *Avicennia*, *Osbornia*, *Bruguiera*, *Excoecaria*, *Aegiceras*, *Aegialitis* and *Acrostichum*. Measurements of the osmotic potential of xylem sap in species which lack salt glands gave values of less than -0.2 MPa (Scholander et al. 1962), indicating that the concentration of soluble salts in the xylem is close to that of many plants from non-saline environments. The osmotic potential of xylem sap in salt-secreting species appears to range from -0.4 to -0.7 MPa (Scholander et al. 1966), showing that they are somewhat less efficient in excluding salt than those species without salt glands. However, Downton (1982) has shown that the osmotic potential of *Avicennia marina* is correlated with the salinity of the growth medium, ranging from -1.6 MPa at zero salinity to -3.6 MPa at full sea water. Nevertheless, salt-secreting species apparently still exclude 80–90 per cent of the salt in sea water (Scholander 1968), although the physical and biochemical basis for this is still poorly understood (Field 1984). Scholander (1968) found that neither chilling nor metabolic inhibitors caused any change in the capacity of the roots to exclude salt, and he concluded that the process was simply a passive function of the differential permeability of membranes in the root. This was supported by the absence of any obvious diurnal variation in the salt concentration of the xylem sap (Scholander et al. 1966), which suggests that the flux of salt into the root is tied closely to water uptake.

Salt Accumulation

Salt-accumulating mangroves (*Excoecaria, Lumnitzera, Avicennia, Osbornia, Rhizophora, Sonneratia* and *Xylocarpus*) often deposit sodium and chloride in the bark of stems and roots and in older leaves (Atkinson et al. 1967; Joshi, Jamale and Bhosale 1975; Clough and Attiwill 1975). Leaf storage of salt is generally accompanied by succulence (Jennings 1968). Joshi, Jamale and Bhosale (1975) have shown that prior to leaf fall in *Sonneratia, Excoecaria* and *Lumnitzera*, sodium and chloride are deposited in senescent leaves. In this way, excess salt is removed from metabolic tissue. For deciduous species such as *Xylocarpus* and *Excoecaria*, annual leaf fall may be a mechanism for the removal of excess salt prior to the onset of a new growing and fruiting season (Saenger 1982).

The movement of salt into viviparous and cryptoviviparous seedlings while still attached to the parent tree appears to be regulated in *Rhizophora, Ceriops, Bruguiera, Aegiceras, Avicennia* and *Acanthus* (Chapman 1944; Loetschert and Liemann 1967; Joshi, Jamale and Bhosale 1975). Seedlings taken from *Avicennia marina* growing on tidal mudflats had osmotic potentials more negative than sea water, yet they contained little sodium or chloride (Downton 1982). It appears that while still attached to the tree, seedlings can control the uptake of sodium and chloride, and adjust osmotically by the accumulation of organic rather than inorganic solutes (Downton 1982), but after falling they rapidly increase their salt content until their root system is capable of ultra-filtering sea water (Chapman 1944; Field 1984).

Although it is clear that the internal salt concentration in mangroves must be maintained if turgor potential is to be constant, the metabolic effects of salt are inadequately known.

Salt may influence the functioning of metabolic enzymes and therefore affect such vital processes as respiration, photosynthesis and protein synthesis. For example, Joshi et al. (1974) and Joshi, Jamale and Bhosale (1975) suggested that high salt concentrations in the cell inhibit ribulose diphosphate carboxylase, an enzyme of the carboxylation process. In addition, activity of the enzyme malic dehydrogenase was significantly lower in mangroves than in other plants, and this was attributed to salt inhibition and/or the unavailability of calcium to the metabolic tissues. Through the use of radioactive CO_2, Joshi et al. (1975) were able to show a rapid (one-hour) build-up of amino acids which was consistent with the inhibition of ribulose diphosphate carboxylase and malic dehydrogenase.

The high content of amino acids and their presence as initial products of photosynthesis suggest a large pool of readily available nitrogen in the leaf (Joshi et al. 1975). It has been known for some time that plants from saline soils have higher carbohydrate and nitrogen contents than plants from non-saline soils and that amino acids accumulate in their tissues (Udovenko and M'Inko 1966; Strogonov et al. 1970). Other experimental work showed that a disturbance of protein synthesis could be related to substrate salt levels (Kahane and Poljakoff-Mayber 1968; Hall and Flowers 1973).

Mizrachi, Pannier and Pannier (1980) tested the response of seedlings of *Avicennia germinans* (as *A. nitida*) and *Rhizophora mangle* to different salt concentrations and simultaneously determined the chloride and nitrogen content (total N, protein N and amino N) and the rate of uptake of the labelled amino acid, leucine, in both leaves and roots. The two species responded differently in some respects, but both showed a reduction in leucine uptake with increasing soil salinity, indicating reduction in protein synthesis. In *R. mangle* the amino N increased with increasing salinity, whereas in *A. marina* there was an initial increase to a salinity of 9.6 ‰ followed by a rapid decline. Amino N accumulated at all salt concentrations in the roots of both species.

These effects of salt on enzyme activity suggest that the enzymes of mangroves and saltmarsh plants do not differ from those of other plants; the enzymes probably would not function if they were directly in contact with the salt levels implied by the overall salt content of the plant. How are enzymes kept out of contact with unfavourable salt levels? One possibility is that most of the salt is contained in the vacuole and that in the cytoplasm, where enzymes are located, low concentrations of salt are maintained. However, the water potential of the cytoplasm and the vacuole must be balanced. Consequently, if partitioning of salt actually occurs, other solutes which do not adversely affect enzyme function must be in the cytoplasm at concentrations sufficient to achieve a water potential equal to that of the vacuole. Several organic compounds have been found in various halophytes which appear to function as such cytoplasmic osmoregulators. For example, glycinebetaine, a quarternary ammonium compound, has been detected in the leaves of *Avicennia marina* (Wyn Jones and Storey 1981), and Downton (1982) has calculated that if this compound occupies a cytoplasmic volume of 5–10 per cent of the cell, then its reported concentration is sufficient to balance total leaf osmotic potential. This would be consistent with the current view that halophytes successfully compartmentalize inorganic ions in a way that salt-sensitive species do not, utilizing ions from the environment to maintain vacuolar osmotic potential lower than that of the external solution, while protecting the salt-sensitive cytoplasm from dehydration and ion excess by the substitution of compatible organic solutes (Downton 1982).

Other compounds with osmoregulatory properties found in various halophytes include choline-O-sulphate, choline-O-phosphate, the amino acid proline and the sugar alcohol sorbitol. The amino N accumulation reported with increasing salt concentrations in *Avicennia germinans* and *Rhizophora mangle* (Mizrachi, Pannier and Pannier 1980) suggests that proline may be involved in these species.

Recently, studies on chloroplasts isolated from *Avicennia marina* showed that they have different and unusual properties compared with chloroplasts from species that are not tolerant of salt (Critchley 1982). The *Avicennia* chloroplasts were found to require chloride for maximal production of oxygen during photosynthesis. Similar requirements were also found in *Avicennia germinans* (Critchley et al. 1982) and in two saltmarsh species (Critchley et al. 1982; Critchley 1983), and it was suggested that these halophytes might preferentially accumulate chloride in the chloroplasts.

It should be apparent from the above discussion that, although the generally adverse effects of salt on whole plants are well documented, the metabolic basis for these effects are still speculative: nitrogen metabolism, protein synthesis, carboxylation enzyme inhibition and stimulation of photosynthetic oxygen production all appear to be involved.

Conserving Desalinated Water

The mangrove environment has frequently been described as "physiologically dry" or "physiologically arid", and this apparent contradiction must be clarified. Clearly, most mangroves have an abundant supply of water around them at all times. However, because this water is saline compared with the internal sap concentration of the mangrove, it must be taken up against an osmotic gradient. An energy cost is coupled to this process, and the real availability to the plant of water of reduced salinity is determined by the amount of metabolic energy the plant can make available for desalination. In other words, it is the high physiological cost of this water that underlies the physiological dryness of the mangrove environment.

Having obtained desalinated water at considerable cost, many mangroves display features, generally associated with plants of arid environments, which tend to conserve or retain that water; these are referred to as xeromorphic features.

Xeromorphic Features

Leaves of most mangroves exhibit a range of xeromorphic features, that is, features normally associated with plants from arid or semi-arid regions (Stace 1966), although this is disputed by some (for example Uphof 1941), and there are few experimental studies in support of the water-conserving function of such leaf characteristics (Miller, Hom and Poole 1975). The major xeromorphic features are discussed below.

All species of mangroves have a thick-walled epidermis which, at least on the upper leaf surface, is covered by a thick waxy cuticle that would seem to retard evaporative loss, or by a layer of variously shaped hairs (*Avicennia*, *Hibiscus tiliaceus*) or scales (*Heritiera*, *Camptostemon*) (figure 7). These usually cover salt glands and stomata (when present) and, presumably, reduce water loss via these apertures.

Stomata are minute pores in the leaf that can open or close. They permit the passage of gases into and out of the leaf, and it is through these that much evaporative loss occurs. With a few exceptions, stomata are restricted to the lower leaf epidermis. In terms of frequency and dimension, mangrove stomata are similar to those of plants of other habitats but many species show stomata sunk beneath the level of the epidermis — for example, *Avicennia*, *Aegiceras*, *Bruguiera*, *Ceriops*, *Lumnitzera* and *Rhizophora*. Substomatal chambers are present in *Avicennia*, *Ceriops* and *Rhizophora* (Sidhu 1975b).

Three types of stomata have been reported from mangroves, but Sidhu (1975b) was unable to attach any ecological significance to them. The stomata of mangroves show considerable variation in behaviour; Joshi et al. (1975) reported that in a number of Indian species the stomata are wide open between 4 a.m. and 10 a.m., closed in the early afternoon, and again slightly open in the evening. On the other hand, photosynthetic studies with *Avicennia* and *Rhizophora* in Australia have shown the stomata to remain open throughout the day (Attiwill and Clough 1980; Clough, Andrews and Cowan 1982). It is possible that high temperature also affects stomatal behaviour which, in turn, is reflected in transpiration rates (Lewis and Naidoo 1970; Lugo et al. 1975; Steinke 1979).

Both Stace (1966) and Sidhu (1975b) concluded that the presence of a thick cuticle, wax coatings, sunken stomata and the distribution of cutinized and sclerenchymatous cells throughout the leaf, including the epidermis, are xeric characters which probably developed in response to the physiological dryness of the mangrove environment.

Succulence, or storage of water in fleshy tissue, is a xeromorphic feature common in mangrove leaves. Based on studies of *Rhizophora* (Bowman 1921) and *Sonneratia* (Walter and Steiner 1936) growing in saline and freshwater conditions, it appears that succulence is a response to the presence of chloride. Anatomical

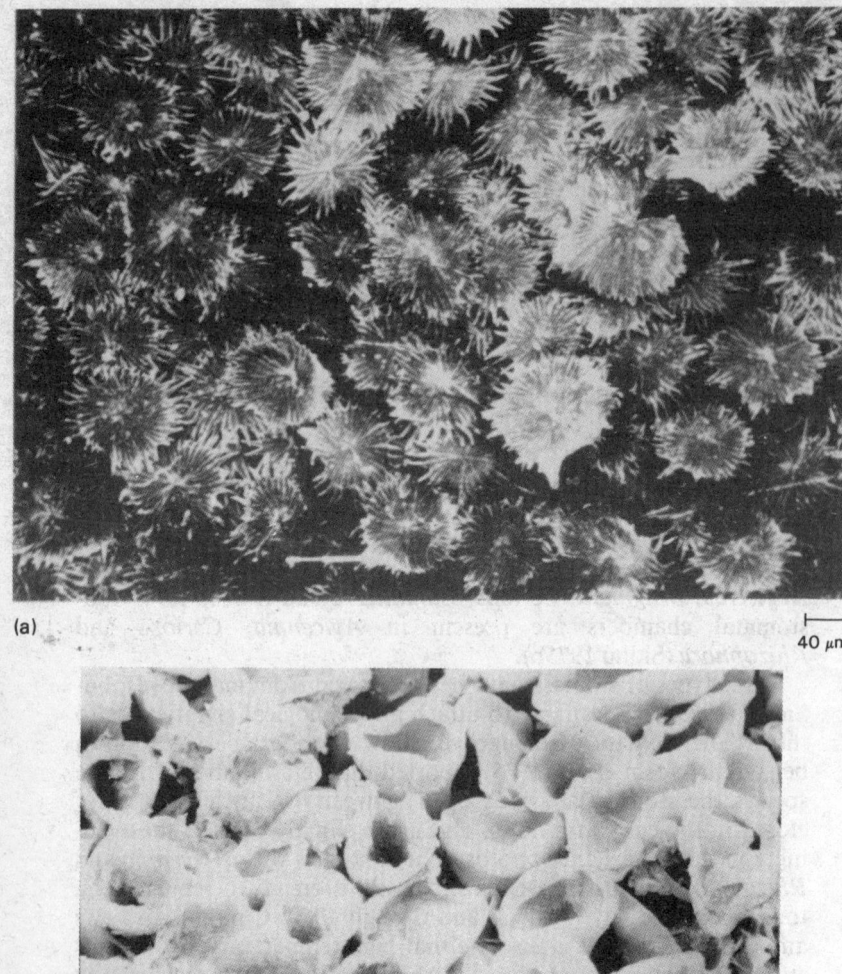

Figure 7 Leaf hairs and scales in mangroves: (a) scales on the underside of leaves of *Heritiera littoralis* (scale bar given); (b) glandular hairs on the underside of leaves of *Avicennia marina* (scale bar given).

factors contributing to succulence in leaves include the presence of a well-developed, large-celled, water-storing hypodermis, a strongly developed palisade mesophyll and generally small intercellular volumes. With the exception of *Ceriops*, those species with isobilateral leaves (upper and lower sides of the leaf the same) do not possess a hypodermis (Saenger 1982), but in several of these, large undifferentiated mesophyll cells form a central water-storing tissue (figure 8). Spongy mesophyll is absent from species with

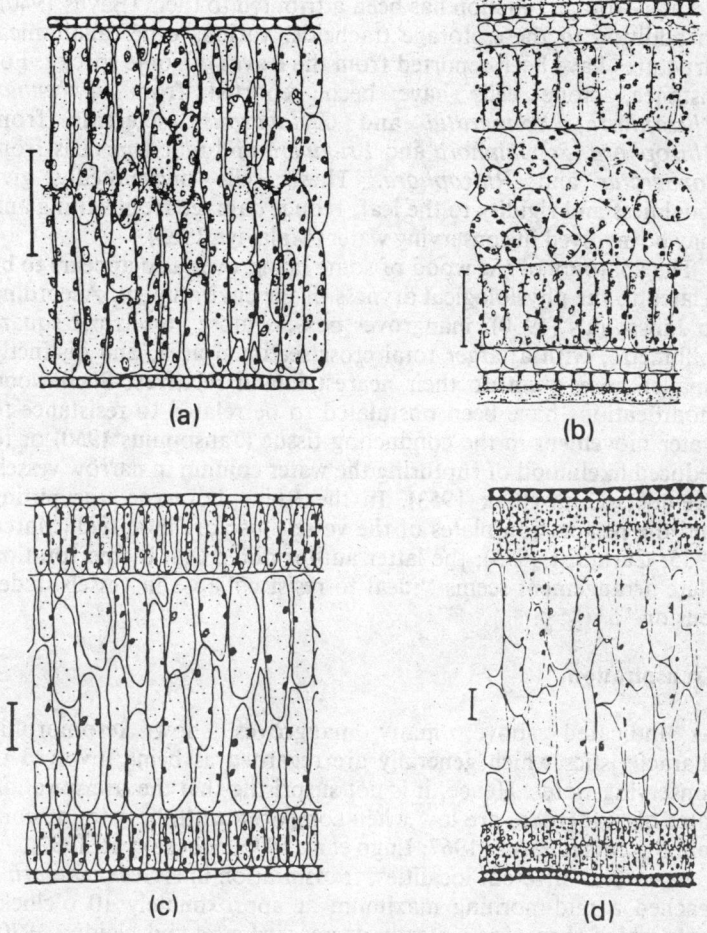

Figure 8 Transverse sections of leaves of Australian mangroves with isobilateral leaves (bar scale is μm in all sections): (a) *Aegialitis annulata*; (b) *Ceriops tagal* — note upper and lower hypodermis and enlarged spongy mesophyll; (c) the mangrove mistletoe *Amyema mackayense*; (d) *Sonneratia caseolaris, Lumnitzera racemosa* and *Osbornia octodonta* — note that all these species have enlarged water-storing cells.

isobilateral leaves with the exception of *Ceriops*, and it generally forms less than 40 per cent of the cross-sectional area of those species with dorsiventral leaves.

In several species, including *Avicennia*, *Bruguiera* and *Ceriops*, the ends of the vascular bundles (tissue conducting water and solutes throughout the plants) are surrounded by irregular groups of tracheid cells which are much larger than the conducting elements. Their walls bear spiral, reticulated or pitted thickenings and, since they possess a flange-like connection to the hypodermis, a water-storage function has been attributed to them (Baylis 1940). In addition to these storage tracheids, various other anatomical structures have been reported from the leaves of some species. For instance, stone cells have been reported from *Avicennia*, *Rhizophora*, *Sonneratia* and *Xylocarpus*, sclereids from *Rhizophora*, *Scyphiphora* and *Bruguiera* and mucilage cells from *Sonneratia* and *Rhizophora*. These cells undoubtedly give toughness and rigidity to the leaf, reduce damage from wilting and may be involved in conserving water (Malaviya 1963).

The anatomy of the wood of some mangroves also appears to be related to the physiological dryness of the environment. According to Janssonius (1950), mangroves possess more vessels per square millimetre, with a larger total cross-sectional area, and distinctly smaller pores than do their nearest inland relatives. Such wood modifications have been postulated to be related to resistance to water movement in the conducting tissue (Jansonnius 1950) or to reduced likelihood of rupturing the water column in narrow vessels (Reinders-Gouwentak 1953). In the Rhizophoraceae, the pitting and the perforation plates of the vessels also are modified (Marco 1935; Carlquist 1975); the latter author noted that the perforation plate arrangement seems "ideal to resist collapse in vessels under tension".

Transpiration

As indicated above, many mangroves show xeromorphic characteristics which generally are regarded as being involved in conserving water. Hence, it is not surprising that the transpiration rates of mangroves are low when compared with the rates of non-saline plants (Gessner 1967; Lugo et al. 1975; Moore et al. 1982).

At several different localities, transpiration in *Avicennia marina* reached a mid-morning maximum at approximately 10 o'clock, after which there was a steady decrease (Lewis and Naidoo 1970; Leshem and Levison 1972; Steinke 1979; Attiwill and Clough 1980). The afternoon decrease in transpiration was not influenced by tidal inundation, and it is assumed that the water potential gradient was so steep that the consequent high rate of transpiration induced an internal water deficit which resulted in the closure of the stomata (Steinke 1979). Once the stomata were closed, the continuing high temperatures prevented stomatal reopening, even when water was freely available (Steinke 1979). Attiwill and Clough (1980), however, found that in a temperate population of *Avicennia* in Westernport Bay, Victoria, there was no change in water stress during the day, even on days of sustained sunlight.

Scholander et al. (1962, 1965) suggested that rates of water loss in mangroves are related to salinity adaptations. Those plants that grow in highly saline situations tend to transpire less than those growing in less saline conditions. This is partly due to the fact that the water is supplied to the leaf at a considerable negative hydrostatic pressure potential, and the demand for water, in terms of the vapour pressure difference at the leaf and air interface, often can be high. In addition, maintenance of the osmotic potential of the xylem sap (although considerably higher than the hydrostatic pressure potential) has a high energy cost because water is taken up against an osmotic gradient. Finally, Gessner (1967) suggested that since the xylem sap of mangroves contains salt, a high transpiration rate would rapidly concentrate salt in the leaf with consequent deleterious effects.

Scholander at al. (1965) discussed some of the physiological cost of these adaptations and indicated that there is a limit to the amount of water a plant can effectively take up and transport against the osmotic gradient of its environment; this limit is reflected in lower transpiration and higher respiration rates.

Energy diverted to the uptake of water or the production of water-conserving structures such as hairs, waxy cuticles and scales is clearly not available for growth and reproduction. Consequently, a number of strategies have developed in relation to this energy cost. In some species, the energy expenditure is reduced (1) by growing in less saline conditions, (2) by reduction of transpirational water loss, or (3) by growing in bursts when fresh water is available and "marking time" during other periods. Other species have increased the efficiency of energy production by such means as leaf orientation and a large photosynthetic surface area.

Even with these strategies in highly saline situations the net productivity is decreased and dwarfing or stunting may occur. Dwarfed mangrove systems apparently use a larger portion of their energy supply for respiration and low-loss recycling mechanisms, and a correspondingly smaller proportion for growth (Lugo et al. 1975).

Root Specializations

Two of the problems affecting mangroves are waterlogged soils that are low in oxygen (anaerobic) and a semi-fluid substrate that provides little mechanical support. The root systems of many species display adaptations which aid in overcoming these problems.

Below the surface, most mangroves possess a laterally spreading cable root system with smaller, vertically descending anchor roots; the latter bear fine nutritive roots. The root system is shallow, generally less than 2 metres deep; tap roots have not been observed. Despite such shallow root systems, the ratio of below- to above-ground biomass is higher for mangroves than for other vegetation types (Saenger 1982), particularly during early developmental stages. This high biomass ratio may be an adaptation to unstable substrate conditions.

Some species do not possess a specialized root system (such as *Aegialitis*, *Excoecaria*) and their roots lie near or on the substrate surface. Since in these species only relatively small surface areas are available for the assimilation of oxygen, they tend to be found on less waterlogged soils (*Excoecaria*) or on coarser, more aerobic sediments (*Aegialitis*). However, *Nypa*, the mangrove palm, grows from an underground rhizome and yet has no specialized aerial root system; it is found in areas of frequent inundation and may occur on waterlogged soils (Tomlinson 1971).

There is an array of above-ground root types displayed by mangroves (figure 9). These include: (1) Pneumatophores — roots arising from the cable root system (for example, *Avicennia*, *Xylocarpus* and *Sonneratia*) and extending upward into the air as small conical projections; (2) Knee-roots — modified sections of the cable root which first grow upward above the substrate and then downward again (for example, *Bruguiera*); (3) Stilt roots — generally branched roots that arise from the trunk and grow into the substrate (*Rhizophora*, *Ceriops*); (4) Buttress roots — similar to stilt roots in origin but expanding into flattened, blade-like structures (*Heritiera*); (5) Aerial roots — generally unbranched roots arising from the trunk or lower branches and descending downward but usually not reaching the substrate (for example, *Rhizophora*, *Avicennia* and *Acanthus*). Most mangrove genera have one or more of these types (see table 2).

Evidence that these root structures are adaptations providing aeration for subterranean roots and which physically anchor the plant comes from a variety of sources. The most apparent is that those mangroves growing at lower tide levels and which are, consequently, more frequently inundated tend to possess the greatest array of above-ground root types. The presence of aerenchymatous tissue and numerous lenticels in most above-ground roots provides further supporting evidence. The mechanism of air uptake, through the development of a negative gas pressure, has been investigated by Scholander, Van Dam and Scholander (1955) in

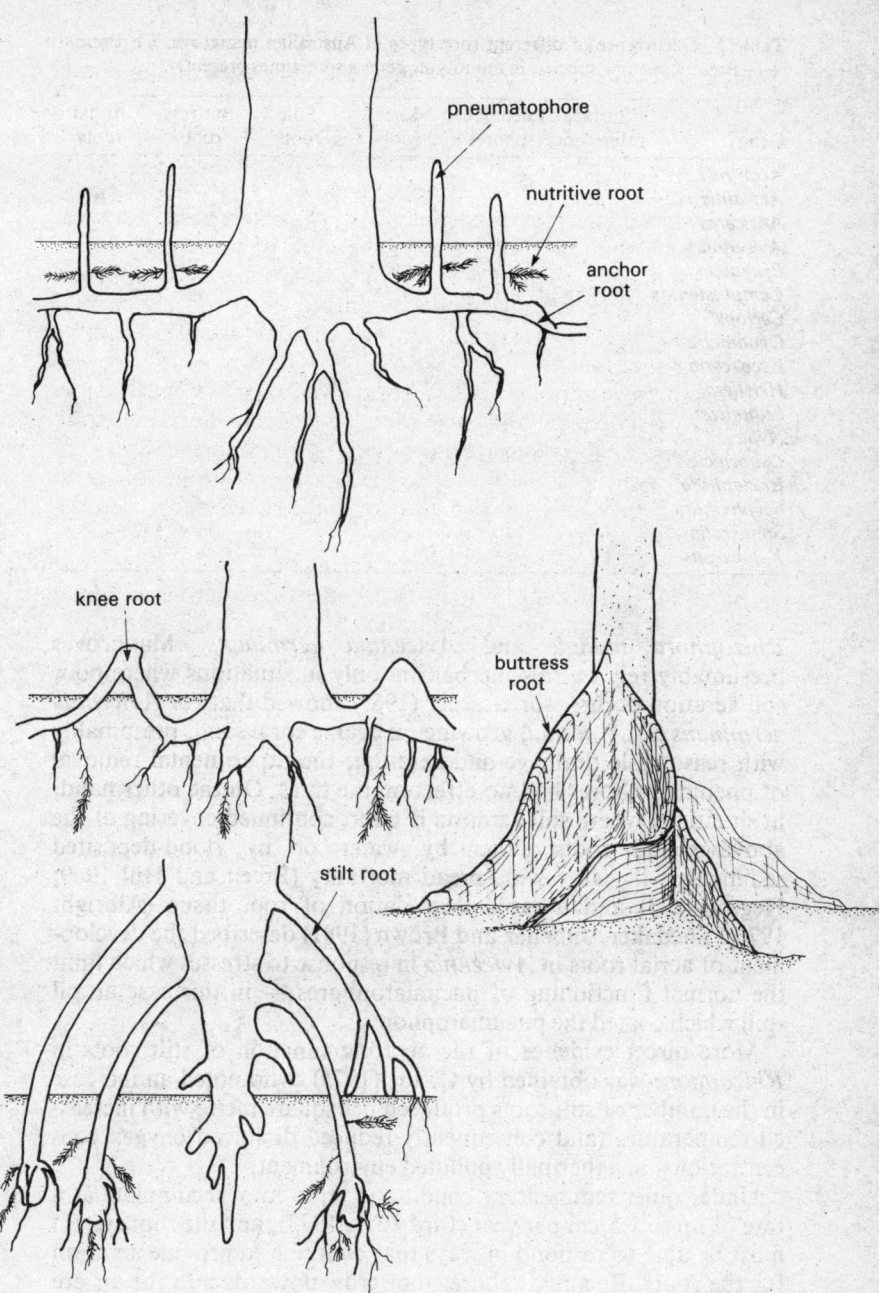

Figure 9 Major morphological root types found in mangroves.

Table 2 Occurrence of different root types in Australian mangroves (+ Present; +/- Present in some species; in monotypic genera sometimes present)

Genus	Surface cable roots	Pneumato-phores	Knee roots	Stilt roots	Buttress roots	Aerial roots
Acanthus	+			+		+
Aegialitis	+					
Aegiceras	+					
Avicennia		+	+/-			+
Bruguiera		+/-	+	+/-	+/-	+/-
Camptostemon		+			+	
Ceriops			+	+	+	
Cynometra					+/-	
Excoecaria	+					
Heritiera					+	
Lumnitzera		+/-	+	+/-		
Nypa	+?					
Osbornia	+	+/-				
Rhizophora				+		+
Scyphiphora	+?					
Sonneratia	+/-	+/-			+/-	
Xylocarpus	+/-	+			+	

Rhizophora mangle and *Avicennia germinans*. Mangroves presumably rely on this mechanism only in situations where poor soil aeration occurs, for Gessner (1967) showed that for *Avicennia germinans* (as *A. nitida*) growing on coarse coral sand, presumably with reasonable drainage and aeration, the experimental removal of pneumatophores had no effect on the trees. On the other hand, in situations where soil aeration is poor, continued covering of the above-ground roots either by water or by flood-deposited sediments will cause widespread mortality (Breen and Hill 1969; Hegerl 1975) and a rapid degradation of root tissue (Albright 1976). Snedaker, Jimenez and Brown (1981) described the development of aerial roots in *Avicennia* in response to stresses which limit the normal functioning of pneumatophores — in this case an oil spill which coated the pneumatophores.

More direct evidence of the aerating function of stilt roots in *Rhizophora* was obtained by Canoy (1975), who noted an increase in the number of stilt roots produced per square metre with increased temperature (and consequently reduced dissolved oxygen concentrations) in a thermally polluted environment.

Under quiet sedimentary conditions, mud may accumulate at a rate of up to 1.5 cm per year (Bird 1971, 1973), and the root system must be able to respond in ways that continue to provide aeration for the roots. Pneumatophores that grow upwards into the air are an adaptation found in *Avicennia*, *Xylocarpus australasicus* and *Sonneratia*. Other species cope with sediment accumulation by forming extra arches of the stilt roots, additional knee roots or the upward secondary thickening of roots or buttresses.

In exposed situations where sediments are removed by erosion and do not accumulate, the problem is reversed and plants must remain firmly anchored in order to survive. Massive development of the various above-ground roots may serve to reduce water movement around the plant, and an extensive cable root system, as in *Avicennia* and *Osbornia* for example, may effectively anchor the plant despite its shallowness. Thom, Wright and Coleman (1975) described the successful colonization of tidal sand flats in the Joseph Bonaparte Gulf, Western Australia, by *Avicennia* despite rigorous conditions imposed by the passage of long-period waves.

The effect of water movement and inundation on above-ground root development in the Japanese mangrove, *Kandelia candel*, was described by Hosokawa, Tagawa and Chapman (1977). In this species of the Rhizophoraceae, the basal part of the stem is buttress-like on plants near creeks with flowing water, but in comparatively still water typical stilt roots are formed.

The anatomical changes occurring in the aerial and stilt roots of *Rhizophora* upon penetrating the substrate have been described by Bowman (1921), Gill and Tomlinson (1971, 1975, 1977) and Karsted and Parameswaran (1976); the external colour of the root changes from tan to white as the thin surface layers lose their chlorophyll, thickened walls are no longer formed and the ground parenchyma contains many gas-filled spaces. The aerial root has approximately 5 per cent gas space before penetration into the substrate compared with about 50 per cent after penetration.

Lenticels are common in the periderm of the stems and roots of most mangroves (Chapman 1947; Roth 1965; Lugo and Snedaker 1975). Outwardly, a lenticel often appears as a vertically or horizontally elongated mass of loose cells that protrudes above the surface through a fissure in the periderm. The dimensions and frequency of lenticels vary between species and with height above the water surface. Baker (1915) described the lenticels occurring on the pneumatophores and stems of *Avicennia marina* as "raised black spots scattered over the surface . . . a section showing these layers of cells to be raised over what is a vacant cavity or air space in direct communication with the ventilating system", and he concluded that they may be secondary organs of ventilation. On the stems, lenticels in this species were noted up to 3 metres above highwater mark.

Little information is available about the physiology and metabolism of mangrove roots (Clough, Andrews and Cowan 1982), and the following comments are highly speculative. The most important aspects requiring investigation are the metabolic adaptations of the root to waterlogging and oxygen deficiency on the one hand and to salt exclusion on the other.

Waterlogging and the accompanying anaerobic conditions in the soil are often associated with the development of aerenchyma (loose tissue containing many air spaces). Such tissue may serve either as an oxygen reservoir, or as a system which allows the maximum volume of root or rhizome per quantity of living tissue, thereby achieving an economy in oxygen consumption per unit volume. However, the facilitation of gaseous diffusion by morphological adaptation is not an entirely satisfactory explanation of waterlogging tolerance in mangroves, and the possible role of metabolic adaptations to anaerobic conditions must be considered.

Experiments with species other than mangroves during waterlogging have demonstrated a range of metabolic responses to anaerobic conditions. These studies suggest that under anaerobic conditions malic acid is more important than ethanol as an end product, that malic, shikimic and quinic acids accumulate in the roots and rhizomes and that succinic acid is a common product of anaerobic respiration. Crawford and Tyler (1969) examined organic acid accumulation in a selection of species ranging from tolerant to intolerant of waterlogging, and found that there was an immediate change in organic acid metabolism with the advent of flooding. Malic acid accumulated in those species tolerant of waterlogging but fell in those species that were intolerant. Species intolerant of waterlogging showed increases in succinic acid and to a lesser extent in lactic acid.

The significance of malic acid accumulation is threefold. First, malic acid, like other organic acids, can accumulate in plant cells in considerable quantities without injury to the plant, and in this respect differs considerably from ethanol. Second, malic acid is an alternative product to ethanol in anaerobic respiration in plants and it can subsequently be metabolized on the return of aerobic conditions. Third, in cation absorption, malic acid accumulates in greatest quantity and thus preserves the electrical neutrality of the cell.

Whether the malic acid respiratory pathway exists in mangroves as it does in other species tolerant to waterlogging is not presently known. If, however, respiration is involved in the salt-excluding mechanism, either directly in providing energy for the process or indirectly in the synthesis and maintenance of a salt-excluding membrane, the respiratory requirements of the roots of mangroves may be higher than those of non-mangrove plants (Clough, Andrews and Cowan 1982). Under such circumstances, the ability to switch to a non-damaging anaerobic respiratory pathway while waterlogged would clearly provide a significant benefit to the plant. If that pathway could, at the same time, assist in the maintenance of ionic neutrality in the root cells, it would seem to be a very important adaptation. Clearly, this is an aspect of the physiology of mangroves which may reward investigation.

Responses to Light

Light and Form

Wylie (1949) found that leaves developing in high light intensity show a higher degree of xeromorphy than those protected from it. Consequently, it is possible that some of the xeromorphic features discussed earlier may be responses to high light intensities rather than (or in addition to) water shortages. Isobilateral leaf anatomy is generally regarded as a xeromorphic character, but when it is combined with a mechanism for orientating the leaf towards the sun (as occurs for example in *Ceriops*) it seems reasonable to assume a light response also may be involved. Other leaf characteristics associated with high light intensities include such xeromorphic features as a high ratio of volume to surface area and a well-developed, highly differentiated, often isobilateral palisade mesophyll. These features are present in many mangrove species, and in conjunction with the leaf pattern of arrangement on the shoots (Tomlinson and Wheat 1979) may constitute an adaptation optimizing exposure to light under varying conditions of light and shade.

Since leaves developing in intense light show a greater degree of xeromorphy (Wylie 1949), one can distinguish between "sun" and "shade" leaves. Several mangrove species showed a marked morphological differentiation between sun and shade leaves, particularly *Lumnitzera*, *Ceriops* and *Aegiceras* (Saenger 1982). In a detailed examination of the leaves of *Ceriops tagal*, it was noted that shade leaves, when compared with sun leaves, are larger, thicker, have a higher volume-to-surface ratio, possess fewer stomata per square millimetre on the lower leaf surface and possess a proportionately thicker tannin-filled hypodermis on both upper and lower surface and a proportionately thinner upper palisade mesophyll, lower epidermis and lower cuticle (Saenger 1982). These characteristics are those associated with xeromorphy except that xeromorphic leaves generally possess a higher volume-to-surface-area ratio than shade leaves (Shields 1950).

Seedling leaves of *Avicennia marina* also can be subdivided into sun and shade leaves on a morphological basis; the shade leaves contained more chlorophyll on both a leaf-area and fresh-weight basis, were enriched in chlorophyll *b* relative to *a*, and had a lower specific weight and greater leaf area than sun leaves (Ball and

Critchley 1982). In terms of gas exchange and photosynthetic characteristics, however, distinction between sun and shade leaves could not be made; both leaf populations were typical of sun leaves. Unfortunately, no comparative photosynthetic data are available for other Australian mangrove seedlings.

The canopy shape of mangroves is determined largely by endogenous growth patterns which lend themselves to a plant architectural analysis (Halle, Oldeman and Tomlinson 1978). However, the canopy seems to assume certain shapes under specific environmental conditions. For example, an "umbrella-type" canopy has been recorded from *Avicennia marina* in certain situations (Macnae 1968; Wester 1967; Saenger and Hopkins 1975), and Baker (1915) has suggested that, since the pneumatophores of this species must be shaded, a flattened canopy may be an adaptation protecting pneumatophores of isolated trees from intense light.

Photosynthesis

Photosynthesis is necessary for production of food by plants, and many plants have adaptations or physiological responses that optimize this process.

In southern Australia, *A. marina* sun leaves have a rate of photosynthesis approximately 4.5 times that of shade leaves (Attiwill and Clough 1980), although at high light intensities both types of leaves can potentially reach the same maximum rate of photosynthesis per unit of leaf surface. In terms of the quantum efficiency of the canopy of *A. marina*, Attiwill and Clough (1980) found a value of 0.0135 mol CO_2 per mol of photosynthetically active radiation; this is low compared with non-mangrove plants (Bjorkman 1970). Furthermore, this quantum efficiency was obtained only at low light intensities; at full midday light intensities the quantum efficiency was reduced to approximately one-tenth of this level. While such a decline in efficiency is shared with many species, it appears to be of greater magnitude than in many non-mangrove species, and it suggests that although the photosynthetic mechanism of *Avicennia* is inefficient it is best adapted to shade conditions (Attiwill and Clough 1980).

Ball and Critchley (1982) investigated the photosynthetic responses of seedlings of *Avicennia marina* and suggested they are best adapted to growing in exposed conditions and appear to have a low capacity to acclimate to low light intensities. However, these seedlings appear to be able to use sunflecks highly efficiently, and consequently are able to survive in the understorey environment (Ball and Critchley 1982).

Joshi et al. (1974, 1975) investigated the photosynthetic carbon metabolism of two Indian species of mangroves and concluded that they were of the aspartate type. They suggested that this type of metabolism may be due to the inhibition of the enzyme malic dehydrogenase in the presence of high sodium chloride concentrations. To test this hypothesis, plants were labelled with radioactive carbon and the early products were identified; aspartate and alanine both had become heavily radioactive, and consequently it was concluded that the C_4 carbon fixation pathway operated in these species.

More recently, however, the occurrence of this biochemical adaptation of photosynthetic carbon fixation has been questioned for mangroves, and other evidence, drawn from leaf anatomy, the $^{13}C/^{12}C$ carbon isotope ratio and gas exchange properties (Moore et al. 1972; Cowan 1978; Clough, Andrews and Cowan 1982), seems to indicate that mangroves possess the more common C_3 photosynthetic carbon metabolism.

Light and Other Physical Factors

With high light intensity also come higher temperatures and increased water losses. Consequently, the adaptations of plants need to strike a balance between photosynthetic advantage and harmful effects of other physical conditions associated with intense radiation.

Different light and shade requirements have been noted in mangroves, and geographic variation among adults, seedlings and saplings are apparent (Saenger 1982). Two groups of mangroves seem to emerge (see table 8): (1) those which are shade tolerant both as seedlings and as adults (*Aegiceras*, *Ceriops*, *Bruguiera*, *Osbornia*, *Xylocarpus*, *Excoecaria*), and (2) those species which are shade intolerant (*Acrostichum*, *Acanthus*, *Aegialitis*, *Rhizophora*, *Lumnitzera*, *Scyphiphora*, *Sonneratia*). *Avicennia* may be shade intolerant in the seedling stage but shade tolerant as a tree.

Isobilateral leaves are found in both of the above groups. Three

Table 3 Early development of *Avicennia marina* seedlings grown under different salinities

Salinity (as % sea water)	Time (days) from planting until:		
	Splitting of pericarp	Separation of cotyledons	Shoot emergence
0	2.2	12.8	18.8
10	3.6	14.1	25.1
25	4.1	16.6	21.4
50	4.6	18.6	20.8
75	7.9	25.1	33.1
100	7.0	37.8	53.0

Source: Downton (1982).

species, including *Ceriops* and *Osbornia* (shade tolerant) and *Lumnitzera* (shade intolerant), possess leaves that point upwards and are orientated towards the sun. By such leaf orientation, effective photosynthesis is increased along with the effective length of the photosynthetic day. At the same time, the heat input per unit leaf area is reduced. For example, measurements of leaf temperatures high in the canopy of *Rhizophora stylosa* in northern Queensland in full sunlight in summer showed that leaves at their natural inclination were often over 5°C cooler than leaves experimentally held horizontal; they also had correspondingly lower rates of water loss and higher photosynthetic rates than the horizontal leaves (Clough, Andrews and Cowan 1982).

Uphof (1941) suggested that as a water storage tissue is present between the epidemis and palisade mesophyll of most mangrove leaves, its function is to protect the mesophyll from excessive heat or from infra-red radiation. Tannin cells on the upper surface of the leaf of *Rhizophora* and *Ceriops*, for example, may protect the leaf from intense visible or ultraviolet radiation. However, the role of tannins is not clear. For example, the hypodermis is relatively reduced in sun leaves compared with the shade leaves; this finding does not support the theory that a tannin-filled hypodermis protects the palisade mesophyll from high levels of visible or ultraviolet radiation. It has been suggested also that tannins may be involved in preventing fungal infestations or in the removal of excess salt.

Lugo et al. (1975) were able to show that net daytime photosynthetic rates in two American species, *Rhizophora mangle* and *Avicennia germinans* (as *A. nitida*), were about twice as high in sun leaves as in shade leaves. At night, the shade leaves had respiration rates that were four times as high as those of sun leaves. The two species behaved differently in terms of transpiration. *Rhizophora* sun leaves had a higher transpiration rate than the shade leaves while in *Avicennia*, and also in *Laguncularia*, the sun leaves had lower transpiration rates when compared with shade leaves. For many Australian species, both leaf orientation and general leaf morphology vary according to the leaf's position in the canopy (Saenger 1982). Although comparable physiological data for Australian species is being collected only now, it is apparent that certain aspects of leaf physiology, such as responses to light and temperature, also vary with the leaf's position in the canopy (Clough, Andrews and Cowan 1982).

Living with Wind, Waves and Frosts

Mangroves have adapted in various ways to physical damage.

When the factor causing damage is of low to moderate intensity, species differ in their degree of tolerance. At high to catastrophic intensities, most species are killed or damaged severely, but various recovery patterns can be observed.

Most mangroves are susceptible to frosts, although the degree of susceptibility varies with species and geographic location. McMillan (1975a) showed that both *Avicennia germinans* and *A. marina*, collected from a range of localities and subjected to frost under identical conditions, have populations selectively adapted to a latitudinal range of habitats, including ones with recurrent low winter temperatures. Leaf scorch seems to be the predominant symptom (Chapman and Ronaldson 1958), often followed by a reduction in the leaf area index (Lugo and Zucca 1977).

Tropical storms are of frequent occurrence in northern Australia. Stocker (1976) classified wind damage caused by cyclone "Tracy" into four types: (1) windthrow, where the tree is felled; (2) crown damage, where leaves and twigs are removed and/or branches are torn off; (3) bole damage, where the trunk is broken, severely fractured or leaning; and (4) death, where the tree remains standing. Because all these damage types also can be caused by wave action (which generally accompanies high winds), no distinction is made between wind and water damage in the following paragraphs.

Windthrow is the severest form of damage and Stocker (1976) found several mangroves to be particularly susceptible, including *Camptostemon schultzii*, *Ceriops tagal*, *Rhizophora stylosa*, *Bruguiera parviflora* and *Excoecaria agallocha*.

Other species such as *Xylocarpus australasicus*, *Aegiceras*, *Aegialitis* and *Lumnitzera racemosa* showed little or no windthrow, and they rapidly developed new crowns. It seems likely that windthrow-susceptible trees are those with weakly developed cable root systems, or whose root system is weakened by erosion or bank-slumping, or by some biological agency such as infestation by isopods or wood-boring molluscs. For most species, windthrow results in death, although for *Sonneratia* and *Avicennia* epicormic shoots will rapidly develop if some root connection remains.

Susceptibility to bole damage varies considerably among species. The anomalous wood structure of *Avicennia*, with its non-concentric, non-annual growth rings of alternating bands of xylem and phloem (Gill 1971), gives the wood unusual qualities: (1) it is extremely strong for its weight; (2) it is extremely difficult to split radially yet it is easy to do so tangentially (hence it was used to make shields by the Aborigines); and (3) the unusual ring structure ensures that, if any part of the trunk is damaged, sufficient intact conductive tissue remains to supply the crown and epicormic shoots — as a consequence of this distribution of xylem and phloem tissue, *Avicennia* cannot be killed by ringbarking, an apparently useful adaptation in minimizing damage from waterborne objects.

The secondary wood anatomy of other mangrove species has been studied slightly (Panshin 1932; Marco 1935; Venkatiswarlu and Rao 1964). In *Ceriops*, thick-walled bast fibres form a mechanical tissue cylinder giving strength and rigidity to the stem (Rao and Sharma 1968). In *Rhizophora*, abundance sclereids occur in non-functional phloem tissue (Karsted and Parameswaran 1976) and stone cells and fibres occur throughout the plant. The wood of *Bruguiera* has been described as extremely strong (Banerji 1958) as has that of *Heritiera*, *Rhizophora apiculata* and *Lumnitzera littorea* (Panshin 1932).

In the case of a broken bole, a few species are able to regrow from the stumps. *Avicennia*, *Sonneratia*, *Xylocarpus* and *Excoecaria* and the western hemisphere *Laguncularia* and *Conocarpus* coppice readily.

Crown damage is the most common type of damage, with the plant being defoliated in extreme cases. Leaves of most mangroves are leathery and strengthened by various sclerenchymatous cells, and in strong winds leaf-bearing twigs appear to be shed rather than individual leaves. Recovery from twig or leaf damage is usually rapid; *Avicennia*, *Excoecaria* and *Sonneratia* have abundant reserve buds in the stem. In *Rhizophora*, buds are present in the stems of saplings but become restricted to thin terminal branches as the tree matures (Gill and Tomlinson 1969). Conditions severe enough to remove or kill all branches possessing viable reserve buds will kill *Rhizophora*.

When the tree is dead but remains standing, a number of causative factors may be involved, including changes in the substrates, fatal root or bole damage caused by wind sway, or stress following the near-total loss of leaves.

Reproductive Adaptations

Flowering and Pollination

Flower primordia develop on young plants when little more than three or four years old. The initiation of flowering seems to be independent of size, but the actual factors involved are largely unknown. Most Australian species begin flowering in spring and continue through the summer months (Jones 1971; Saenger 1982; Duke, Bunt and Williams 1984); the predominance of summer flowering in central Queensland species is shown in figure 10.

Pollination in most mangroves occurs through the agency of wind, insects and birds (Clifford and Specht 1979; Saenger 1982),

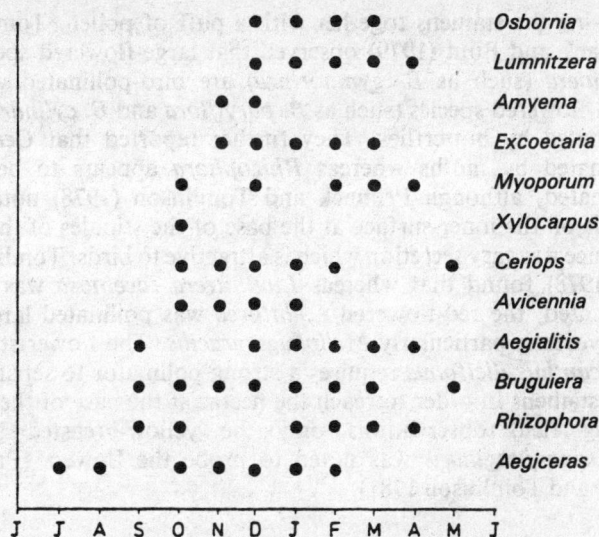

Figure 10 Flowering times of mangroves and mangrove associates in Port Curtis, Queensland, during 1975 to 1979.

and most species possess small, non-sticky pollen grains (Wright 1977) which are distinctive for each species (Muller and Caratini 1977) and are even recognizable as fossils (Muller 1964; Churchill 1973). In *Nypa* the pollen is sticky and pollination probably occurs via the many insects that visit its inflorescences (Uhl 1972), particularly drosophilid flies (Essig 1973). The vasculature, histology and growth patterns of the flowers of *Nypa* appear to be directly related to insect pollination (Uhl and Moore 1977). *Aegiceras*, *Cynometra* and probably *Avicennia*, with their scented flowers, are predominantly bee-pollinated (Blake and Roff 1972; Chanda 1977; Clifford and Specht 1979); the western mangrove *Avicennia germinans* appears to be exclusively pollinated by the bee *Apis mellifera* (Percival 1974). *Excoecaria* is dioecious, bears flowers in catkins and possesses a two-celled pollen grain (Venkateswarlu and Rao 1975) and it can be presumed that it is wind-pollinated. *Sonneratia* releases copious amounts of dry or slightly sticky pollen at dusk when the flower opens (Muller 1969), and it is dispersed by bats (Faegri and van der Pijl 1971; Semeniuk, Kenneally and Wilson 1978) and moths (Primack, Duke and Tomlinson 1981). In South Africa, *Bruguiera gymnorhiza* is pollinated by insects and sunbirds, and the petals of this species are peculiarly adapted to this method of pollen dispersal (Davey 1975); *Bruguiera* petals possess a heavily cutinized epidermal region which, on the application of gentle pressure, causes the petal lobes to spring apart, thereby releasing the stamens together with a puff of pollen. Tomlinson, Primack and Bunt (1979) observed that large-flowered species of *Bruguiera* (such as *B. gymnorhiza*) are bird-pollinated whereas small-flowered species (such as *B. parviflora* and *B. cylindrica*) are pollinated by butterflies. They further reported that *Ceriops* is pollinated by moths whereas *Rhizophora* appears to be wind-pollinated, although Primack and Tomlinson (1978) noted that glands on the inner surface at the base of the stipules of the latter produce a sugary secretion which is attractive to birds. Tomlinson et al. (1978) found that whereas *Lumnitzera racemosa* was insect-pollinated, the red-flowered *L. littorea* was pollinated largely by honeyeaters, particularly *Meliphaga gracilis*. The flower structure of *Acanthus ilicifolius* requires a strong pollinator to separate the four stamens in order to reach the nectar at the base of the ovary; during field observations only the yellow-breasted sunbird *Nectarina jungularis* was noted to probe the flowers (Primack, Duke and Tomlinson 1981).

Propagule Production

Most mangroves on the eastern Australian coast bear mature propagules in the summer months (February to March) (Jones 1971; Graham et al. 1975; Saenger 1982; Duke, Bunt and Williams 1984); the occurrence of mature propagules in mangroves on the central Queensland coastline is shown in figure 11. Similarly, on the

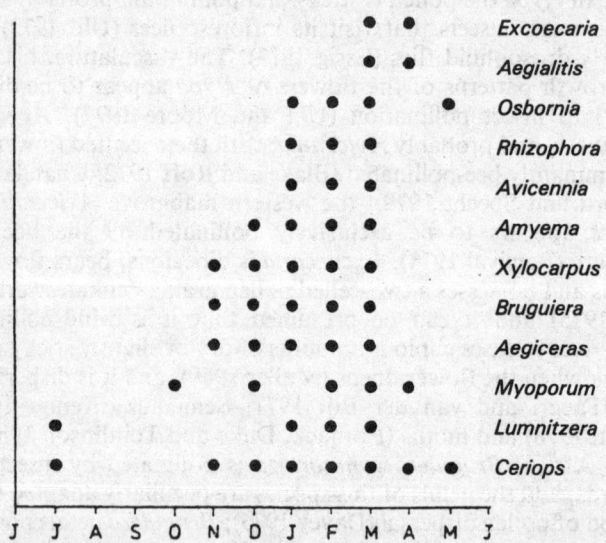

Figure 11 Fruiting times of mangroves and mangrove associates in Port Curtis, Queensland, during 1975 to 1979.

eastern coast of South Africa, *Avicennia marina* has its main fruiting period in March–April (Steinke 1975).

In some species, the time from flower primordium to mature propagule is considerable: 3 years in *Rhizophora apiculata* (Christensen and Wium-Anderson 1977); 1–1.5 years in *Bruguiera gymnorhiza*, *Ceriops tagal* and *Rhizophora stylosa* (Duke, Bunt and Williams 1984); 6 months from open flower to mature propagule in *Rhizophora mangle* (Guppy 1906); 12 months from flower buds to mature propagules in *Kandelia candel* (Nishihira and Urasaki 1976), *Rhizophora mangle* (Gill and Tomlinson 1971b) and *Aegiceras corniculatum* (Carey and Fraser 1932). In spite of this lag, both flowering and fruiting largely occur in the summer months, and it would seem that some common environmental parameter is involved in controlling both processes. Since leaf production in many species is also seasonal with the maxima for many species during summer (Saenger and Moverley 1985), it seems that fruiting, like flowering, is timed for the period most favourable for growth.

Considerable mortality has been reported for developing seedlings still attached to the tree. Gill and Tomlinson (1971b) showed that for *Rhizophora mangle* only between 0 and 7.2 per cent of flower buds produced mature seedlings, although the number of flowers produced may be increased markedly by an increase in nutrients (Onuf, Teal and Valiela 1977). Lugo and Snedaker (1975) followed the development of selected seedlings of *R. mangle* while still attached to the parent tree in Florida in the United States, and they found a mortality of 9 per cent, 13.4 per cent and 20.9 per cent for the months of January (winter), April and May (spring) respectively. Similar figures were noted in *Rhizophora apiculata*, for which Christensen and Wium-Andersen (1977) reported that only 7 per cent of flower buds formed flowers, and only 1–3 per cent formed fruits. In *Kandelia candel* less than 30 per cent of the flower buds ultimately developed into mature propagules (Nishihira and Urasaki 1976). For the Australian *Sonneratia alba*, young fruit developed from forty-one of forty-six flower buds (Primack, Duke and Tomlinson 1981). As in non-mangrove species, much of this pre-dispersal mortality can be attributed to fungal and insect attack on the fruit and to such inherent factors as albinism (Handler and Teas 1983) and other morphogenetic malfunctions.

Vivipary and Cryptovivipary

Various types of fruits are found among the mangroves and those of Australian genera are listed in table 4. In several genera, the fruits contain seeds which develop precociously; the seed germinates while still attached to the parent tree. In these species, the

Table 4 Reproductive units of Australian mangrove and associated genera, together with references to detailed descriptions of their embryology and/or seedling development

Genus	Fruit	Description of embryology and/or seedling development
Acanthus	Capsule with several flat seeds	
Acrostichum	Spore	Stokey and Atkinson 1952; Lloyd 1980
Aegialitis	Indehiscent nut	
Aegiceras	Fleshy capsule, shed with calyx attached	Haberlandt 1896; Carey and Fraser 1932; Collins 1921
Amyema	Baccate, with viscous seeds	
Avicennia	Fleshy capsule with single seed	Treub 1883; Collins 1921; Padmanabhan 1960, 1962a, 1962b; Butler and Steinke 1976
Bruguiera	Fleshy single-seeded berry, shed with calyx attached	
Camptostemon	Capsule with 2 to several woolly seeds	
Ceriops	Fleshy berry, usually single-seeded, shed with calyx attached	Carey 1934
Cynometra	Wrinkled one-seeded pod	
Excoecaria	Trilobed exploding capsule, each lobe one-seeded	Venkateswarlu and Rao 1975
Heritiera	Clusters of woody, keeled carpels	
Hibiscus	Hairy capsule splitting into 5 locules; many-seeded	
Lumnitzera	Indehiscent woody fruit with thin outer fleshy layer	Clifford and Specht 1979
Lysiana	Ovoid with single viscous seed	
Nypa	Aggregate head of one-seed fruits	Tomlinson 1971
Osbornia	Capsule	
Rhizophora	Ovoid fleshy berry, with single seed	Carey 1934; Cook 1907; Gill and Tomlinson 1969
Scyphiphora	Axillary clusters of ribbed fruits, surmounted by calyx	
Sonneratia	Many-celled, many-seeded capsule	Venkateswarlu 1935
Xylocarpus	Several-seeded capsule	Percival and Womersley 1975

embryo develops into a seedling without any dormant period (Gill and Tomlinson 1969), although a form of seedling dormancy may be induced by low water content (Sussex 1975). In *Bruguiera*, *Ceriops*, *Rhizophora*, *Kandelia* and *Nypa* the embryo ruptures the pericarp and grows beyond it, sometimes to considerable lengths, while still attached to the parent tree. This condition is known as vivipary. In *Aegialitis*, *Acanthus*, *Avicennia*, *Aegiceras*, *Laguncularia* and *Pelliciera*, the embryo, while developing within the fruit, does not enlarge sufficiently to rupture the pericarp. This condition is termed cryptovivipary. In the remaining species, the seeds, like those of most plants, pass through a resting stage prior to germination, and do not germinate while still on the parent tree.

Vivipary and cryptovivipary frequently have been cited as an adaptation to some aspect of the mangrove environment. Its adaptive significance could include rapid rooting (Macnae 1968), salt regulation (Joshi 1933), ionic balance (Joshi et al. 1972), development of buoyancy (Gill 1975) and prolonged attainment of nutrients from the parent (nutritional parasitism) (Pannier and Pannier 1975; Bhosale and Shinde 1983). In the viviparous seagrasses, *Amphibolus* and *Thalassodendron*, vivipary appears to be an adaptation assisting rapid root attachment of the plant (Ducker and Knox 1976). However, the occurrence of apparently successful mangroves without viviparous fruits (such as *Osbornia*, *Sonneratia*, *Lumnitzera*, *Xylocarpus* and *Excoecaria*) makes it doubtful whether the possession of vivipary *per se* is of any real adaptive advantage. Tidal buffeting and waveborne objects pose a threat to establishing seedlings, and it would be expected that the smaller the seedling, the larger the threat. Because of this, vivipary in a mangrove may simply be a means of producing a large seedling which is less likely to be damaged by water movements (Saenger 1982). It is interesting to note in this respect that many of the non-viviparous genera also possess large seeds (such as *Xylocarpus*, *Heritiera*, *Cynometra*), which similarly may be a means of alleviating damage by water movement.

Propagule Dispersal and Establishment

The seeds of the mangrove mistletoes are dispersed by the mistletoe bird, *Dicaeum hirundinaceum*, and they are capable of withstanding passage through the alimentary canal of this species. The spores of the mangrove fern *Acrostichum* appear to be wind-dispersed since they do not float; the prothalli of this species, however, float in sea water but mostly sink in fresh water. The dispersal unit of mangroves may be a single seed (*Excoecaria*), a one-seeded fruit (*Cynometra iripa*), a several-seeded fruit (*Sonneratia, Xylocarpus*), a multiple fruit (*Heritiera*), an aggregated fruit (*Nypa*) or a precociously developed seedling (*Avicennia, Aegiceras, Rhizophora, Ceriops, Bruguiera*). The propagules of all mangroves trees are buoyant and are adapted to dispersal by water (Saenger 1982).

Few data are available on the periodicity of propagule dispersal but Clarke and Hannon (1971) found that dispersal of *Aegiceras* coincided with unusually high tides whereas that of *Avicennia* coincided with low tides.

Buoyancy of mangrove propagules may be due to the radicle as in *Rhizophora*, the pericarp and cotyledons as in *Avicennia* (Steinke 1975; Butler and Steinke 1976), the endoderm (for example, *Xylocarpus*) on the cotyledon as in the Panamanian *Pelliciera*. Changes in any of these features can alter the buoyancy. For example, Steinke (1975) showed that propagules of *Avicennia marina* sink after losing their pericarp, generally within four days. Subsequent investigation of the rate of pericarp shedding showed that high and low salinities decreased the rate at which they were shed when compared with the rate in water of intermediate salinity. Consequently, propagules in brackish water will disperse less than those in water of high or low salinity.

High temperatures also increase the rate of pericarp shedding and consequently shorten the potential dispersal distance (Steinke 1975). Using Australian material of *A. marina*, Downton (1982) showed that the time required for the splitting of the pericarp and the separation of the cotyledons increased with increasing salinity and, in this respect, appeared to differ from South African examples of the same species.

A buoyant propagule appears to be an efficient means of widespread, water-based dispersal. Among the seagrasses, however, only a few genera (*Posidonia*, *Thalassodendron* and *Enhalus*) have buoyant fruits and, paradoxically, these species have restricted distributions (Den Hartog 1970). It would appear from this that the role of buoyancy in effecting widespread dispersal needs experimental evaluation.

Rabinowitz (1978a) investigated the parameters affecting dispersal of six Panamanian mangroves, including longevity and vigour, period of floating, period required for establishment and the period of obligate dispersal. Two contrasting dispersal patterns were observed, one for small and another for large propagules.

In contrast to those of *Avicennia marina*, the propagules of *A. germinans* always float, and this species seems to have an absolute requirement for a period of stranding in order to establish itself. This species is restricted to higher ground where inundation is less frequent and where it is free of tidal disturbances. The time required for this species to root is approximately seven days whereas *A. marina*, whose propagules sink after approximately four days, becomes firmly rooted in five days (Clarke and Hannon 1970). *Laguncularia*, whose propagules sink after approximately twenty days, also requires a period of stranding of five days or more in order to become firmly rooted (Rabinowitz 1978a).

The two genera that have large propagules (*Rhizophora* and *Pelliciera*) tolerate tidal disturbance better than do either *Avicennia* and *Laguncularia*; the propagules of the former two are capable of taking root in water of various depths because their weight affords resistance to tidal buffeting, and growth continues under water. Longevity of propagules ranged from thirty-five days in *Laguncularia* to a year or more in *R. mangle* (Rabinowitz 1978a).

These findings led Rabinowitz (1978b) to suggest that the seedling populations of mangroves with smaller propagules turn over annually whereas those with larger ones are made up of overlapping cohorts. In other words, two reproductive strategies are involved: mangroves with small propagules pepper the swamp annually with short-lived seedlings which may establish in gaps that have arisen during the previous year; those with larger propagules form a persistent seedling bank which can maintain itself until a gap in the canopy occurs (if shade-intolerant), or grow in the shade to reach the canopy (if shade-tolerant).

It is doubtful, however, that such a simple scheme operates in Australian mangroves. The ability to utilize sunflecks efficiently, as can *Avicennia marina* seedlings (Ball and Critchley 1982), blurs the boundary between the shade-tolerant and shade-intolerant species with large propagules. Furthermore, newly arrived seedlings tagged in permanent study areas at Port Curtis, central Queensland, survived as two- to four-leaved seedlings for up to eight years whether from small (*Lumnitzera*), medium (*Avicennia*, *Aegiceras*, *Aegialitis*) or large (*Rhizophora*, *Ceriops*) propagules (Saenger, unpubl. data). Nevertheless, a persistent seedling bank appears to be an important colonization strategy in Australian mangroves.

The number of mangrove propagules establishing along 30 metres of intertidal shoreline in permanent study areas (Saenger and Robson 1977) at Port Curtis, a semi-enclosed bay in central coastal Queensland, is given in table 5 together with comparative data from Repulse Bay, near Proserpine. The number of propagules establishing per adult of the same species is also given. These figures are low in view of the apparently high numbers of propagules borne by most species. However, considerable mortality occurs prior to dispersal, and crabs (particularly species of *Sesarma*) and insects damage many propagules after they have fallen. Further mortality occurs during dispersal, including stranding on unfavourable substrates, injury by boring or decomposing marine organisms, and sinking as a result of the attachment of fouling organisms such as barnacles and tubeworms (serpulid polychaetes). Once the propagules are stranded, physical damage by waveborne objects frequently occurs. Among those that establish successfully — that is, become firmly rooted and possess at least one leaf — mortality rates are variable and site-dependent (see table 5). In Queensland, at Port Curtis, mortalities in the first year ranged from 72 per cent in *R. stylosa* to 0 per cent in *L. racemosa* and *A. annulata*. At Repulse Bay, mortality rates during the first year were much more equable. The main factors determin-

Table 5 Production, establishment and mortality rates for propagules of Queensland mangroves

Gladstone

Genus	No. of propagules establishing per 30 m of shoreline line during 4 years	No. of established propagules per adult of same species	% mortality of established propagules during first year	Mean mortality (%) of adults during one year
Rhizophora	276	1.64	71.7	2.98
Aegialitis	3	1.50	0	0
Avicennia	199	1.47	22.1	5.97
Lumnitzera	9	1.00	0	2.78
Aegiceras	27	0.18	14.8	1.51
Ceriops	52	0.13	36.5	1.01

Source: Data from permanent plots, 1975–1979 (after Saenger 1982).

Proserpine

Genus	No. of propagules establishing per 80 m of shoreline line in 1 year	No. of established propagules per adult of same species	% mortality of established propagules during first year	Seedling growth (% height increase/year)
Avicennia	28	1.58	38.1	18.7
Rhizophora	35	0.77	29.2	18.4
Excoecaria	9	0.21	25.0	41.6
Aegiceras	20	0.17	26.7	50.6
Ceriops	11	0.15	12.5	30.3
Lumnitzera	3	0.13	20.0	27.5

Source: Data from permanent plots, 1980–1982 (Saenger, unpubl. data).

ing post-establishment mortality are physical such as waveborne objects, biological such as crab damage, and physiological such as water stress, insufficient light and high soil salinities. The mortality rates from Queensland (table 5) show trends different from those reported by Rabinowitz (1978b), who found that mortality rate was inversely correlated with initial propagule weight.

3. Mangroves and Their Environment

If the broad ecological prerequisites outlined in the Introduction are fulfilled at a particular locality, a mangrove community is likely to develop. However, such communities are not uniform structurally, floristically or functionally when compared one with another, and even within any one community, considerable heterogeneity is apparent.

Differences in and among mangrove communities are due to a number of environmental factors — abiotic, biotic and fortuitous — which act differentially on individual mangrove species. These factors lead to three types of interactions: (1) those between the physico-chemical environment and the plants, (2) those among the plants themselves, and (3) those between plants and animals. The ultimate structure and function of a particular mangrove community is the outcome of all these interactions.

Physico-Chemical Environment-Plant Interactions

A number of physico-chemical factors, arising out of the broad mangrove environmental prerequisites, have been recognized as primary determinants of mangrove growth and development (figure 12). These can operate to modify one or more of the essential life processes within the mangrove community (table 6), and consequently determine whether a species is able to survive and grow at that particular locality.

In addition to these factors, there are others who are prerequisites for the normal growth of all plants such as gravity and the availability of carbon dioxide. However, as a specialized group of plants is under consideration, only those factors are discussed which either are specific to the mangrove environment, or to which mangroves show an interesting or unusual response.

Temperature

Temperature, because of its critical effect on both photosynthetic and respiratory processes, regulates a large number of internal energetic processes. Perhaps the most important of these are salt regulation and excretion, and root respiration.

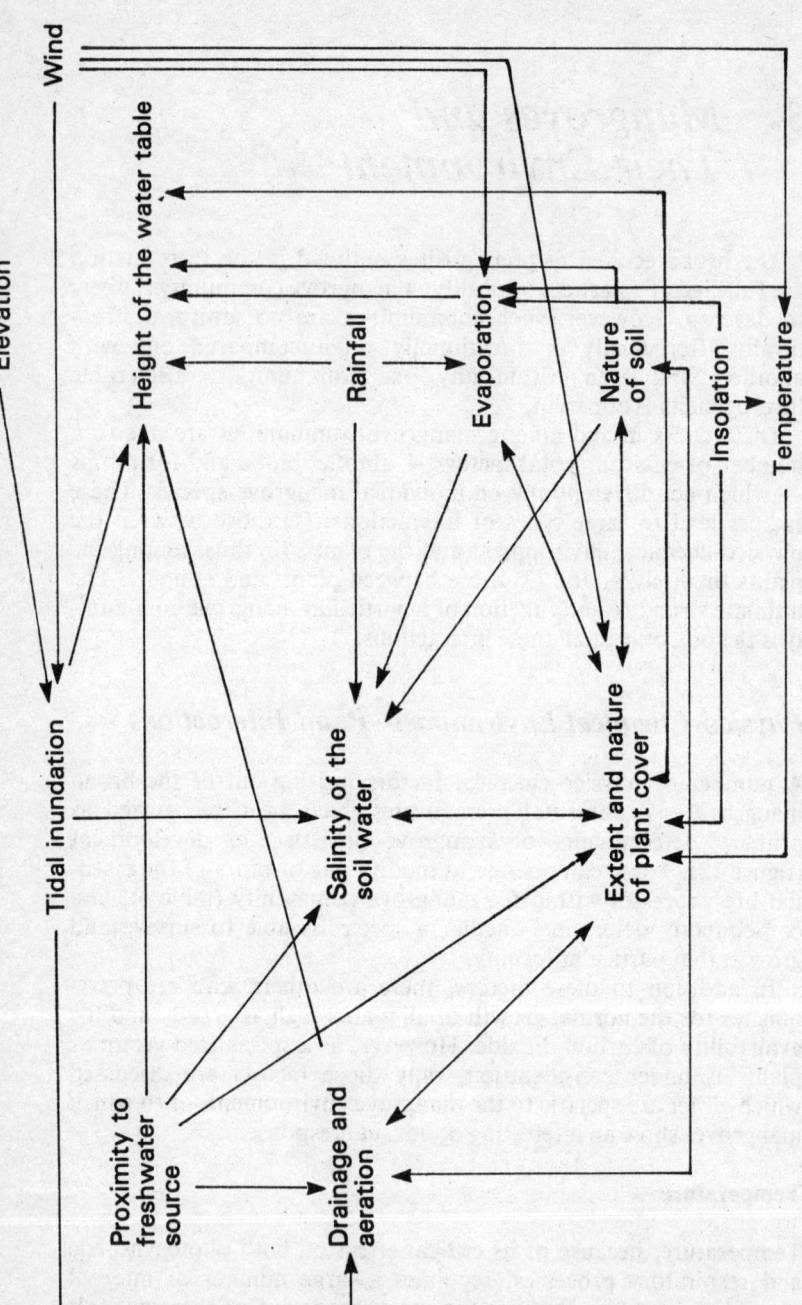

Figure 12 Interrelationships between major physico-chemical factors and the extent and nature of the mangrove plant cover.

Table 6 Relationship between physico-chemical factors and the essential life processes of mangroves

Indirect determinants	Direct determinants	Essential life process affected
Location	Evaporation	Photosynthesis
Rainfall	Wind	
	Temperature	
	Insolation	
	Height of watertable	
Tidal regime	Drainage/aeration	Respiration and growth
	Nature of soil	
	Proximity of freshwater source	
Elevation	Salinity of soil water	Water balance/transpiration
Coastal configuration		

Specht (1981a) recognized three thermal groups in the Australian vegetation, based both on species distribution and on the threshold temperature at which shoot growth is initiated. In the tropical–subtropical group, shoot growth is initiated when the mean air temperature rises above 25°C; the warm–temperate group shows shoot growth between 15 and 25°C; and the cool–temperate group shows shoot growth when the mean air temperature rises above 10°C.

Based on species distributions, mangroves belong predominantly to the tropical–subtropical group (refer to figure 4, chapter 1), although some species extend considerably southwards of the subtropics.

Very few shoot-growth data are available for mangroves; consequently, leaf-growth data are used instead. The monthly production of new leaves (Saenger and Moverley 1985) for nine species of mangroves from Gladstone (lat. 24°S) are shown in relation to air temperature in figure 13. The data for *Avicennia* show an approximately linear increase in leaf production with increasing temperature up to 20°C, followed by a decline at higher temperatures. Extrapolating the line suggests that leaf production ceases at 12°C and that *Avicennia* belongs to the cool–temperate group.

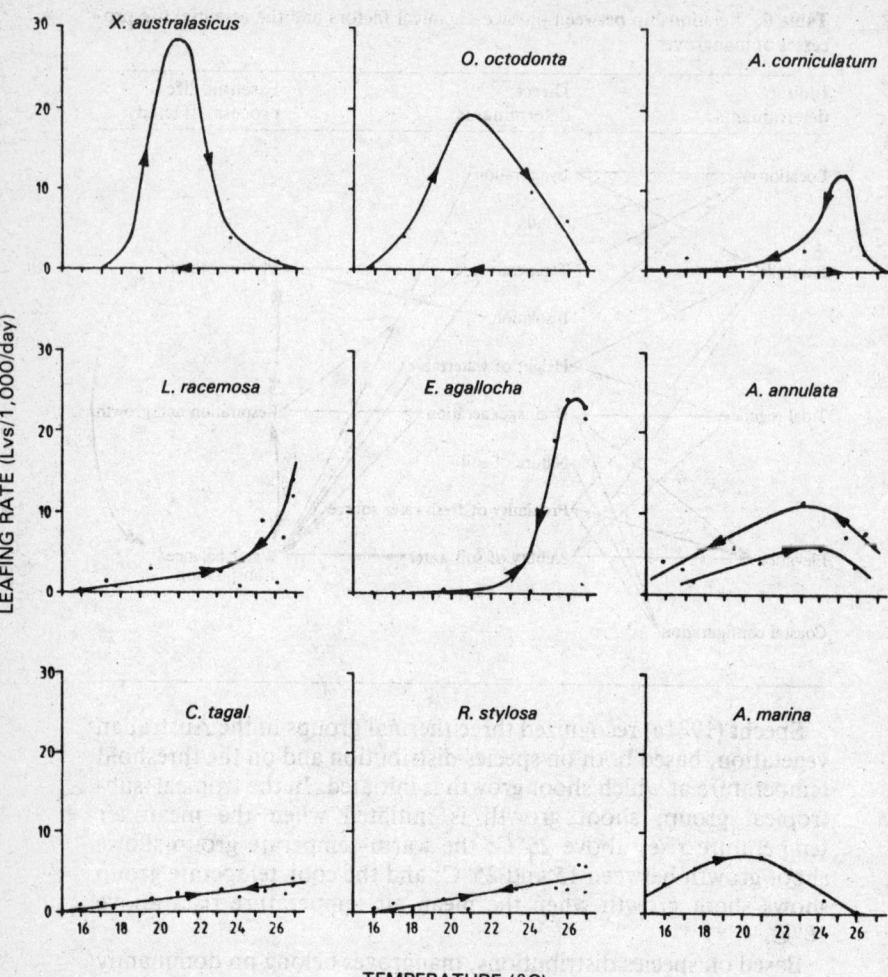

Figure 13 Rates of leaf formation in nine species of mangroves at different mean monthly temperatures at Gladstone, Queensland (after Saenger and Moverley 1985).

The data for *Rhizophora stylosa* suggest that leaf formation would cease below 16°C. Working with *R. mangle*, Miller (1975) demonstrated, by measuring leaf resistance, that the stomata of this species are only fully open above 18°C (figure 14), thereby restricting transpiration and photosynthetic gas exchange at low temperatures. Consequently, it appears that both *R. stylosa* and *R. mangle* fit well into the warm–temperate group.

Osbornia octodonta shows a slightly different pattern in that it

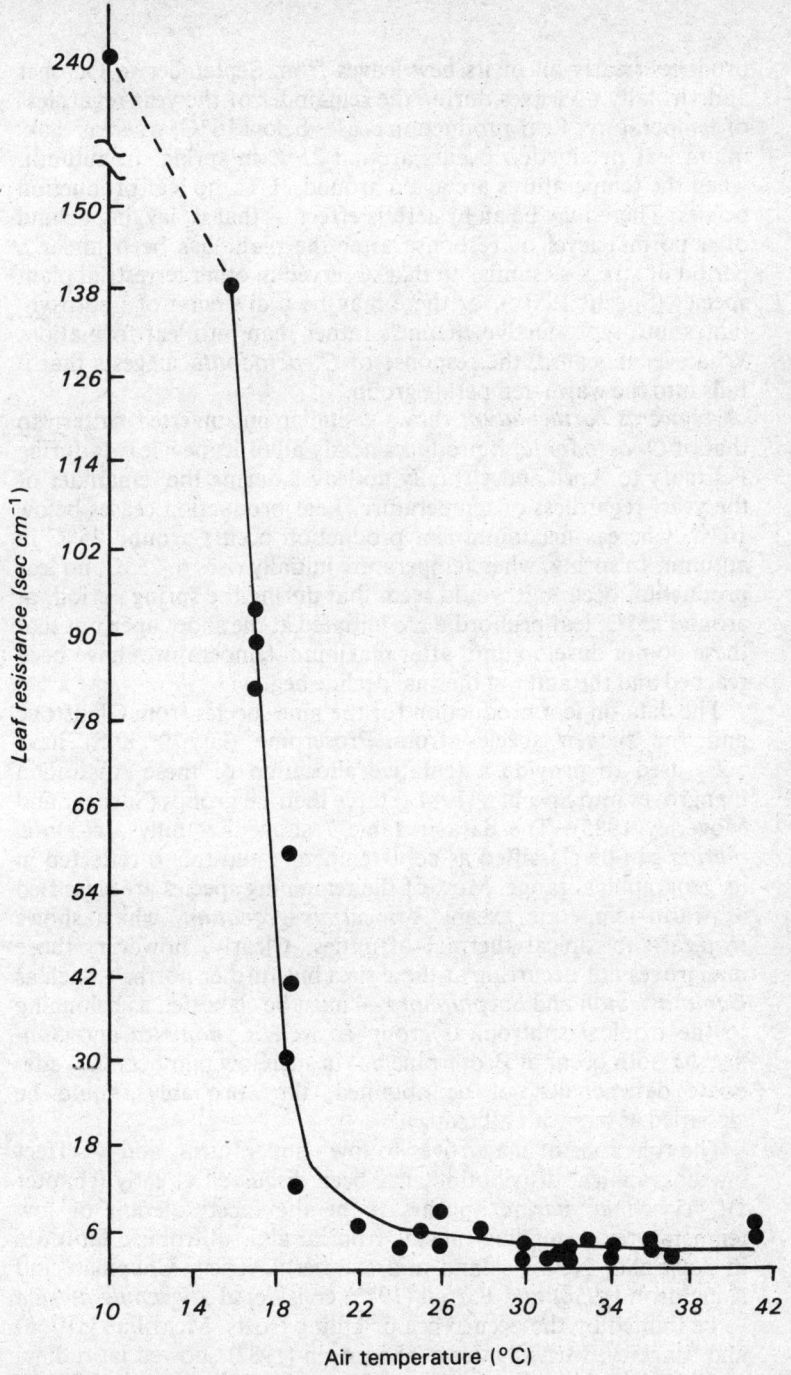

Figure 14 Stomatal opening (measured as decreasing leaf resistance) in *Rhizophora mangle* in response to various air temperatures (redrawn from Miller 1975).

produces nearly all of its new leaves from September to October and virtually no leaves during the remainder of the year regardless of temperature. Leaf production ceases below 16°C, whereas maximum leaf production occurs around 21°C in spring. In autumn, when the temperatures are again around 21°C, no leaf production occurs. There may be an hysteresis effect — that is, lagging behind of a normal level of response after the plant has been under a period of stress — similar to that observed in other terrestrial plant species (Specht 1981c), or there may be a diversion of photosynthates into reproductive channels rather than into leaf formation. Whatever its cause, the response of *O. octodonta* suggests that it falls into the warm-temperate group.

Aegiceras corniculatum shows a similar but inverted pattern to that of *O. octodonta*; it produces nearly all of its new leaves during February to April and virtually no leaves during the remainder of the year, regardless of temperature. Leaf production ceases below 16°C, whereas maximum leaf production occurs around 25°C in autumn. In spring, when temperature initially rises to 25°C, no leaf production occurs; it would seem that during the spring period, at around 25°C, leaf primordia are initiated at the shoot apex but that these do not develop until after maximum temperatures have been reached and the autumn thermal decline begins.

The data on leaf production for the nine species from Gladstone and for sixteen species from Proserpine (lat. 20° 30'S) have been used to provide a tentative allocation of these Australian mangroves into Specht's (1981a) three thermal groups (Saenger and Moverley 1985). The data in table 7 show that only *Avicennia marina* can be classified as cool-temperate and this is reflected in its geographical range. Most of the remaining species are classified as warm-temperate except *Xylocarpus granatum*, which shows tropical-subtropical thermal affinities. Clearly, however, those mangroves not occurring at these sites but further north — such as *Camptostemon* and *Scyphiphora* — must be classified as belonging to the tropical-subtropical group as well. *Cynometra* and *Sonneratia* both occur at Proserpine but in such low numbers that adequate data could not be obtained; they probably should be classified as tropical-subtropical.

The tolerance of mangroves to low temperatures, and its effect on geographical distribution, has been discussed already (chapter 1); *Avicennia marina* appears to be the most tolerant of low temperatures, extending outside tropical and subtropical latitudes in Australia, New Zealand and southern Africa. Chapman and Ronaldson (1958) and Farrell (1973) considered *Avicennia marina* to be limited by the occurrence of killing frosts. McMillan (1975a) and Markley, McMillan and Thompson (1982) showed latitudinal

Table 7 Classification of Australian mangroves into thermal groups (data from Port Curtis and Repulse Bay; where differences occur, the Repulse Bay data are given in parentheses)

Species	Temperature (°C) at which: Leaf production ceases	Leaf production is maximal
Cool-temperate		
Avicennia marina	12	20
Warm-temperate		
Aegialitis annulata	14(15)	23(27)
Ceriops tagal	15	27
Aegiceras corniculatum	16	25(27)
Rhizophora stylosa	16	28
Osbornia octodonta	16(17)	21
Xylocarpus australasicus	17(23)	21(26)
Lumnitzera racemosa	17(18)	28
Excoecaria agallocha	18(16)	26(28)
Bruguiera exaristata	17	27
Bruguiera parviflora	17	27
Bruguiera gymnorhiza	16	15
Heritiera littoralis	24	28
Acanthus ilicifolius	17	26
Hibiscus tiliaceus	22	27
Tropical-subtropical		
Xylocarpus granatum	26	>28

variation in response to chilling in a range of mangrove species from Australia and America.

The response of Australian mangroves to high temperature is not well known. However, mangroves growing in the discharge areas of coastal power stations such as Torrens Island, Adelaide, and the Gladstone and Howard power stations, central Queensland, show no visible effect, although some thermal effects may be masked by related conditions associated with circulation and chlorination of cooling water, and the discharge of airborne pollutants (Saenger, unpubl. data).

Canoy (1975) showed that *Rhizophora mangle* in Puerto Rico developed more stilt roots per unit area where it was subjected to a 5°C temperature increase from a cooling water discharge point, and that in temperature-stressed areas this species formed more, but significantly smaller, leaves (Canoy 1975; Lugo and Snedaker 1974). McMillan (1971) reported that young seedlings of *Avicennia germinans* were killed by water temperatures of 39°C to 40°C, although established seedlings and trees were not damaged.

Based on modelling studies of the bioclimate, leaf temperatures and primary production of *R. mangle* in southern Florida, Miller (1972) suggested that production is decreased by increasing air temperature and increasing humidity above optimum levels. Moore et al. (1972, 1973) found that the optimum temperature for photosynthesis by Florida mangroves was subject to some seasonal variation, but for all species the optimum temperatures were below 35°C with little or no photosynthesis occurring at 40°C. These findings in American mangroves emphasize the need for studies of community metabolism in Australian species under the influence of elevated water and air temperatures. Preliminary data for the Australian *A. marina, R. apiculata* and *R. stylosa* indicate that they, too, show a sharp decline in photosynthesis above 35°C (Clough, Andrews and Cowan 1982).

Smillie (1984) investigated the cold and heat tolerances of Australian mangroves by measuring the decline in induced chlorophyll flourescence following application of a cold or heat stress to the leaf tissue, a technique initially developed for crop plants (Smillie and Hetherington 1983). Susceptibility to cold injury was assessed by the decrease in the rate of induced rise of chlorophyll flourescence in dark-adapted leaves kept at 0°C; heat tolerance was determined by the decrease in chlorophyll flourescence after heating in water to 49°C for ten minutes.

Cold tolerance was measured in twenty-seven species of mangroves. Certain species such as *Bruguiera exaristata* and *Ceriops decandra* were very intolerant, accounting for their confinement to the tropics and the warmer subtropics. Overall, a wide range of cold tolerances was found, but within genera the cold tolerances of species were correlated with their latitudinal distribution. In other words, the further south the species occurred, the greater its cold tolerance (Smillie 1984).

For *Avicennia marina*, the most southerly extending mangrove, there was considerable cold adaptation in the southern populations compared with the more northerly ones, a finding consistent with that described by McMillan (1975) from seedling growth studies.

Twenty mangrove species from tropical areas were assessed for their heat tolerance. It was concluded (Smillie 1984) that all the species showed a very high degree of heat tolerance compared with other plants tested by the same technique, and that mangroves appear to be at the extreme high end of the heat tolerance range for non-arid tropical plants. The most heat-sensitive species were the mangrove fern (*Acrostichum speciosum*), *Acanthus ilicifolius* and *Rhizophora stylosa*. Both *Acrostichum* and *Acanthus* grow in sun-flecked shade and are subject to short periods only of solar heating. *Rhizophora stylosa*, on the other hand, most commonly grows in full sunlight, a situation difficult to reconcile with its apparent heat sensitivity.

In contrast to the finding of cold adaptation, there was no evidence for any latitudinal differentiation of heat tolerance in

Avicennia marina and *Aegiceras corniculatum*, the only two species tested (Smillie 1984).

Insolation

The physiological and morphological adaptations of mangroves to high levels of incident solar radiation have been discussed in chapter 2, and it remains now to examine the ecological role of light.

Attiwill and Clough (1980) examined the relationship between light and photosynthesis in branches of *Avicennia marina*; for branches both within and at the top of the canopy, they found that photosynthesis became light saturated at a total short-wave radiation between 200 and 400 Wm^{-2} (figure 15). The photosynthetic efficiency of this species was low compared with non-mangrove plants. Moreover, at full midday light intensities (approximately 1,000 Wm^{-2}), this quantum efficiency was greatly reduced, and Attiwill and Clough (1980) suggested that the photosynthetic mechanism of *Avicennia marina* is relatively inefficient and best adapted to shade conditions.

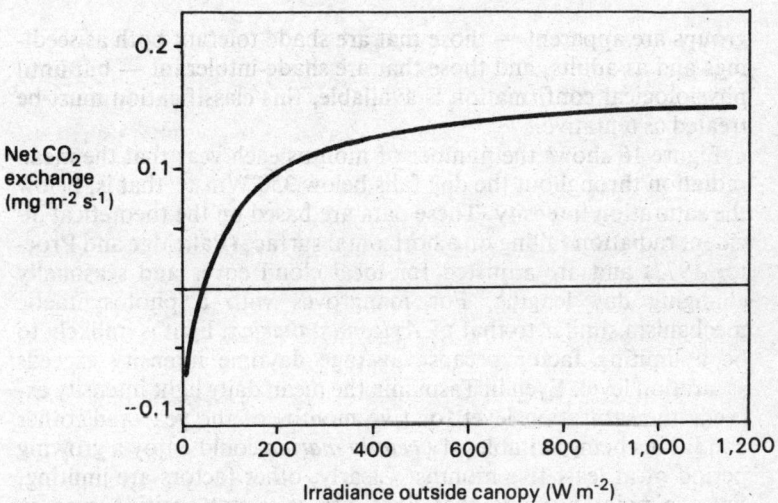

Figure 15 Light-saturation curve for *Avicennia marina* at Westernport Bay, Victoria (redrawn from Attiwill and Clough 1980).

Unfortunately, comparative data for other Australian mangroves are not available. However, based on observations in the field and from culture experiments, various authors have noted the light and shade requirements of different species (table 8). Two

Table 8 Shade tolerance of Australian mangroves

Genus	Shade tolerant	Shade intolerant
Acanthus		Macnae 1966, 1968
Acrostichum		Macnae 1966
Aegialitis		Macnae 1966, 1968
Aegiceras	Clarke and Hannon 1971; Thom, Wright and Coleman 1975	
Avicennia	Clarke and Hannon 1971; Attiwill and Clough 1980	Macnae 1963, 1966, 1968; Thom, Wright and Coleman 1975
Bruguiera	Macnae 1966; Macnae and Kalk 1962; Watson 1928	
Ceriops	Macnae and Kalk 1962; Thom, Wright and Coleman 1975	Macnae 1966, 1968
Excoecaria	Saenger 1982	
Lumnitzera		Macnae 1966, 1968
Osbornia	Saenger 1982	
Rhizophora	Macnae 1966	Macnae 1968
Sonneratia		Macnae 1968
Scyphiphora		Macnae 1966
Xylocarpus	Saenger 1982	

groups are apparent — those that are shade-tolerant both as seedlings and as adults, and those that are shade-intolerant — but until physiological confirmation is available, this classification must be treated as tentative.

Figure 16 shows the number of months each year that the mean radiation throughout the day falls below 350 Wm^{-2}, that is, below the saturation intensity. These data are based on the theoretical incident radiation falling on a horizontal surface (Paltridge and Proctor 1977) and are adjusted for local cloud cover and seasonally changing day lengths. For mangroves with a photosynthetic mechanism similar to that of *Avicennia marina*, light is unlikely to be a limiting factor because average daytime intensity exceeds saturation level. Even in Tasmania the mean daily light intensity exceeds the saturation level for five months of the year and, other conditions being suitable, *Avicennia marina* could enjoy a growing period of at least five months. Clearly, other factors are limiting, but the point is made that light *per se* is not limiting even in Tasmania. At increasingly lower latitudes, light is present above saturation levels for more and more of the time and has led to various morphological adaptations in the mangroves. Even so, Attiwill and Clough (1980) showed that at Westernport Bay in Victoria a decrease in photosynthesis of this species occurred on days of sustained high radiation levels, and they ascribed this to a photochemical inhibition of the photosynthetic mechanism.

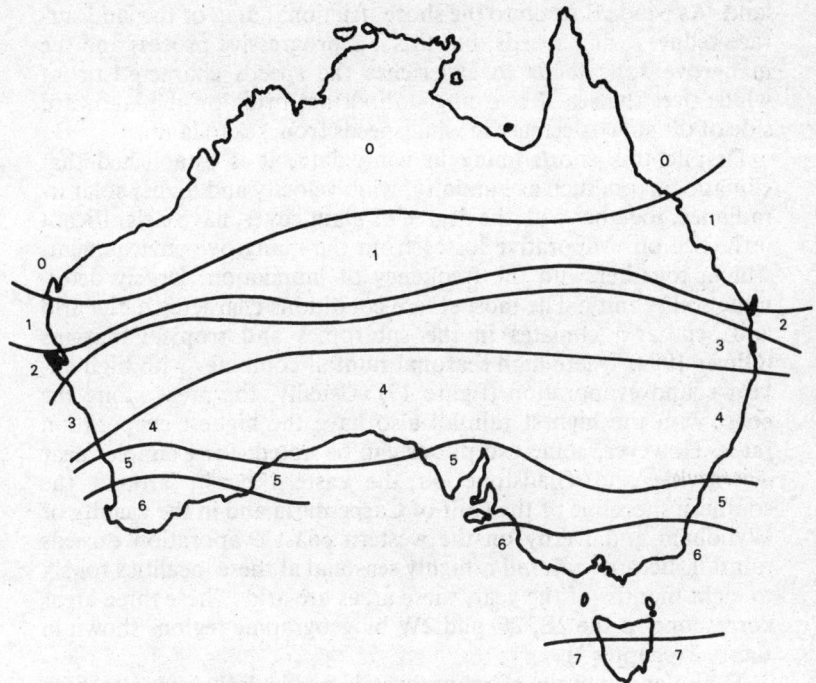

Figure 16 Number of months per year during which the mean radiation level between sunrise and sunset falls below 350 watts m^{-2}.

Wind and Evaporation

Wind affects mangroves in several separate ways. Coastal water drift and tidal currents are modified by wind direction and speed. Wave action is accentuated, especially at high tides, by stormy conditions. Both waves and water movement affect sediment transport. Wind has a major part to play in causing evaporation and in increasing salinity. In addition, it can cause physical damage to canopies and desiccate foliage. On the positive side, it facilitates pollination and the dispersal of propagules in a number of species.

There are, however, three aspects of wind that impinge directly on the physiological performance of mangroves: its evaporative capacity, its effect on sea-level and its role in regulating evapotranspiration from leaves.

Oliver (1982) pointed out that most of the standard wind data do not relate to the mangrove environment because wind recording is usually done close to the ground and some distance from the coast. The sea surface causes less mechanical and thermal obstruction to air flow and therefore wind speeds are greater over water than over land. As winds flow on to the shore, frictional drag of the land surface reduces wind speeds, but this is a progressive process and the mangrove zone tends to experience the speeds characteristic of winds over the sea. Recording stations are often on the landward side of the sharp decrease in wind speeds from sea to land.

Despite this shortcoming in wind data, it is established that climatic factors such as humidity, wind velocity and higher solar irradiance, together with the degree of plant cover, have a significant influence on evaporative losses from the mangrove environment, which, together with the frequency of inundation, largely determine soil salinity. The most severe conditions characterize the arid and semi-arid climates in the subtropics and tropical margins (Oliver 1982) where high seasonal rainfall combines with high all-year-round evaporation (figure 17). Usually, the areas along the coast with the highest rainfall also have the highest evaporation rates. However, some exceptions can be noted; for example, near Townsville and Gladstone on the eastern coast, around the southern shoreline of the Gulf of Carpentaria and in the vicinity of Wyndham and Derby on the western coast evaporation exceeds rainfall. Because rainfall is highly seasonal at these localities for six to eight months of the year, these areas are arid. These three areas correspond to the 2E, 2G and 2W biogeographic regions shown in figure 2 (chapter 1).

The inland margin of mangroves is particularly prone to high evaporative losses and drying out of the substrate. Often, an edge effect is noticeable where mangroves abut salt flats. The evaporative build-up of soil salinity results in mangrove dieback and gradual expansion of the salt flats. Similarly, where breaks occur in the canopy, especially in the mangroves towards the landward margin, evaporation may lead to increased soil salinities which, in turn, may prevent the regeneration of mangroves (Spenceley 1976). In the more humid tropics, on the other hand, rain wetting of leaf surfaces, cloud cover and high humidities reduce evaporative losses and the tendency towards salt flat formation is not so great.

Wind affects evapotranspiration from mangroves by the same mechanism as in other plants. However, because mangroves are at the land–sea interface, they tend to be more consistently exposed to windy conditions, and wind probably assumes a greater importance in relation to evapotranspiration in mangroves than in other plant communities.

As transpiration occurs, there is a tendency for a moist layer of air to form next to the leaf surface. This layer, termed the boundary layer, is variable in thickness, but in those mangroves with epidermal hairs or scales it is thicker than around those leaves with untextured surfaces. Wind conditions also affect the thickness of

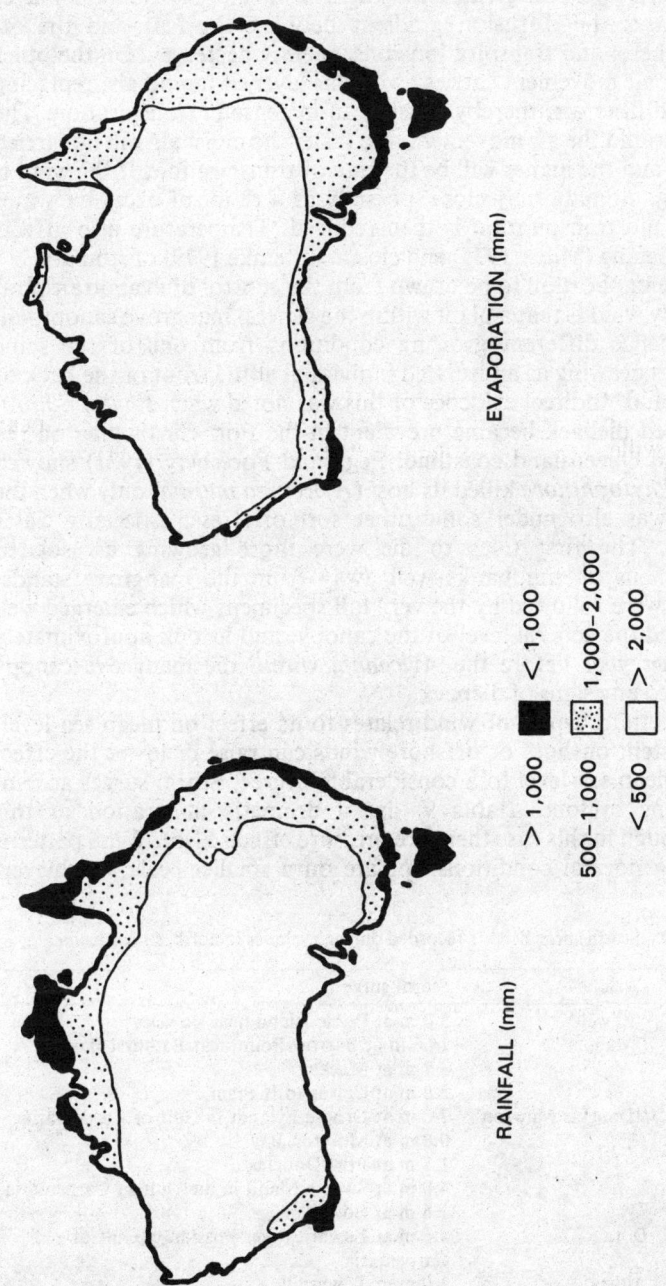

Figure 17 Annual rainfall and evaporation distribution in Australia.

this layer, with the greatest thickness in still air. The boundary layer decreases the diffusion gradient between the leaf and the atmosphere, and transpiration consequently decreases. On the other hand, air movement carries away this layer of humid air, replacing it with drier air, thereby causing an increase in transpiration. The more rapid the air movement, the faster the moist air will be carried away and the higher will be the rate of transpiration. If the wind is strong, stomata may close, possibly as a result of excessive water loss, and transpiration is then reduced. Temperature also affects the opening (Miller 1975) and closure (Steinke 1979) of stomata.

The implication to be drawn from the control of evapotranspiration by wind is that a plant within the general mangrove canopy will experience different growing conditions from one of the same species growing as an isolated individual at the front or the back of the stand. Indirect evidence of this was noted when *Phytophthora*-induced dieback became prevalent in the Port Curtis area on the central Queensland coastline. Pegg and Foresberg (1981) showed that *Phytophthora* killed its host (*Avicennia marina*) only when the host was also under some other sort of stress, especially water stress. The first trees to die were those growing as isolated specimens on mudbanks well away from the mangrove stands; these were followed by the very tall specimens which emerged well beyond the general level of the canopy, and it took approximately another year before the *Avicennia* within the mangrove canopy showed any signs of dieback.

The third aspect of wind relates to its effect on mean sea-level. Persistent onshore or offshore winds can raise or lower the effective mean sea-level to a considerable degree. Storm surges accompanying cyclones (table 9) are a dramatic illustration of this (although in this case there are pressure effects also). Wind patterns during normal conditions operate on a smaller scale, but never-

Table 9 Storm surge heights recorded during cyclones in northern Australia

Date	Cyclone	Storm surge
1884	"Bowen"	3.0 m at Poole Island near Bowen
1899	"Mahina"	14.6 m at Barrow Point near Bathurst Bay
1918		3.7 m at Mackay
1918		3.0 m at Cairns to Ingham
1923	"Douglas Mawson"	7.0 m at Groote Eylandt in Gulf of Carpentaria
1931		0.8 m at Moreton Bay
1934		1.8 m at Port Douglas
1948		4.0 m at Sweers Island in the Gulf of Carpentaria
1958		1.6 m at Bowen
1964	Dora	4.6 m at Edward River Mission in Gulf of Carpentaria
1971	Althea	3.0 m at Townsville
1976	Ted	2.0 m at Karumba in the Gulf of Carpentaria

theless have physiological implications for the mangrove community.

Munro (1973) described seasonal changes of mean sea-level of up to 1 metre in the southeastern part of the Gulf of Carpentaria owing to seasonally changing wind patterns (south to southeast in winter and mainly north to northeast in summer). Similar but less pronounced sea-level changes were detected from yearly tide recordings at Port Curtis on the central Queensland coastline (figure 18) where winter winds from the south to southwest are offshore and summer–autumn winds from the north to northeast are onshore.

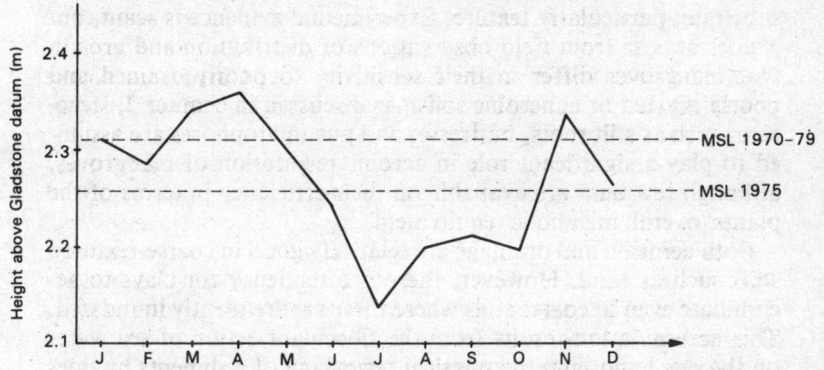

Figure 18 Seasonal mean sea-level changes at Gladstone, Queensland, during 1975. Note that the long-term mean sea-level obtained from continuous tide records during 1970 to 1979 is also shown.

Even changes of 30 cm on a relatively flat coastline represent a significant increase in depth and frequency of tidal inundation and in terms of the area subject to tidal effects. The season during which such sea-level changes occur seems to be ecologically important. For example, in the Gulf of Carpentaria, mean sea-level is raised during the summer months when the river discharges are at their maximum. Consequently, flooding is frequent and freshwater inundation of the mangroves and salt flats aids in the leaching of salt from these communities. At Port Curtis, however, sea-level is raised during the dry autumn season and, rather than remove salt, may in fact contribute salt to those communities at or near high-water spring levels, particularly when winter evaporation is high. Occasional wetting of the salt flats during the dry autumn months at Port Curtis allows temporary development of filamentous algal mats which contribute to overall productivity. In contrast, the salt flats of the southeastern part of the Gulf of Carpentaria in the dry season are dried to the point of cracking, and are covered in salt crystals and are virtually lifeless (Saenger and Hopkins 1975).

Although it is too early to state categorically that seasonal changes in sea-level are important, the examples discussed above suggest that this is an area worthy of detailed investigation.

Drainage/Aeration

Soil aeration is important in mangrove environments in supplying oxygen for respiration. Aeration is directly related to soil drainage and is therefore highly variable. It depends upon elevation, steepness of the topography and the physical characteristics of the substrate, particularly texture. Experimental evidence is scant, but it does appear from field observations of distribution and growth that mangroves differ in their sensitivity to poorly drained and poorly aerated or anaerobic soils. As discussed in chapter 2, structures such as stilt roots, buttresses and pneumatophores are assumed to play a significant role in aerobic respiration of mangroves, although few data are available on their efficiency in terms of the plants' overall metabolic requirements.

Both aeration and drainage are relatively good in coarse-textured soils such as sand. However, there is a tendency for clays to accumulate even in coarse soils where these are frequently inundated. This accumulation results from the flocculant action of sea water on the one hand, and the physical reworking of sediments by tides and waves on the other. In consequence, drainage in the lower to middle regions of the intertidal zone may be reduced except in areas of turbulence owing to wave action or where there are high current velocities.

In those areas where the parent material of the catchment is predominantly argillaceous — that is, leading to the formation of clays during erosion — the entire estuarine system may consist of finely textured soils. Port Curtis on the Queensland coastline is an example (Jardine 1925; Conaghan 1966; Saenger and Robson 1977), with the major area of coastline consisting of clayey deposits, predominantly quartzite and albite. Even there, however, subsurface soil horizons of greater coarseness can sometimes provide good drainage and replenishment of oxygen through subsurface drainage. Finely textured sedimentary soils can act as plugs preventing drainage from higher elevations. Drainage problems at such sites may last for extended periods and impose stresses on the plants growing there.

Clays and sands are not the only soil types on which mangroves will grow. For example, Teas (1979) described mangroves growing on karst limestone formations with no sediment accumulation, and

Macnae (1968), Thom (1975) and Stoddart (1980) reported them as growing on coral rubble on islands and cays in the northern Barrier Reef Province. However, in terms of aeration and drainage, sandy and clayey soils usually can be considered as the extremes in which mangroves will grow. The properties of these two kinds of soils are compared in table 10.

Table 10 Comparison of various soil characteristics between sand and clay

	Sand	Clay
Infiltration	High	Low
Porosity	High	Low
Permeability	High	Low
Seepage velocity	High	Low
Water-holding capacity	Low	High
Salt retention	Low	High
Leaching	Rapid	Slow
Capillarity	Low	High

Tidal inundation combined with specific drainage and aeration properties generally lead to mangrove soils characteristically having a high water content, low oxygen content, and often high levels of salinity, free hydrogen sulphide, Eh values between −100 and +400 mv and pH values ranging from 4.9 to 7.2. In addition, these soils are often semi-fluid and poorly consolidated.

The high water content (expressed on a wet-weight basis) of mangrove soils from Port Curtis ranges from 38 to 46 per cent in clay to 20 to 37 per cent in sand (figure 19). Clarke and Hannon (1967) reported higher values from Sydney mangrove soils, but their values were expressed on a dry-weight basis. When their values are converted to a wet-weight basis, the Sydney values for sandy mangrove soils range from 22.2 per cent to 58.9 per cent. These values are still significantly higher than those from Port Curtis, but the Sydney soils have a high content of roots and other organic material. Hesse (1961), working on West African mangroves, reported a moisture content (wet-weight basis) of 37.5 per cent in *Rhizophora* soils with 11.9 per cent fibrous organic matter, and only 33.9 per cent in soils with 5.5 per cent fibrous organic matter. Naidoo (1980), using South African mangrove soils, experimentally determined the moisture content at saturation using the methods of Bower and Wilcox (1965); saturation values ranged from 42.5 per cent in clay to 25.8 per cent in sand. In addition, a direct relationship was found between saturation water content and the organic content of the soil; the latter ranged from 2.5 to 6.3 per cent.

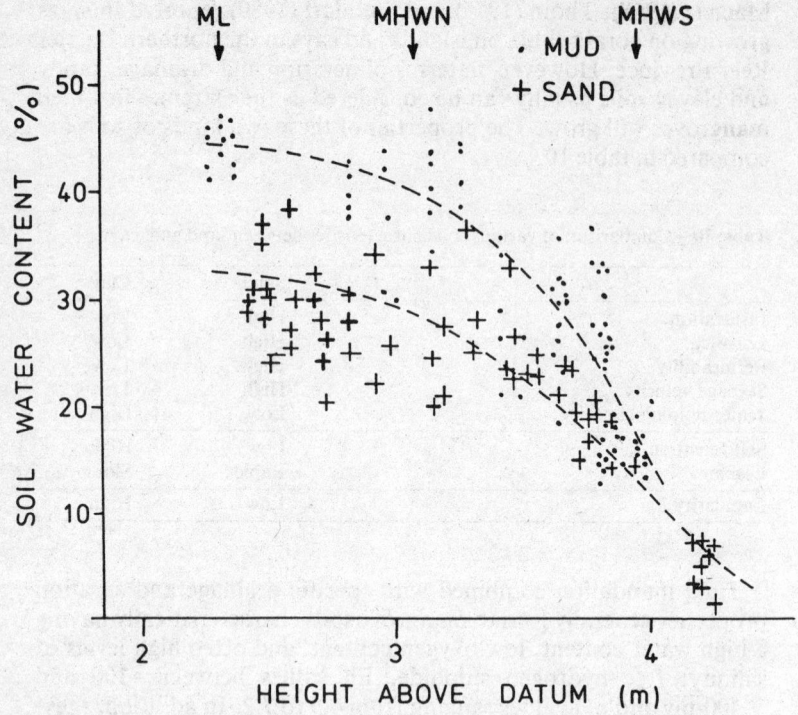

Figure 19 Soil water content (expressed as wet weight) from three study areas at Gladstone, Queensland, 1975-80.

Because many of the mangrove soils are nearly saturated most of the time, spaces in the soils are water-filled and consequently the penetration of oxygen, an essential requirement for active root growth and development, is reduced. Low soil oxygen levels generally lead to highly reducing conditions in the soil, a decrease in the soil pH, high soil Eh values and the formation of free sulphides or pyrites (FeS_2) from the anaerobic breakdown of organic matter.

Low rates of water infiltration into many mangrove soils can compound the already adverse soil conditions described above. Clarke and Hannon (1967) measured the infiltration rates into mangrove and saltmarsh soils in Sydney (table 11) and, with the exception of rates measured near crab holes, infiltration rates were low despite the sandy texture of these soils.

Table 11 Soil infiltration rates of mangrove soils in Sydney, NSW

Habitat	Replicate no.	Infiltration rate (cm/hour)	Mean
Mangrove	1	0	
	2	0.33*	
	3	0.08	
	4	0.02	
	5	0.36*	
	6	0.04	0.04
Salt marsh	1	0.30	
	2	0.67	
	3	0.25	
	4	0.02	
	5	0.42	
	6	0.06	0.29

* Near crab holes; values not included in means.
Source: Clarke and Hannon (1967).

With impeded drainage and little or no internal water movement, the interstitial water bathing the roots could quickly become exhausted of major plant nutrients. Low infiltration and the generally mediocre cation exchange capacity of the soils mean that little replacement of such nutrients occurs around the roots. Furthermore, in the immediate vicinity of the roots, the development of anaerobic conditions can lead to pH changes which, in turn, can change the availability of nutrients. For example, phosphorus becomes unavailable at low pH values.

The major effect of waterlogged soils, however, appears to be the induction of a root oxygen stress which, in turn, leads to decreased permeability of the root membranes, thereby inducing water stress on the plant. Two strategies have been adopted in terrestrial plants and it is likely that these are also of relevance to mangroves (Hook and Scholtens 1978).

In one strategy, auxin accumulates in the stem of the waterlogged plant, and the high levels of this growth hormone result in the formation of adventitious roots. For example, Snedaker, Jimenez and Brown (1981) reported intense aerial root formation in *Avicennia germinans* growing in areas subject to oil spills and prolonged waterlogging. It seems that by this strategy partial restoration of root function occurs, and water stress in the plant is reduced.

The second strategy also brings about a reduction in plant water stress by lowering the water loss through transpiration. Decreased cytokinin export from the waterlogged root system (Itai, Richmond and Vaadia 1968) and the accumulation of abscissic acid in the leaves lead to stomatal closure, rapid leaf senescence and shedding and a retardation of shoot development and elongation. All of these responses reduce the transpirational water loss. Growth rates of the shoots may also be reduced by the altered giberellin balance or the accumulation of ethylene in the plant; this may also contribute to a reduction of water losses through transpiration (Crawford 1978).

The root adaptations described in chapter 2 facilitate survival in the generally adverse soil environment of mangroves. Species vary, however, in their tolerance to soil conditions. For example, Hesse (1961) and Naidoo (1980) found that *Avicennia, Rhizophora* and *Bruguiera* grow on soils with different characteristics (table 12). Spenceley (1983) examined soil characteristics in various mangrove and saltmarsh zones at two Queensland localities; rather inconsistent patterns were found. Adequate characterization of Australian species of mangroves as far as tolerances to soil conditions are concerned awaits more complete data.

Table 12 Comparison of soils supporting stands of *Rhizophora*, *Avicennia* and *Bruguiera*

Soil parameter	Relative soil concentrations				
pH	Rhizophora	>	Avicennia	>	Bruguiera
S⁻	Rhizophora	>	Avicennia	=	Bruguiera
N	Rhizophora	>	Avicennia		?
P	Rhizophora	>	Avicennia		
Organic carbon	Rhizophora	>	Avicennia	>	Bruguiera
Cation exchange capacity	?		Avicennia	>	Bruguiera
Exchangeable bases	?		Avicennia	>	Bruguiera
Exchangeable acidity	?		Avicennia	<	Bruguiera
Clay content	?		Avicennia	<	Bruguiera
Al^{+++}	?		Avicennia	<	Bruguiera

Source: After Hesse (1961) and Naidoo (1980).

The height of *Avicennia marina* appears to depend on drainage properties of the soil, with the tallest trees growing on well-drained banks close to streams (Chapman and Ronaldson 1958; Macnae 1966). Macnae (1966) maintained that *Ceriops tagal* in Australia is found only on well-drained soils, and he suggested that its virtual absence from areas of high rainfall may be as much due to drainage irregularities as to rainfall. Measurements of soil water content among *Ceriops tagal* stands along the Queensland coastline (Saenger and Robson 1977) do not support his suggestion but it is possible that considerable geographic variation occurs.

Macnae (1966) cited two examples of extreme variability in response to soil drainage: (1) *Rhizophora stylosa* grows on well-drained soils in Malaysia (Ding Hou 1958), whereas in Australia it grows on a range of substrates, but with the tallest trees occurring on soft, waterlogged muds; (2) in Australia, *Lumnitzera racemosa* is recorded from well-drained sandy soils on the landward fringe (Macnae 1966), but in southern Africa, however, Macnae (1966: 96) records it "as a true mangrove extending down to almost high water neaps".

Field data on the soil water content in which various species grow were collected from Proserpine from four sites over one-and-a-half years; the upper and lower limits are presented in table 13. Three groups, based on the water content of the soil on which they grow, can be recognized. *Osbornia octodonta* and *Bruguiera parviflora* grow on soils that have low water contents, either because of good drainage (*B. parviflora*) or because of their location on the landward margins of the mangroves where tidal inundation is infrequent (*O. octodonta*). The second group contains those species growing on soils with an intermediate water content and includes *Bruguiera gymnorhiza*, *B. exaristata*, *Clerodendron inerme* and the mangrove fern *Acrostichum*. The third group, containing eleven species, grows in soils that have high water contents, either because of frequent tidal inundation (such as *R. stylosa* and *A. marina*), or because high freshwater run-on occurs (such as *Heritiera littoralis* and *Cynometra iripa*).

Table 13 Soil water content at which various species occurred at Proserpine, Queensland, October 1980–May 1982 (water content is expressed as percentage of wet weight of soil)

Species	% soil water content
Osbornia octodonta	9–17
Bruguiera parviflora	17–21
Bruguiera gymnorhiza	21–27
Bruguiera exaristata	22–29
Clerodendron inerme	22–28
Acrostichum speciosum	22–28
Heritiera littoralis	26–28
Lumnitzera racemosa	27–30
Xylocarpus australasicus	28–33
Xylocarpus granatum	28–30
Cynometra iripa	28–38
Rhizophora stylosa	29–37
Ceriops tagal	29–32
Acanthus ilicifolius	30–32
Excoecaria agallocha	31–32
Avicennia marina	31–34
Aegiceras corniculatum	32–38

Although the above grouping is tentative, it provides some indication of the soil water regime under which the various species grow. It provides no information on the tolerances of these species under extreme conditions.

Salinity of the Soil Water

Salinity of the interstitial soil water has long been recognized as an important factor regulating growth, height, survival and zonation of mangroves (Bowman 1917; Macnae and Kalk 1962; Mogg 1963; Macnae 1968; Cintron et al. 1978; Teas 1979; Semeniuk 1983). The physiological importance of salinity has been investigated using culture experiments and these have provided many useful results (Connors 1969; Sidhu 1975a). However, in the field, the response to salinity is more variable, and mangroves have been found at salinity levels that exceed those suggested by laboratory experimentation. For example, Macnae (1968) showed that *Avicennia marina* and *Lumnitzera racemosa* can tolerate salinities of up to 90‰ in the soil whereas *Rhizophora mangle* is probably limited by soil salinities above 65‰ (Cintron et al. 1978; Teas 1979). *Avicennia germinans* was reported to become dwarfed and gnarled in Florida when soil salinities approached 60–80‰ . In culture, *Avicennia* and *Aegiceras* showed maximum growth at 25 per cent sea water (figures 20, 21).

Soil salinity is regulated by a number of factors, including tidal inundation, soil type and topography, depth of impervious subsoils, amount and seasonality of rainfall, freshwater discharge of rivers, run-on from adjacent terrestrial areas, run-off and evaporation. However, in tidally inundated situations, evaporative losses and the frequency of flooding are the major factors determining soil salinity (Oliver 1982). Other climatic factors, such as humidity, wind velocity and high solar radiation together with the extent of plant cover, have a significant influence on evaporative losses from the mangrove community.

Particularly where the clay content is high, soils have a high resistance to internal salt and water movement (Clarke and Hannon 1967; Blackmore 1976). As a result, tidal inundation, rainfall and evaporation principally affect the soil surface, although with time an equilibrium with the soil at considerable depths will be reached. However, as an approximation, an initial understanding of the regulation of soil salinities can be made considering only the surface processes. At any particular point in the intertidal gradient the soil salinity can be directly related to:

- salinity of the tidal water
- time interval between inundations
- rainfall
- evaporation rate
- retention properties of soil
- run-on minus run-off.

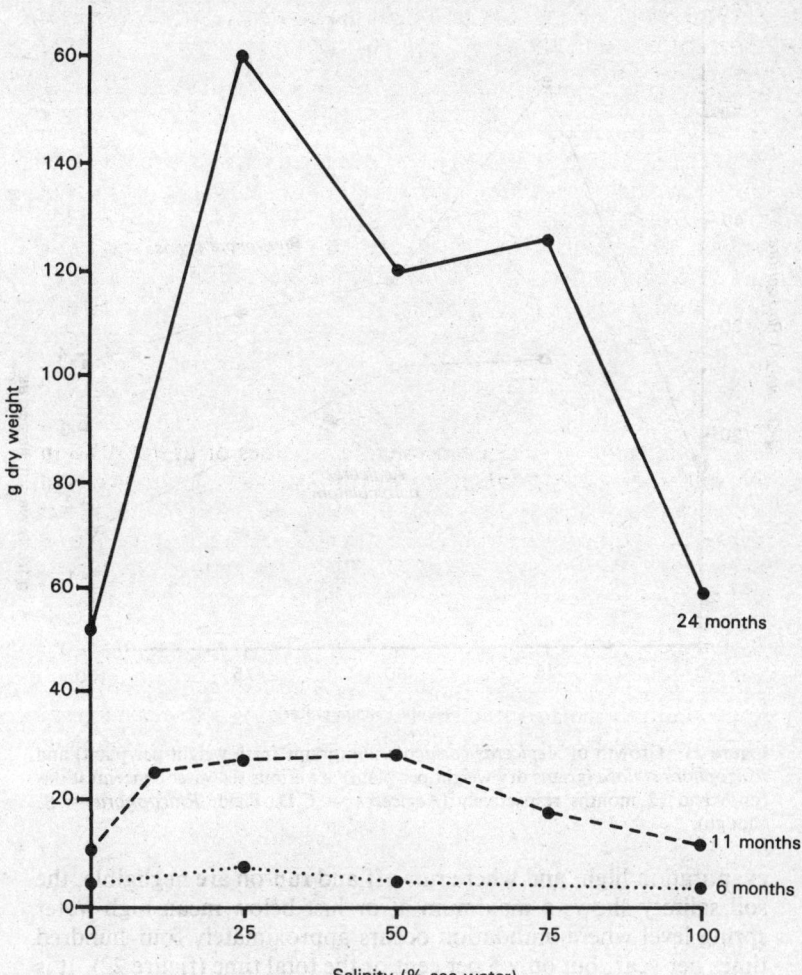

Figure 20 Growth (measured as grams of dry weight per plant) of *Avicennia marina* at various seawater concentrations over varying periods (24 months – B. Clough; 11 months – Downton 1982; 6 months – C.D. Field).

At any one locality, several of these processes are more or less constant, including evaporation, rainfall and run-on/run-off. Salinity of the tidal water and soil properties also can be relatively constant at a particular locality, and in this instance the soil salinity along the intertidal gradient is determined more or less by the time interval between inundations. For example, on the central Queensland coastline, where clayey soils predominate, where rainfall is low and

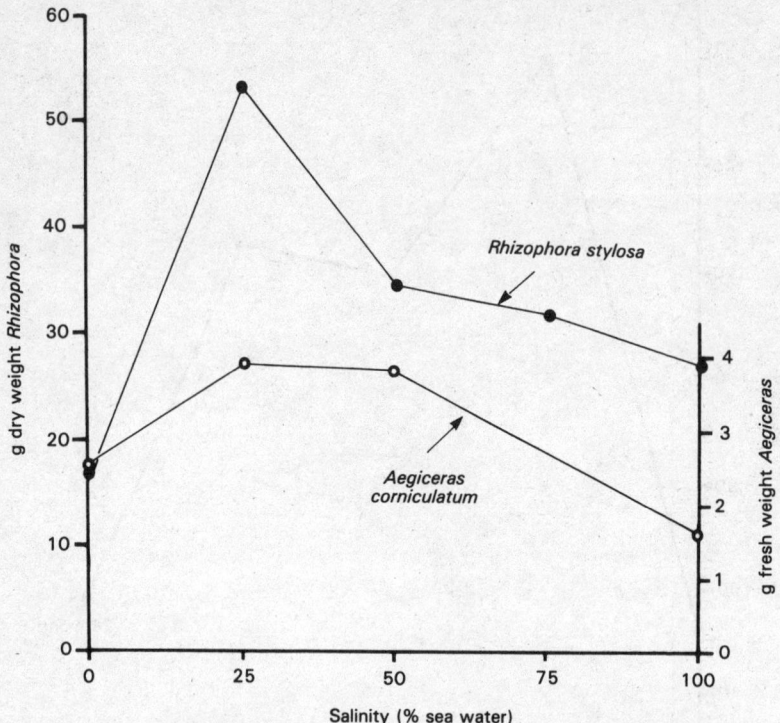

Figure 21 Growth of *Aegiceras corniculatum* (grams fresh weight per plant) and *Rhizophora stylosa* (grams dry weight per plant) at various seawater concentrations for 6 and 12 months respectively (*Aegiceras* – C.D. Field; *Rhizophora* – B. Clough).

evaporation high, and where run-off and run-on are negligible, the soil salinity shows a maximum at or just below mean high-water spring level where inundation occurs approximately four hundred times per year, but only 5 per cent of the total time (figure 22). It is clear, however, that at another locality with lower rainfall and higher evaporation, the soil salinity maximum may be broader and located lower down in the intertidal gradient. On the other hand, where rainfall greatly exceeds evaporation, no soil salinity build-up occurs, and the soil salinity will simply show an approximately linear decrease from seawards to landwards.

Baltzer and Lafond (1971) listed two extremes of coastal environment with respect to interstitial soil salinity: (1) areas of high rainfall that are permanently wet, where the soil salinity decreases progressively from the sea towards the inland and the vegetation ranges from mangroves to inland vegetation without discontinuity; and (2) areas of low rainfall where dry and wet seasons alternate,

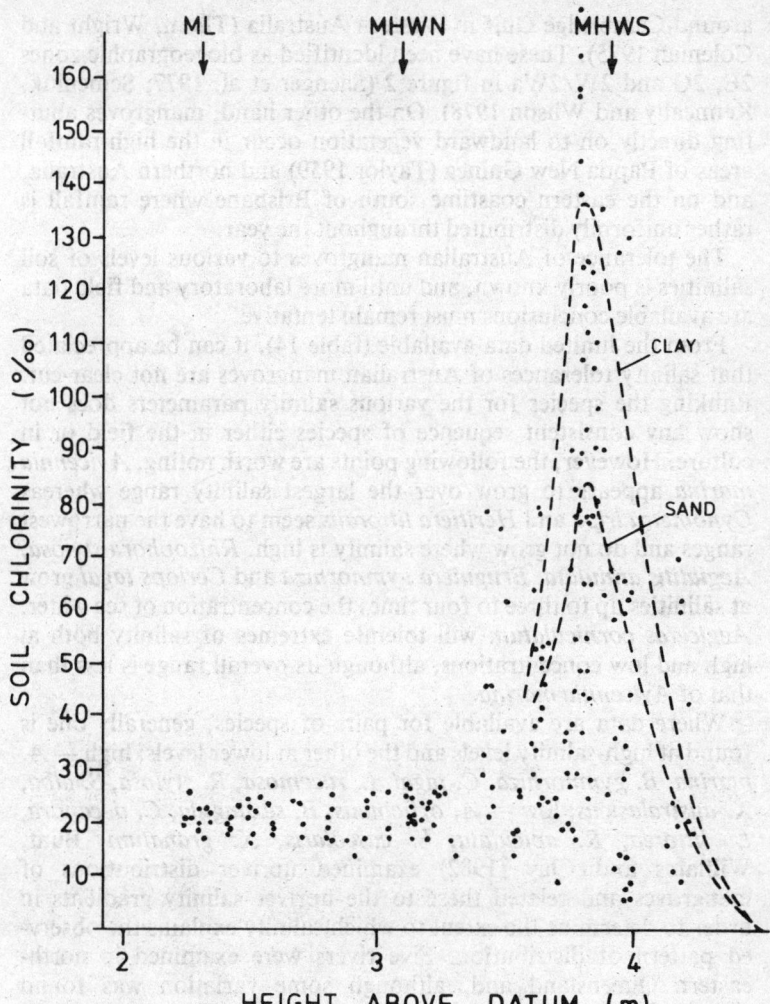

Figure 22 Relationship of soil chlorinity to tidal levels at four study areas at Gladstone, Queensland, 1975-80.

and where salt flats and/or saltmarshes appear between the mangrove community and the inland vegetation.

Both of these extremes are well represented in Australia: extensive mangrove communities with landward saltflats occur from Gladstone to Townsville and at Princess Charlotte Bay in Queensland (Fosberg 1961; Macnae 1966; Saenger and Robson 1977; Saenger et al. 1977; Elsol and Saenger 1983), in the southeastern Gulf of Carpentaria (Saenger and Hopkins 1975) and around Cambridge Gulf in Western Australia (Thom, Wright and Coleman 1975). These have been identified as biogeographic zones 2E, 2G and 2W/2Wa in figure 2 (Saenger et al. 1977; Semeniuk, Kenneally and Wilson 1978). On the other hand, mangroves abutting directly on to landward vegetation occur in the high-rainfall areas of Papua New Guinea (Taylor 1959) and northern Australia, and on the eastern coastline south of Brisbane where rainfall is rather uniformly distributed throughout the year.

The tolerance of Australian mangroves to various levels of soil salinities is poorly known, and until more laboratory and field data are available conclusions must remain tentative.

From the limited data available (table 14), it can be appreciated that salinity tolerances of Australian mangroves are not clear-cut. Ranking the species for the various salinity parameters does not show any consistent sequence of species either in the field or in culture. However, the following points are worth noting. *Avicennia marina* appears to grow over the largest salinity range whereas *Cynometra iripa* and *Heritiera littoralis* seem to have the narrowest ranges and do not grow where salinity is high. *Rhizophora stylosa*, *Aegialitis annulata*, *Bruguiera gymnorhiza* and *Ceriops tagal* grow at salinities up to three to four times the concentration of sea water. *Aegiceras corniculatum* will tolerate extremes of salinity both at high and low concentrations, although its overall range is less than that of *Avicennia marina*.

Where data are available for pairs of species, generally one is found at high-salinity levels and the other at lower levels: high — *A. marina, B. gymnorhiza, C. tagal, L. racemosa, R. stylosa, S. alba, X. australasicus*; low — *A. officinalis, B. sexangula, C. decandra, L. littorea, R. apiculata, S. caseolaris, X. granatum*. Bunt, Williams and Clay (1982) examined upriver distributions of mangroves and related these to the upriver salinity gradients in order to determine the extent to which salinity explains the observed pattern of distribution. Five rivers were examined in north-eastern Queensland and, although some variation was found among river systems, nine species showed significant correlation with upriver salinity gradients. Of these nine species, four showed positive correlations, that is, they occurred at the high-salinity areas and were absent at lower salinities. These were *Rhizophora stylosa, R. apiculata, Sonneratia alba* and *Ceriops tagal*. Five species (*Heritiera littoralis, Excoecaria agallocha, Acrostichum* sp., *Aegiceras corniculatum* and *Rhizophora mucronata*) showed negative correlations in the above ranking, indicating their presence in the upstream, low-salinity areas. Three further species (*Avicennia, Bruguiera gymnorhiza* and *Xylocarpus granatum*) showed no significant correlations suggesting that they can grow over almost the complete salinity range from fresh water to sea

Table 14 Salinity data for Australian mangroves

Species	Salinity Parameters (°/oo)			
	Max. salinity of tidal water (Wells 1982)	Soil salinity range over 1.5 years, Repulse Bay, Qld	Mean soil salinity and range over 7 years, Port Curtis, Qld	Salinity for optimal growth in culture
Acanthus ilicifolius	65	40–44		8[e]
Acrostichum speciosum		61–90		
Aegialitis annulata	85		33–83–114	
Aegiceras corniculatum	67	34–36	16–85–148	8–15[g]
Avicennia marina	85	24–34	11–126–300	8–15[a, c, d, e, f]
Avicennia officinalis	63			
Bruguiera exaristata	72	53–69		8[e]
Bruguiera gymnorhiza	37	59–62	33–79–85	8[e]–34[b]
Bruguiera parviflora	66	61–90		8[e]–17[b]
Bruguiera sexangula	33			
Camptostemon schultzii	75			
Ceriops decandra	67			17[b]
Ceriops tagal	72	35–39	49–110–300	34[b]
Clerodendron inerme		60–63		
Cynometra iripa		8–15		
Excoecaria agallocha	85	33–44		
Heritiera littoralis		9–16		
Lumnitzera littorea	35			
Lumnitzera racemosa	78	36–41	11–47–110	
Osbornia octodonta	56	92–99		
Rhizophora apiculata	65			8[e]
Rhizophora stylosa	74	33–35	43–92–148	8[e]
Scyphiphora hydrophyllacea	63			
Sonneratia alba	44			
Sonneratia caseolaris	35			
Xylocarpus australasicus	76	30–34		8[e]
Xylocarpus granatum	34	25–29		8[e]

[a] Connors (1969)
[b] Sidhu (1975)
[c] Clarke and Hannon (1970)
[d] Farrell (1973)
[e] Clough (unpubl. data)
[f] Downton (1982)
[g] Field (unpubl. data)

water, and that their upriver distributions are determined by factors other than salinity.

Bunt, Williams and Clay (1982) concluded that, to the extent generalizations are possible from their analysis, the species can be ranked in order of decreasing tolerance of or adaptation to sea water: *Rhizophora stylosa*; *R. apiculata*; Sonneratia alba; *Ceriops tagal* > *Aegiceras corniculatum*; *Bruguiera parviflora* > *Excoecaria agallocha*; *R. mucronata*; *Acrostichum* sp.; *Heritiera littoralis*; *Nypa fruticans* > *Barringtonia* sp.; *B. sexangula*; *Sonneratia caseolaris*; *Hibiscus tiliaceus*. Although this ranking shows some similarities to those in table 14, particularly that of Wells (1982), it must be remembered that salinity of the tidal water is only one of the variables determining soil salinities and that the salinities immediately adjacent to the roots of mangroves — be they maxima, minima, means or ranges — will ultimately determine growth and success of mangroves in that particular situation. As Bunt, Williams and Clay (1982) have emphasized, tidal water salinity is only one of the factors affecting the distributions of even those species with significant correlations. In the case of those species not showing significant correlations, factors other than salinity appear to be more important.

Interaction of soil texture with salinity tolerance of mangroves was suggested by experimental studies on the effects of hypersalinity (McMillan 1975b). Two American mangroves — *Avicennia germinans* and *Laguncularia racemosa* — were experimentally subjected to hypersaline conditions while growing in a range of soils with differing clay contents (McMillan 1975). Seedlings of various ages up to three-and-a-half years were subjected to salinities up to five times that of sea water for forty-eight hours, and their responses noted. In soils with a high sand content, whether coarse- or fine-grained, the plants failed to survive this treatment. In soils with a clay content of 7–10 per cent, the hypersaline exposure was tolerated. *Avicennia* seedlings tested over a broad range of salinity survived forty-eight-hour exposure to 60‰ in sand and water culture but failed to survive at higher salinities (McMillan 1975b).

It was suggested by McMillan (1975b) that, although the actual mechanisms underlying interaction of soil texture and salinity tolerance is not understood, depression of the pH in all the experimental soils indicated the involvement of cation exchange. It was suggested that in clay soils the exchange of Na^+ and H^+ ions may reduce the salinity of the interstitial water immediately around the roots, and that the adsorption of Na^+ ions would cause the clay particles to deflocculate. In turn, this would reduce the contact of roots and hypersaline water, thereby facilitating the uptake of water and simultaneously reducing the uptake of salt; wilting and salt excretion by the experimental plants suggested that this took place (McMillan 1975b).

Whether such a mechanism operates in the field has not been investigated, but some supportive evidence suggests that it does. Core samples of clayey soils from salt flats on the seasonally arid coastline of Queensland were compared with clayey soils from adjacent monospecific stands of *Ceriops tagal*. The results obtained (figure 23) indicate a higher water content in the soils of the salt flats, particularly at depth, than in those under *Ceriops*. The soil salinity is also higher on the salt flats than among the *Ceriops*, but it should be noted that in the *Ceriops* soil only surface salinities are high and there is a decrease with depth. X-ray fluorescence analyses of these soils showed that concentrations of two major soil components — calcium and iron — were twice as high in the *Ceriops* soils as in those of the salt flats.

Blackmore (1976) has shown that calcium is involved in reversing the "salt-sieving" ability of clay soil aggregates, a process whereby salt is constrained within the pores of clay particles by anion exclusion. He also found that when sodium ions are present but salt is at relatively low concentrations, salt is trapped and held within the microfabric of the clay. In the presence of calcium ions and when salt was at high concentrations this retention could be reversed, so that salt moved out of the clay much more easily.

In other words, in highly saline situations, the tendency for salt retention by clays appears to be reduced by high calcium ion concentrations, or by the addition of gypsum (Blackmore 1976). Consequently, the concentration of exchangeable calcium in the clay will modify the salinity of the pore spaces with which the roots are in direct contact. The possibility thus exists, that the often sharp boundary between salt flats and *Ceriops* stands is influenced not only by water content and salinity *per se*, but also by the levels of exchangeable calcium.

Height of the Watertable

The subsurface height of the watertable is one of the complex of factors determining the water status of the soil. At the outset, however, it should be emphasized that a "watertable" — the upper surface of free-moving interstitial ground water — is expressed only in those soils whose texture allows such a water surface to develop. In clays, with their low infiltration, low lateral and vertical seepage velocities, low porosity, low permeability, high water-holding capacity and high capillarity, such an internal water surface rarely develops. Clays may be saturated beyond a certain depth, above which they might be below their water-holding capacity. Nevertheless, a hole drilled to the saturated clay level will not fill with water. This is in sharp contrast to sandy or loamy soils in which an actual internal water surface may exist (Ericksen 1970; Lanyon, Eliot and Clarke 1982), and where a hole drilled to an adequate depth will fill with water at a rate depending on permeability and on the lateral seepage velocity.

Despite this reservation, the concept of a watertable, even in

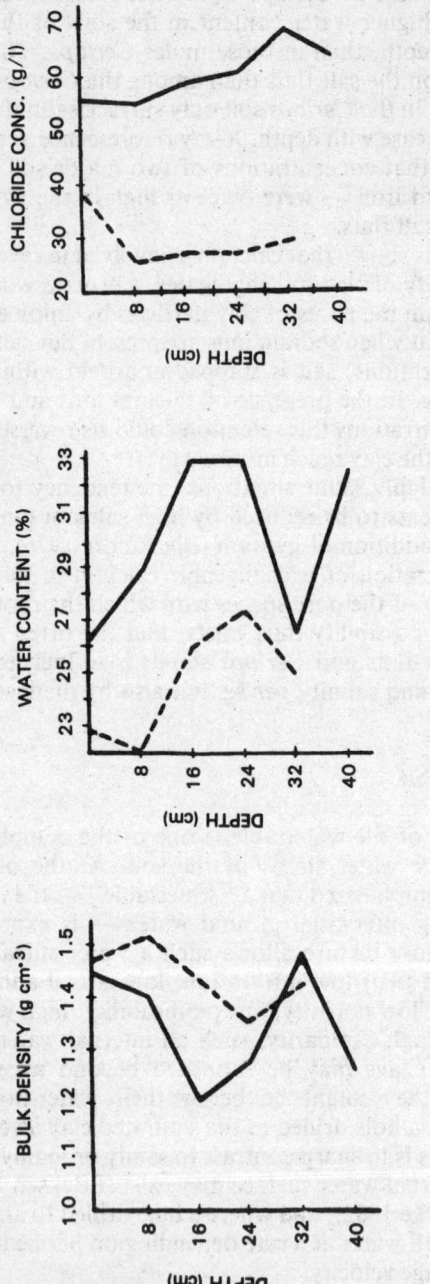

Figure 23 Characteristics of the soils from the vegetation-free salt flats and adjacent mangrove (*Ceriops*) communities at Gladstone, Queensland.

clays, is a useful one, and in the field, plugs or layers of coarser materials or crab burrows can facilitate subsurface flows in clay and result in considerable groundwater movement.

The importance of the subsurface height of the watertable to mangrove and saltmarsh communities has received little attention except by Chapman (1938) and more recently by Clarke and Hannon (1969) and by Semeniuk (1983).

Chapman (1938) found an almost linear relationship between tide and watertable level for non-flooding tides, whereas during flooding tides, even when the surface was inundated, the watertable rarely rose to the surface. This was considered to be due to an aerated layer in the mud immediately below the surface.

Clarke and Hannon (1969) studied the movement of the watertable in Woolooware Bay, Sydney, by inserting perforated plastic tubes into the ground and measuring the level of water that accumulated in them. They found that tides were the most important factor in determining watertable height, and that watertable fluctuations were greatest in the mangroves, followed in descending order by the saltmarsh and *Casuarina glauca* zones. They also found that heavy rainfall caused a rise in watertable level, but this effect was diminished by the fact that water may lie on the soil surface without infiltrating.

The physical properties of the soils in Woolooware Bay suggest that rapid drainage should occur, but the infiltration rates showed such not to be the case (table 11). Clarke and Hannon (1969) observed that both rainwater and sea water may lie on the soil surface for some days when the watertable is several centimetres below the soil surface. This appears to support Chapman's (1938, 1960) hypothesis that even when the soil is covered by a flooding tide, air is available to plant roots because it is trapped just below the soil surface in small pockets. On the other hand, Clarke and Hannon (1969) found that the water level did rise to the surface especially during spring tides, and they did not substantiate Chapman's (1938) hypothesis in its entirety. They concluded that an aerated layer may be present in poorly drained areas, but in their study area such a layer did not exist throughout.

The fact that the amount of water lying on the surface is not necessarily an accurate indication of watertable depth (Clarke and Hannon 1969) has certain implications of relevance to clayey soils. The relationship between tidal and watertable movement may be the result of lateral seepage owing to saturated flow rather than to actual inundation. This could be due to the permeability of the subsurface soil being greater than that of the surface soil, or possibly because the forces behind tidal movement are more effective in causing lateral seepage than in causing vertical seepage. As pointed out earlier, in clayey soils, layers of shell material and crab burrows are important in facilitating lateral flows (Wolanski and Gardiner 1981).

Additional implications are that (1) heavy rainfall is relatively ineffective in leaching salt down into the soil, (2) evaporation of water lying on the soil surface is important in overall salt balance, and (3) if an area can be flooded by sea water without infiltration, tidal inundation may not always add fresh salt but may, in fact, dissolve salt crystallized on the soil surface. These points have been discussed under the section on soil salinity, and at least for clayey soils the assumption has been made that soil salinity is largely determined by surface processes.

The ecological role of watertable heights and fluctuations is poorly understood. Clarke and Hannon (1969: 232) suggested:

> The various components of the physiographic factors are closely interrelated, but that the overall governing element is that of the tide-elevation complex which control plant distribution through the frequency of inundation and exposure, mechanical action of tidal water and salinity. The movement of the watertable and drainage and aeration in the soil are less important components of the tide-elevation complex in explaining plant distribution.

By following salinity changes in a small mangrove creek on Hinchinbrook Island, Queensland, Wolanski and Gardiner (1981) were able to postulate that considerable salt was removed from the mangroves bordering the creek by groundwater flow during periods of non-inundation, and that once back in the creek, tidal flushing removed salt from the creek. This finding is at odds with that of Clarke and Hannon (1969), who suggested that even heavy rainfall is ineffective in leaching salt through the soil. At Hinchinbrook Island, however, Wolanski and Gardiner (1981) observed that rainwater percolated directly into the mud with no surface run-off; in other words, the Hinchinbrook muds have extremely high infiltration rates. In this system, groundwater flows were observed to be the dominant water-transport process, at least when inundation was minimal, and it was greatly accelerated near the surface of such biological disturbances as crab holes (Wolanski and Gardiner 1981).

Another potential role of watertable movement was postulated by Hicks and Burns (1975) from their study of the effects of drainage canals which intercept the overland sheet flow in the mangroves of southwestern Florida. They suggested that nutrients regenerated by the breakdown of detritus may be transported into the root zone of the mangroves by the vertical motion of the watertable; unfortunately, no evidence to affirm or negate this suggestion is presently available.

It thus seems that, although the ecological role of watertable levels (or groundwater flows) is not clear, fluctuations in watertables are relatively unimportant in explaining plant distributions within the mangroves and littoral complex (Clarke and Hannon 1969); they may be important in salt flushing (Wolanski and Gardiner 1981) and nutrient recycling (Hicks and Burns 1975). However, in view of the apparent array of adaptations to growing in waterlogged and oxygen-deficient conditions displayed by mangroves as a group, it seems that, along with drainage and waterlogging, watertables close to the surface give mangroves a competitive edge over other plants lacking such adaptations which might otherwise invade this environment.

Nature of the Soil

The nature of the soil in a mangrove community is largely determined by a range of geological and geomorphological processes. Some of these (such as sea-level change or erosion) may affect the mangroves directly, but more often they change certain characteristics of the sediment which, in turn, renders it more (or less) suitable for mangrove growth and development.

Mangrove communities develop best in sheltered depositional environments where, in the absence of drastic resculpturing of the coastline, there is a steady accretion of sediments. A gradual elevation of the sediment surface in relation to sea-level results, and with it a gradual change in soil water characteristics. However, such gradual and directed changes rarely occur, for the coastal environment is a dynamic one where erosion occasioned by storms or flooding can rapidly reverse the biologically mediated depositional phase (Bird 1971, 1972; Spenceley 1982). The likelihood of such disruptive change depends on the geography of the coast and on its geomorphological history.

Two major types of intertidal landforms can be recognized: those which contain a veneer of transported or trapped sediment over a consolidated parent material, and those which are the result of sedimentary accretion, producing prograding shorelines (Thom 1982). The latter is the more common and includes many fringing substrates and deltas. The former type may be important regionally, for example, where sediments accumulate over fossil coral platforms. This type also includes the comparatively narrow terrigenous beaches which occur along sunken river valleys.

Stability of the landform is strongly influenced by differing geomorphological origins. Accretion or erosion may be a continuing, seasonal or periodic process in depositional substrates. Modification of the landforms of more consolidated shores may be intermittent and arise from catastrophic events such as severe storms.

Thom (1967, 1975) and Thom, Wright and Coleman (1975)

studied the ecology of mangroves in terms of the response of the plants to habitat change induced primarily by geomorphic processes. They considered that given the climatic-tidal environment and a pool of mangrove species, each of which possesses a certain physiological response to habitat conditions, the history of the land surface and contemporary geomorphic processes jointly determine the nature of the soil surface on which mangroves grow (Thom 1982). Such attributes of the substrate as moisture content, texture, salinity, redox potential and chemical composition are, to a large extent, a function of past and present geomorphic processes. The mangroves reflect each of these geomorphic situations by responding to the environmental gradients of elevation, drainage, stability, soil characteristics and nutrient input, which each of these situations produces. According to the physiological response of species to moisture and/or salinity stress, for example, there will be more, or less, favourable plant growth in a particular habitat. Thus, landform properties and geomorphic processes find expression in the variation in growth, morphology and metabolism of mangroves along environmental gradients.

Although mangrove development is bound historically to the geomorphic processes of a region, it is an expression of the resultant properties of the soils that occur there. From an ecological viewpoint, a study of the soil relationships of the mangroves will provide more direct information on mangrove growth and development than will historical (geomorphological) analysis. This is not to deny the importance of geomorphological studies, for these place mangrove soil characteristics into broader, more causally related contexts. For example, Spenceley (1983) showed that there are differences in elemental concentrations between open accreting shores and estuarine coastlines, and that the temporal and spatial behaviour of the elements also differ.

Understanding mangrove-soil relationships is complicated by the ability of most mangrove species to grow on a variety of substrates and because they often alter the substrate through peat formation or by changing the pattern of sedimentation. Mangrove trees are found on a wide variety of substrates including muds, silts, peat, sand, and even rock and coral shingle, provided there are sufficient crevices for root attachment. Mangrove ecosystems, on the other hand, appear to be best developed only on muds and fine-grained sand (Butler et al. 1977a; Galloway 1982); these muds are often highly saline and gypseous, with soft loose surfaces showing neither seasonal cracking nor change in texture with depth. These physical characteristics are important in terms of the drainage and aeration of the soils.

The composition and texture of the soil also can affect its salinity on the one hand, and the response by mangroves to that salinity on the other.

Nutrient availability in mangrove soils is another important consideration of mangrove-soil relationships. Although the level of nutrients in a particular mangrove soil will reflect the chemistry of its parent material and of the surrounding waters, the availability of nutrients will depend largely on the type of soil and its microbial characteristics.

For the two major plant nutrients — nitrogen and phosphorus (N and P) — microbiological processes are the main determinants of their release in a form available for plant growth. In the mangrove environment, nitrogen becomes available through microbial fixation of atmospheric N_2 and through the biological decomposition of organic matter in the soil. Nitrogen bound up in proteins is converted to ammonia by numerous proteolytic bacteria and fungi; this process is termed "ammonification". Ammonia can serve directly as a source of nitrogen but, more importantly, it provides energy for nitrite bacteria which, in the presence of oxygen, oxidize the ammonia to nitrite. As a rule, the nitrite is further oxidized to nitrate by another group of nitrifying bacteria; this whole process is termed "nitrification". These various pathways are summarized in table 15.

Table 15 Microbial reactions involved in the availability of nitrogen for plant growth

Process	Facilitating organisms	Reaction
1. N_2 fixation	N_2 fixers e.g. bacteria and blue-green algae	$8 H^+ + N_2 \rightarrow 2 NH_4^+ \rightarrow$ Organic N
2. Ammonification	Most proteolytic bacteria and fungi	Organic N $\rightarrow NH_4^+$
3. Nitrification		
(a) Nitrition	Nitrite bacteria	$NH_4^+ + 1½ O_2 \rightarrow NO_2^- + H_2O + 2 H^+$
(b) Nitration	Nitrifying bacteria	$NO_2^- + ½ O_2 \rightarrow NO_3^-$

Studies of nitrogen fixation in mangroves have been limited (Zuberer and Silver 1975, 1978; Kimball and Teas 1975; Potts 1979; van der Valk and Attiwill 1984). In the mangroves of Florida, Zuberer and Silver (1978) found that the order of nitrogen-fixing activity was: plant-free muds < plant-associated muds < root tissue of *Rhizophora mangle, Avicennia germinans* and *Laguncularia racemosa*. Most of the nitrogen fixation appeared to be carried out by photosynthetic bacteria.

Zuberer and Silver (1975) suggested that the establishment of a nitrogen-fixing bacterial population around the roots may be a critical factor in mangrove establishment during early stages of stand development when trapped detritus is minimal and nitrogen is limiting. Furthermore, once such microbial populations are established, the rates of nitrogen fixation are sufficient to supply much of the nitrogen requirements for these Floridian mangrove communities (Zuberer and Silver 1978).

The importance of nitrogen fixation to Australian mangrove communities is questionable; preliminary studies of northern Queensland mangroves have not revealed any potential areas of high nitrogen-fixing activity (Boto 1982), and van der Valk and Attiwill (1984) found low rates of nitrogen fixation associated with the root zone and sediments of the mangroves in Westernport Bay, Victoria. Consequently, the process of ammonification and nitrification may be more important for maintaining nitrogen levels in Australian mangrove soils.

With the seasonal discharge of fresh water, a short-term increase in phosphate occurs in Australian estuaries (Rochford 1951; Spencer 1956). This phosphate does not remain in the water column but becomes adsorbed on the sediments and suspended particulate matter as insoluble ferric phosphate (Rochford 1951). Under reducing conditions resulting from an oxygen deficiency in the overlying water or through bacterial activity, ferric phosphate is reduced to ferrous phosphate and may subsequently be leached into the water column. Evidence in support of the bacterial mediation of such phosphate adsorption and release (Mee 1978) comes from two studies which demonstrated that (1) this process was temperature dependent and (2) poisoning the sediments with formalin impeded phosphorus exchange between the sediments and the water column. Thus, the sediments from mangrove areas are able to remove dissolved phosphates from the overlying water and bind them. In the reduced mangrove sediments, they would then be available for uptake by mangrove rootlets, and the generally clayey nature of the soils, with their low porosity, would prevent leaching and loss of these soluble phosphates to the overlying waters. As phosphate retention is more efficient in fine sediments than in coarse ones, fine sediments generally possess a greater store of available phosphates; this may partly explain the better development of mangrove communities on fine sediments such as alluvial muds in deltaic environments.

Work on Hawaiian mangroves by Walsh (1967) demonstrated that there is considerable removal of nitrate and phosphate from the tidal water entering the mangrove community. Carter et al. (1973) and Lugo, Sell and Snedaker (1976) confirmed these observations for mangrove communities in Florida. Comparative data from Australian mangrove communities are presently not available (Boto 1982), but it seems likely that the removal of nutrients from overlying tidal waters (sea water and land drainage) together with the release of NO_3 from organic material in the soil are the major mechanisms for maintaining nutrient availability in Australian mangrove soils.

The response of an entire mangrove community to nutrient input and cycling was studied by means of a mathematical model (Lugo, Sell and Snedaker 1976) based on a mangrove forest in southern Florida. In this model, the incoming radiation interacts with nutrients and the mangrove plants to form organic matter through the process of photosynthesis. Some of this gross production is respired by the forest, some is stored as a net increase in forest biomass, and some is deposited on the forest floor as detritus. The detritus may be exported from the forest to the estuary by tidal action. Some of it is grazed *in situ* by mangrove consumers, and some decomposes or accumulates as peat. Decomposition may occur under the influence of oxygen-saturated waters of incoming tides, or of atmospheric oxygen when the forest floor is exposed to the air. Decomposition of detritus within the mangrove system represents a source of nutrients for photosynthesis. Other nutrient sources are from terrestrial drainage, tidal waters, rainfall and sediment storage. Of these, terrestrial drainage is the most significant. In the model, they are all grouped as a single source. Some nutrients are not used and are lost from the system; the rest are sequestered through plant photosynthesis, thus completing the cyclic loop in the model. This model was validated using field data for the various forcing functions, state variables and for the flows within the system, including nutrient input, nutrient uptake and nutrients either not used or exported in the detritus.

The model was highly sensitive to nutrient availability, and when nutrient input to the system was set at zero, mangrove biomass decreased steadily. The decrease in mangrove biomass, gross photosynthesis and nutrient storage with three levels of nutrient input are shown in figure 24. These simulations indicate the dependency of mangrove communities on nutrient inputs derived from the land. Decomposition within the forest and inputs from sea water do not seem to be enough to maintain the observed rates of metabolism because of losses via detrital export. Lugo, Sell and Snedaker (1976) found that the contribution of nutrients from land drainage was ten to twenty times that from sea water during the dry and wet seasons respectively.

Lugo, Sell and Snedaker (1976) concluded that gross photosynthesis appears to be sensitive to terrestrial nutrient input and that the development of mangrove biomass is dependent on the quantity of nutrients and the efficiency of nutrient uptake. In addition, they concluded that during succession, mangroves exert significant control over the amount of nutrients in adjacent water but, if terrestrial nutrient input is reduced, they do not have the capacity to maintain themselves at the same level of production. This is due to

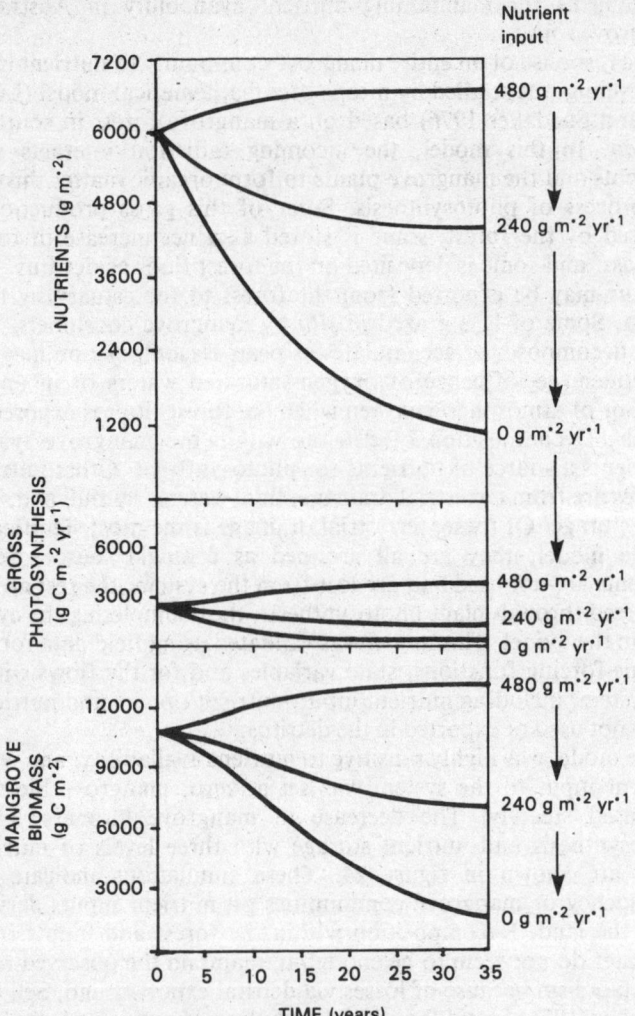

Figure 24 Results of model simulation of Florida mangrove ecosystem. Rates of gross photosynthesis and level of nutrients in system with initial conditions of high nutrient level, mean rates of metabolism and three rates of nutrient input. The response in the mangrove biomass of the system is shown for the same conditions (after Lugo, Sell and Snedaker 1976).

a loss of nutrients via export from the mangrove community, which suggests that there must be selective pressure for mechanisms of recycling.

Onuf, Teal and Valiela (1977) studied the effects of nutrient enrichment of *Rhizophora mangle* islands in Florida. Two islands were compared: one (high-nutrient) with, and one (low-nutrient) without a breeding colony of pelicans and egrets. Their data indicate higher growth rates at the high-nutrient site; trees at the high-nutrient site showed greater additions of leaves, reproductive parts and new lateral branches and larger increases to existing stems. Growth in the fertilized stand began earlier in the year and had a second peak of growth not shared by mangroves at the low-nutrient site.

The nutrient response found in the field by Onuf, Teal and Valiela (1977) appears to verify the more theoretical predictions regarding the effect of nutrients on a mangrove community made by Lugo, Sell and Snedaker (1976).

Comparable data for Australian mangroves are not available; however, the luxuriance of mangroves along the Queensland coastline correlates with those areas where high rainfall results from the proximity of coastal mountain ranges in excess of 700 metres elevation (figure 5). While this may be ascribed in broad terms to high rainfall, these areas are also areas of high-nutrient run-off owing to the erosion of the ranges, and therefore are rich sources of nutrient input for the mangrove communities on the coast. On the other hand, Boto's (1982) preliminary data from selected rivers in northeastern Queensland show that just prior to the onset of the wet season, fresh water flowing into mangrove communities is not a major supplier of dissolved nutrients.

In more detailed studies at Hinchinbrook Island, however, Boto (1983) reported that mangrove standing crop biomass was significantly correlated with (1) extractable phosphorus, (2) soil salinity (negative correlation) and (3) the redox potential or reducing conditions of the soil. Mean ranges of phosphate and ammonium concentrations were 5–20 μg P/g dry soil and 4.6–7.3 μg N/g dry soil respectively.

Experimental fertilization, conducted over a year, showed a significant growth response to soil ammonium enrichment, suggesting that nitrogen limitation was common to all sites (Boto 1983; Boto and Wellington 1983). On the other hand, phosphate limitation was indicated only in those areas where phosphates were less than 5 μg P/g dry soil — the areas where higher elevations occurred in proximity to the coast.

Preliminary data from six Cape York river systems indicate that all sites had low to extremely low phosphate concentrations (0.1–5 μg P/g dry soil), whereas ammonium levels were similar to those at Hinchinbrook Island (Boto 1983). These preliminary findings strongly suggest that phosphorus status is a dominant influence on mangrove primary productivity, at least in northern

Australia, and is consistent with the generally low phosphorus level of Australian soils and its influence on terrestrial plant communities (Beadle 1954).

Proximity of Freshwater Source

There is little doubt that the availability and proximity of fresh water affects mangrove development. For example, along the Queensland coastline the more luxuriant mangroves occur in high rainfall areas (figure 5). Similarly, the biogeographic regions, based in part on floristic data, correlate to a large extent with rainfall. Bunt, Williams and Duke (1982) refined these analyses by examining species assemblages within fifty-six coastal systems between Rockhampton and Cape York and relating these to prevailing hydrological conditions. Analysis of their data included classification of species and sites into species-groups and site-groups respectively and an examination of the relationship between these categories. Nine species-groups and nine site-groups were recognized (tables 16, 17), each of which could be characterized by their ecological tolerances or hydrological characteristics. One site-group comprised members whose shared characteristic was that they were unlike any other members of the system, including one another. These members were allocated to their closest other group (table 17).

The data suggest that species distribution is strongly influenced by the extent of freshwater influence either from rainfall or from

Table 16 Species groups and their characteristics as derived from classification of floristic data from northeastern Australian coastal systems

Species group	Species	Groups characteristics	Common in site groups
1.	Aegiceras corniculatum Avicennia marina Bruguiera gymnorhiza Excoecaria agallocha Osbornia octodonta Rhizophora stylosa	Species of wide ecological amplitude, i.e., tolerant of wide range of salinity and temperature; ubiquitous and often dominant	1, 2, 3, 4, 5, 7 and 8
2.	Acanthus ilicifolius Acrostichum speciosum Heritiera littoralis Xylocarpus granatum	Species associated with freshwater influence and characteristics of the middle and upper reaches of rivers	1, 2, 3, 4, 5 and 6
3.	Rhizophora apiculata Rhizophora lamarckii Xylocarpus australasicus	Species often growing together behind frontal stands of R. stylosa	1, 2 and 3

Table 16 (cont'd)

Species group	Species	Groups characteristics	Common in site groups
4.	*Aegialitis annulata* *Bruguiera exaristata* *Ceriops tagal* var. *tagal* *Lumnitzera racemosa*	Species associated with mid to inner mangrove zones	3 and 5
5.	*Bruguiera parviflora* *Ceriops decandra* *Sonneratia alba* (apetalous)	A degree of freshwater influence seems to be important and limits the distribution of this group	4 and 5
6.	*Cynometra iripa* *Lumnitzera littorea* *Rhizophora mucronata* *Scyphiphora hydrophyllacea*	Species rarely or never found near river mouths or close to seawater influence	1

7.	*Barringtonia racemosa*
	Bruguiera sexangula
	    (daintree)
	Sonneratia caseolaris
	    (johnstone)
	S. caseolaris (tully)
		Species associated with freshwater influence and characteristic of middle and upper reaches of in restricted area	4

8.	*Bruguiera cylindrica* *Bruguiera sexangula* (Jacky Jacky) *Camptostemon schultzii* *Ceriops tagal* var. *australis* *Diospyros ferrea* *Nypa fruticans* *S. alba* (semipetalous)	Mostly species of limited distributions or rare or uncommon within their distribution; no unifying ecological features	
9.	*Barringtonia acutangula* *Dolichandrone spathacea* *Pemphis acidula* *S. alba* (petalous) *S. caseolaris* (claudic) *S. caseolaris* (olive) *S. caseolaris* (mcivor/morgan)		

Source: After Bunt, Williams and Duke (1982).

"wet" rivers, that is, rivers that flow reliably for most of the year (Bunt, Williams and Duke 1982). In contrast, Macnae (1966) was convinced that rainfall was the more important factor, stating that run-off from most Australian rivers is too variable.

Table 17 Site groups and their group characteristics as derived from classification of floristic data from northeastern Australian coastal systems

Site group	Drainage basin	Coastal system	Group characteristics	Common species groups
1.	Jardine Olive/Pascoe Olive/Pascoe Jardine	Jardine River Olive River Pascoe River Cowal Creek	Far northern sites, extensive in areas and covering wide environmental range; all are subject to freshwater influence	1, 2, 3 and 6
2.	Tully Olive/Pascoe Jacky Jacky Daintree Jacky Jacky Murray Barron * Jacky Jacky * Proserpine	Hull River Kangaroo River Barnia Creek Coopers Creek Harmer Creek Deluge Inlet Trinity Inlet Capt. Billy Creek Saltwater Creek	Either rivers receiving limited or only sporadic fresh water from the land and/or medium to large inlets strongly influenced by the sea; intermediate between high- and low-saline areas	1, 2, 3, 4 and 6
3.	Lockhart Jacky Jacky Lockhart Hinchinbrook Jacky Jacky	Claudie River Escape River Lockhart River Missionary Bay Jacky Jacky Creek	Except for the Claudie River, all are extensive estuaries under well-sustained marine influence; the Claudie River is a low-salinity river but, floristically, it appears to resemble the geographically closer but more saline Lockhart River	1, 2, 3, 4, 5 and 6
4.	Daintree Endeavour Jeannie Mulgrave/Russel Murray Tully * Daintree * Johnstone	Daintree River Endeavour River McIvor/Morgan Rivers Mulgrave River Murray River Tully River Bloomfield River Moresby River	Large rivers with strong freshwater influence	1, 2, 3, 5, 6 and 7
5.	Endeavour Normanby Lockhart * Jacky Jacky * Ross * Hinchinbrook	Annan River Annie River Nesbit River Mew River Alligator Creek Fisherman's Point	Relatively dry environments with extensive salt flats and only a narrow mangrove fringe	1, 2, 4 and 5

Table 17 (cont'd)

Site group	Drainage basin	Coastal system	Group characteristics	Common species groups
6.		Bloomfield River Mew River Moresby River Alligator Creek Captain Billy Creek Haydock Island Fisherman's Point Saltwater Creek	Non-conformist group; members respectively reallocated to site-groups 4, 5, 4, 5, 2, 8, 5 and 2	
7.	Jacky Jacky Stewart Stewart Lockhart Haughton Hinchinbrook Ross Jacky Jacky Shoalwater Hinchinbrook	Macmillan River Rocky River Steward River Lloyd Island Chunda Bay Zoe Bay Crocodile Creek Round Point Creek Port Clinton Scraggy Point	Locations under a balance of salt and freshwater influence with neither extreme saline or freshwater conditions developing	1, 3 and 4
8.	Jeannie Stewart Herbert * Hinchinbrook	Jeannie River Cliff Island Sunday Creek Haydock Island	Under predominantly marine influence but with parts under sustained freshwater supply	1 and 4
9.	Jeannie Jacky Jacky Jacky Jacky Jardine Jardine Jacky Jacky Jeannie Jardine Jardine	Flinders Island Haggerstone Island Halfway Island Horn Island Prince of Wales Islands Sir Charles Hardy Is. Stanley Island Tuesday Island Wednesday Island	Islands which, apart from rain, have little or no freshwater influence; mangrove development restricted to sheltered bays regularly influenced by tides	1

* Reallocated from site-group 6.
Source: After Bunt, Williams and Duke (1982).

Bunt, Williams and Duke (1982) doubted that rainfall adequately explained the observed distributions, since the islands, dependent entirely on rainfall, have relatively depauperate mangrove floras.

After dismissing the suggestion that catchment size might be involved, they suggested that the mangrove communities of northeastern Australia had been floristically richer in the past, but with increasing aridity of Australia in recent geological time, the less adaptable species began dying out (Mepham 1983). However, a reasonable number of species survived, even though the long-term future of the rarities may be doubtful (Bunt, Williams and Duke 1982).

This approach raises the question as to what aspect of "increasing aridity" is reduced in certain river systems. Bunt, Williams and Clay (1982) and Bunt, Williams and Duke (1982) examined rainfall, catchment size and the effects of upriver salinity gradients, and although these go far in explaining mangrove floristics, there appears to be some inadequacy. Whereas temperatures and evaporation do not vary sufficiently over the study area to allow for such differential floristic development, catchment characteristics appear to do so, but these have not been examined in detail. Rainfall and flow reliability also may be involved as may their seasonality. However, insufficient data are available, particularly for areas in northern and northeastern Australia, to make an adequate assessment.

Catchment characteristics are sufficiently well known, at least at the level of drainage basins, for some comments to be made. Given identical rainfalls, two catchments may, through their characteristics of geology, topography, soils and vegetation cover, effectively absorb and utilize different proportions of that rainfall, with the residual proportion remaining after evaporation becoming part of the surface run-off from that catchment. For example, the continental islands examined by Bunt, Williams and Duke (1982) and shown to be depauperate have small catchments largely of rock, steep slopes, skeletal soils and sparse vegetative cover. Their run-off coefficients, primarily the ratio of run-off to rainfall, are high because little of the rain is able to infiltrate into the soil where it can be retained and utilized. Those catchments, on the other hand, with extensive swamps, overflow basins, dense vegetation or sandy landscapes, have low run-off coefficients and allow considerable retention of water with more regulated and sustained release to their drainage river systems. However, some catchments without these characteristics may also have low run-off coefficients because of low rainfall and high evaporation or transpiration over the catchment.

Insufficient data are available on rainfall, run-off and catchment characteristics to examine in detail each of the coastal systems treated by Bunt, Williams and Duke (1982). However, a broad picture can be obtained using data (Australian Water Resources Council 1976) on the respective drainage basins to which they belong.

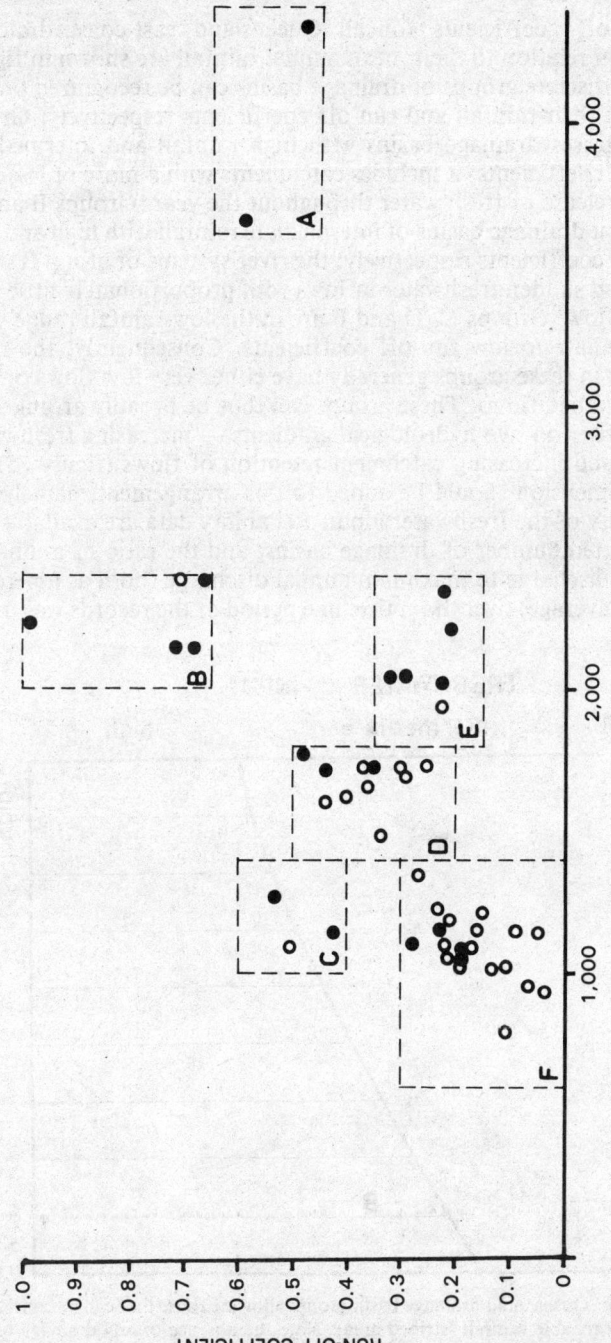

Figure 25 Classification of Queensland drainage basins into six groups on the basis of their run-off coefficients and mean rainfall.

Run-off coefficients for all Queensland east-coast drainage basins in relation to their mean annual rainfall are shown in figure 25. Six discrete groups of drainage basins can be recognized on the basis of their rainfall and run-off coefficients respectively. Group A comprises drainage basins with high rainfall and intermediate run-off coefficients; it includes catchments with a more or less sustained release of fresh water throughout the year. Groups B and E represent drainage basins of intermediate rainfall with high and low run-off coefficients respectively; the river systems of group B show large and sudden freshwater inflows with proportionately little sustained flow. Groups C, D and F are in the low-rainfall range with intermediate to low run-off coefficients. Consequently, the river systems in these groups generally have either very low flows or low but sustained flows. These groups can thus be broadly arranged in some order on two hydrological gradients — increasing freshwater inputs and increasing catchment retention of flows (figure 25). A third dimension should be added to this arrangement, namely the reliability of the freshwater input. Reliability data are available for a restricted number of drainage basins, and the ratio of minimum annual discharge to maximum annual discharge (both as a percentage of average) over the entire time period of the records was used.

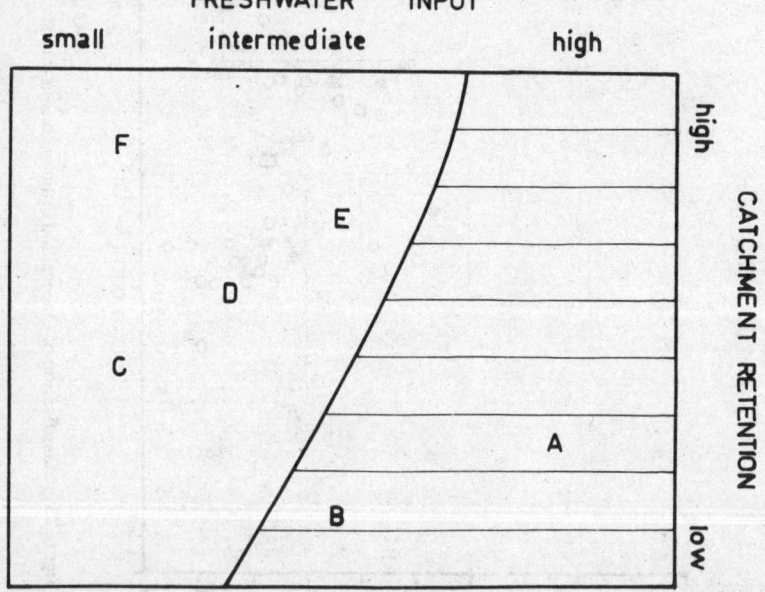

Figure 26 Queensland drainage basin groups showing those that can be considered as having reliable rainfall (striped area). Note the absence of any drainage basins with high and reliable rainfall and high catchment retention.

The more reliable and the less variable the annual flow, the more closely this ratio approaches unity. Using these data, the drainage basins can be subdivided tentatively into those of high or low reliability (figure 26).

Using the above data, a classification of the east-coast drainage systems can be attained based only on hydrological data (table 18). These groupings show some similarity to the detailed classification based on floristic data (Bunt, Williams and Duke 1982) and suggest that catchment characteristics and, to a lesser extent, flow reliabili-

Table 18 Hydrological classification of the Queensland east-coast drainage basins

Group A.	High freshwater input; intermediate catchment retention; high flow reliability:
	+ Johnstone River + Tully River*
Group B.	Intermediate freshwater input; low catchment retention; high flow reliability:
	Mossman River + Mulgrave-Russell Rivers Murray River
	+ Hinchinbrook Island + Daintree River
Group C.	Low freshwater input; intermediate catchment retention; low flow reliability:
	Black River + Jeannie River + Stewart River
Group D.	Intermediate freshwater input; intermediate catchment retention; low flow reliability:
	+ Jardine River + Jacky Jacky Creek + Endeavour River
	O'Connell River Pioneer River Plane Creek
	Fraser Island Noosa River Maroochy River
	Stradbroke Island South Coast Rivers
Group E.	Intermediate freshwater input; high catchment retention; low flow reliability:
	+ Olive-Pascoe Rivers + Lockhart River + Barron Rivers*
	+ Herbert River + Proserpine River Whitsunday Islands
Group F.	Low freshwater input; high catchment retention; low flow reliability:
	+ Normanby River + Ross River + Haughton River
	+ Shoalwater Creek Burdekin River* Fitzroy River*
	Water Park Creek Burnett River* Baffle Creek
	Burrum River Curtis Island Calliope River
	Boyne River Styx River Mary River*
	Kolan River* Don River Pine River
	Logan-Albert Rivers	

+ Rivers from these drainage basins were examined by Bunt, Williams and Duke (1982).
* More than 100 m^3 x 10^6 diverted for irrigation purposes.

ty are about equal in importance to average rainfall of the catchment area in shaping the floristic composition of the mangroves in the various coastal regions.

The additional comment should be made that northeastern Australia does not have any drainage basins which show both high and reliable rainfall and high catchment retention (figure 26, table 18). Although this probably reflects the general aridity and low relief of the continent, it results in the absence of such river basins as the New Guinean Fly River or Purari River deltas with their extensive mangrove development and vast *Nypa* forests in the reduced-salinity reaches (Womersley 1975; Percival and Womersley 1975).

Plant–Plant Interactions

As in any other plant community, the constituent plants of the mangrove community interact with one another, often in specific or defined ways. Many of these interactions are subtle, and most are poorly studied and little understood. There is, however, a gradually increasing awareness that plant–plant interactions within the mangrove community must be important because the distribution and success of the mangroves cannot be adequately explained solely in terms of their interaction with the physico-chemical environment.

Several categories of plant–plant interactions can be recognized as important in determining the structure and/or function of the mangrove community. They are: parasitic, antagonistic, mutualistic and competitive interactions.

Parasitism

Parasitic relationships are those in which the parasite obtains food from its host, which may or may not suffer harm as a result. Many such relationships occur in the mangrove community, but two of them suffice as examples.

Mistletoes (family Loranthaceae) are parasitic plants which, although capable of photosynthesis, tap into the host's vascular system to obtain water and mineral nutrients (see chapter 4). They are relatively benign parasites and rarely kill the host plant. Nevertheless, they deprive the host of desalinated water and nutrients, both of which may be scarce resources for mangroves, as well as causing growth modification and shading of the affected branches.

Being relatively benign and uncommon, the mistletoes exert little influence on the mangrove community as a whole, although individual host plants may suffer considerable stress.

Parasitic fungi, on the other hand, can have devastating effects. Many parasitic fungi occur in the mangroves from the canopy to the root (see chapter 4). Usually some equilibrium is established with the host, but sometimes equilibrium is disturbed and considerable mortality results. For example, on the central Queensland coastline, a species of *Phytophthora* has caused considerable mortality in *Avicennia marina* (Pegg, Gillespie and Foresberg 1980). This fungus is normally a leaf litter decomposer and, as such, it occurs throughout Australian mangrove communities. However, it does have the capacity to become pathogenic, attacking the roots of its mangrove host. As a result, the roots cease to function, or function inefficiently, and severe water stress is induced, leading ultimately to death of the mangrove.

So far, wherever outbreaks of *Phytophthora* and high mortality have been recorded, only one host — *Avicennia marina* — has been involved, and it appears that its susceptibility to this parasite is considerably higher than that of other mangrove species (figure 27).

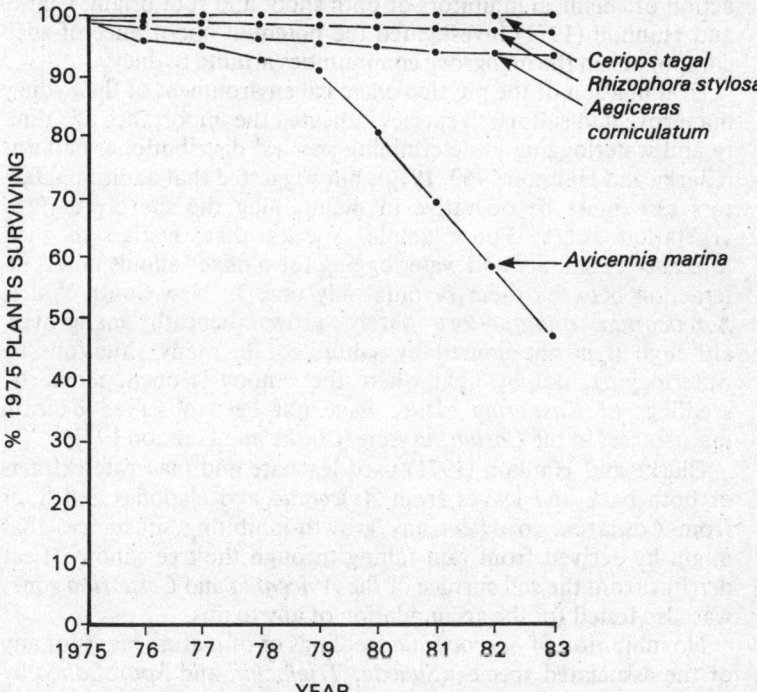

Figure 27 Depletion curves for mangroves at Gladstone, Queensland, showing the high *Phytophthora*-induced mortality in *Avicennia marina*, and the unaffected mortality of the other mangrove species present in the area.

The resultant selective mortality has led to a dramatic change in the species composition at particular localities. For example, in Port Curtis, where almost pure stands of *Avicennia* once occurred, the mangrove community contains virtually no mature *Avicennia marina* at present, and *Rhizophora stylosa* appears to be rapidly filling the gap. Hence, the change from an *Avicennia marina* community to one dominated by *Rhizophora stylosa* has occurred as a result of a fungus over a relatively short period — less than four years.

Antagonism (Ammensalism)

Antagonistic relationships are those in which the growth of a particular plant is inhibited or interfered with through the creation of adverse conditions by another plant, generally through the production and secretion of toxic or inhibitory substances (Garb 1961; Woods 1960; Muller 1966).

Numerous examples have been reported in which vegetational composition and species distributions have been attributed to the action of chemical inhibitors of both shoot and root origin. Clarke and Hannon (1971) investigated the potential importance of such compounds in the mangrove communities around Sydney.

Examination of the physico-chemical environment of the Sydney mangrove and saltmarsh species indicated the importance of salinity and waterlogging in determining species' distributional patterns (Clarke and Hannon 1969, 1970), but suggested that additional factors also must be operative in maintaining the sharply defined vegetation zones. For example, species that overlap in their tolerance of salinity and waterlogging form mixed stands unless interaction between them favours only one. In New South Wales, *Sarcocornia quinqueflora* rarely grows beneath mangroves, although it is not limited by salinity or in many situations by waterlogging, nor by light where the canopy is open. Similarly, seedlings of *Casuarina glauca* have not been observed beneath mature trees in the *Casuarina* zone (Clarke and Hannon 1971).

Clarke and Hannon (1971) used leachate and macerate extracts of both bark and leaves from *Avicennia* and cladodes and litter from *Casuarina* to detect any growth-inhibiting substances that might be derived from rain falling through the tree canopy. Leaf detritus from the soil surface of the *Avicennia* and *Casuarina* zones was also tested for the accumulation of any toxins.

No inhibition of *Sarcocornia* seedlings or of mature plants of any of the associated species (*Suaeda*, *Triglochin* and *Sporobolus*) by mangrove extracts was found. *Juncus* plants were healthy only in the control treatment (tap water), and yellowing of the shoots was common where leachates and macerates were included. Survival and growth of seedlings of both *C. glauca* and *J. maritimus* were inhibited by the presence of a layer of *Casuarina* litter on the soil surface, but no significant differences in growth were found using *Casuarina* extracts.

Clarke and Hannon (1971) concluded that in these Sydney communities it appears highly unlikely that phytotoxic exudates influence the establishment and maintenance of zonation patterns but that *Casuarina* litter creates physical difficulties for germination and seedling growth.

It seems likely, however, that in the more tropical and species-rich mangroves, toxic and inhibitory exudates are potentially of greater significance. Catechol-type tannins, as well as a range of more exotic compounds, are abundant in the bark, wood and leaves of many Queensland mangroves (Brunnich and Smith 1911; Hogg and Gillan 1984). For example, brugine has been recorded from stem and bark extracts of *Bruguiera sexangula*, *B. exarista* and *B. cylindrica* (Loder and Russell 1969; Kato 1975) and a triterpenoidal saponin has been recorded from the roots of *Acanthus ilicifolius* (Minocha and Tiwari 1981). The fish-poisoning properties of the bark and stems of *Barringtonia*, *Thespesia* and *Derris*, three common associates of northern mangroves, were well known and exploited by the Aborigines (Everist 1974). All of these are physiologically active compounds, capable of regulating or inhibiting growth; however, at present, no evidence is available that they do so under field conditions although this would seem to be an aspect worthy of detailed investigation.

Mutualism

Mutualism is the association of individuals of different species such that their ability to survive and reproduce is greater when together than when apart (Roughgarden 1975; Margulis 1981; Lewin 1982).

Numerous mutualistic associations have been documented (Boucher, James and Keeler 1982; Henry 1966, 1967; Trench 1979; Law and Lewis 1983), but the few that are known from mangroves mostly involve interactions of plants and micro-organisms.

Some of the epiphytes occurring in mangroves (see chapter 4) may have a mutualistic association, but whereas the benefit to the epiphyte is easily discernible, the benefit, if any, to the mangrove partner remains questionable.

Four types of mutualistic interactions between mangroves and micro-organisms can be identified, although details of their frequency of occurrence are not available.

Probably the most widespread interaction occurs in the rhizosphere, that zone immediately surrounding the fine roots which is characterized by an enhanced microbial activity (Smith and

Delaune 1984). Although generally not intimately connected with root cells, the fungi and bacteria modify the micro-environment around each root through their metabolic activity, releasing nutrients and altering the pH of the soil. In turn, this microbial flora probably depends on the leakage of organic material from the roots, which can be utilized by the micro-organisms as a source of energy. A number of soil fungi are characteristically associated with mangrove roots (see chapter 4) and form part of this rhizosphere flora.

The second type of interaction is more intimate and involves fungi which form a direct association with roots. These fungal associations are termed mycorrhizae (Harley 1969). In some cases, the fungi are unicellular and live within individual root cells ("endomycorrhizae"), but in many cases the fungi cover the root tips in a thick mat and penetrate the intercellular spaces of the cortex ("ectomycorrhizae"). Although no mycorrhizae have been reported specifically from mangroves so far, they are frequently found in forest and swamp soils that are rich in organic matter. Mycorrhizae also have been reported from sand-dunes with low organic matter, where they appear to facilitate the availability of phosphorus from the mineralized coating on sand grains (Jehne and Thompson 1981). In view of their habitat diversity, it would seem likely that some mangrove mycorrhizae do occur. Like the rhizosphere flora, mycorrhizae facilitate the movement of phosphorus, potassium and calcium into the roots and the movement of metabolites from the roots to the fungus.

Root nodules comprise a third type of mutualistic interaction. The bacterial genus *Rhizobium* forms nitrogen-fixing nodules almost exclusively on the roots of the angiosperm family Leguminosae (Allen and Allen 1981). Nodulated legumes grow more vigorously than do non-nodulated ones in nitrogen-deficient soils. In view of the low nitrogen status of most mangrove soils (Boto 1983), root nodules may be important to the two legumes commonly found in mangroves — that is, the mangrove *Cynometra iripa* and the climbing associate *Derris trifoliata*; critical investigation of these two species is desirable.

Another kind of micro-organism that forms nitrogen-fixing root nodules with higher plants is the actinobacterial genus *Frankia*. A number of unrelated genera of flowering plants form nodules with *Frankia*, including *Casuarina* (Bond 1956, 1963); plants with nodules grow much better in nitrogen-deficient media than those without nodules (Bond 1963). For *Casuarina glauca*, a common inhabitant of the landward margins of mangroves where they abut freshwater swamps, these nodules may be of considerable ecological significance.

The fourth type of mutualistic interaction involves bacterial leaf nodules, and is common in over four hundred species of the angiosperm families Rubiaceae and Myrsinaceae (Lersten and Horner 1976). The bacteria are maintained as a colony in the closed shoot tips of the plant and enter developing leaves through the stomatal pores, ultimately forming chambers along the leaf margin (Miller, Gardner and Scott 1983). These bacterial leaf nodules have been shown to be capable of nitrogen fixation (Van Hove 1976), and they may be involved in synthesis of cytokinin (Miller, Gardner and Scott 1983).

Bacterial leaf nodules are present in the American mangrove *Laguncularia racemosa* (Humm 1944). The Australian mangroves *Lumnitzera racemosa* and *L. littorea* (Combretaceae) possess a small "gland" or domatium at the apex of the leaf, with occasionally smaller glands in the axils of the lesser veins (Jones 1971). These glands are elliptical chambers immediately below the epidermis and are similar in appearance to the leaf nodules of genera that have received detailed study. The species of *Lumnitzera* undoubtedly would repay detailed investigation; similarly, *Aegiceras corniculatum* and *Scyphiphora hydrophyllacea* (which belong respectively to the Myrsinaceae and Rubiaceae), in which bacterial leaf nodule formation is extremely common (Lersten and Horner 1976), would be worthy of detailed study.

Although the information available on mutualistic interactions in the Australian mangrove flora is extremely limited, there is sufficient to suggest that the study of mutualism may be a potentially productive line of investigation. The fact that most of the interactions involve the availability of nitrogen, which is otherwise in short supply in Australian mangroves, suggests that considerable ecological significance may attach to an understanding of the relationships between mangroves and micro-organisms.

Competition

Competition between plants has been defined as the tendency of neighbouring plants to utilize the same quantum of light, ion of mineral nutrient, molecule of water, or volume of space (Grime 1973). According to this definition, competition refers exclusively to the capture of resources and is only one of the mechanisms whereby a plant may inhibit the growth of a neighbour by adversely modifying its environment. In this sense, competition is strongly contrasted with antagonism, two interactions which are often lumped together in the more traditional usage of the term "competition" (Milne 1961; Harper 1961).

The competitive ability of a plant is a function of the area, activity and distribution in space and time of the plant surfaces through which resources are absorbed and, as such, it depends upon a combination of plant characteristics including storage organs, height, lateral spread, phenology, growth rate, response to stress and response to damage (Grime 1979). Several of these characteristics have been discussed already (chapter 2) under the heading of adaptation — the selective change of a particular set of characteristics in a way suited to a particular environment.

Stated in another way, plants will tend to disperse as widely as possible. This may take them into habitats where their physiological optima are exceeded. If they encounter other individuals better suited to the prevailing environment, differences in growth potential, either above or below ground, will result in the suppression of the less-suited individual.

Within the mangrove environment, most plant species are widely dispersed. However, large differences in the environmental conditions also occur, particularly in relation to water, salt, nutrients and light. It seems clear from the experimental work of Clarke and Hannon (1971) that the sharp boundaries between communities dominated by different species are often the direct result of competition.

Even within communities, species composition may be determined, or at least influenced, by competitive interactions between component species. For example, in a detailed transect of the mangroves of the open shoreline at Princess Charlotte Bay, it was found that the distributions of *Ceriops* and *Avicennia* overlapped to a large extent (Elsol and Saenger 1983). Their relative importance values (figure 28) indicate that two broad bands are recognizable: (1) the landward 55 metres in which the importance values of the two species lie on a negatively sloping line, that is, one varies inversely with the other, and (2) from 60 metres seawards with a positive slope, where both species vary in proportion to each other. This suggests that from 0 to 55 metres in the transect the environmental conditions are favourable for both species and they compete with each other. From 60 metres onwards, the conditions are no longer so favourable, and both species together decline in importance (Elsol and Saenger 1983). The similarity of these species, in terms of their salinity (see table 14) and waterlogging (see table 13) tolerances, supports the notion of such a competitive interaction. Undoubtedly, other examples of competition between various mangrove species occur, although little work has been done on this aspect.

Grime (1973, 1979) argued that competition must be viewed in the context of major adaptive strategies which have evolved in plants, and it is important to relate these strategies to the processes which determine the structure and species composition of vegetation.

Two categories of external factors limit the amount of living and

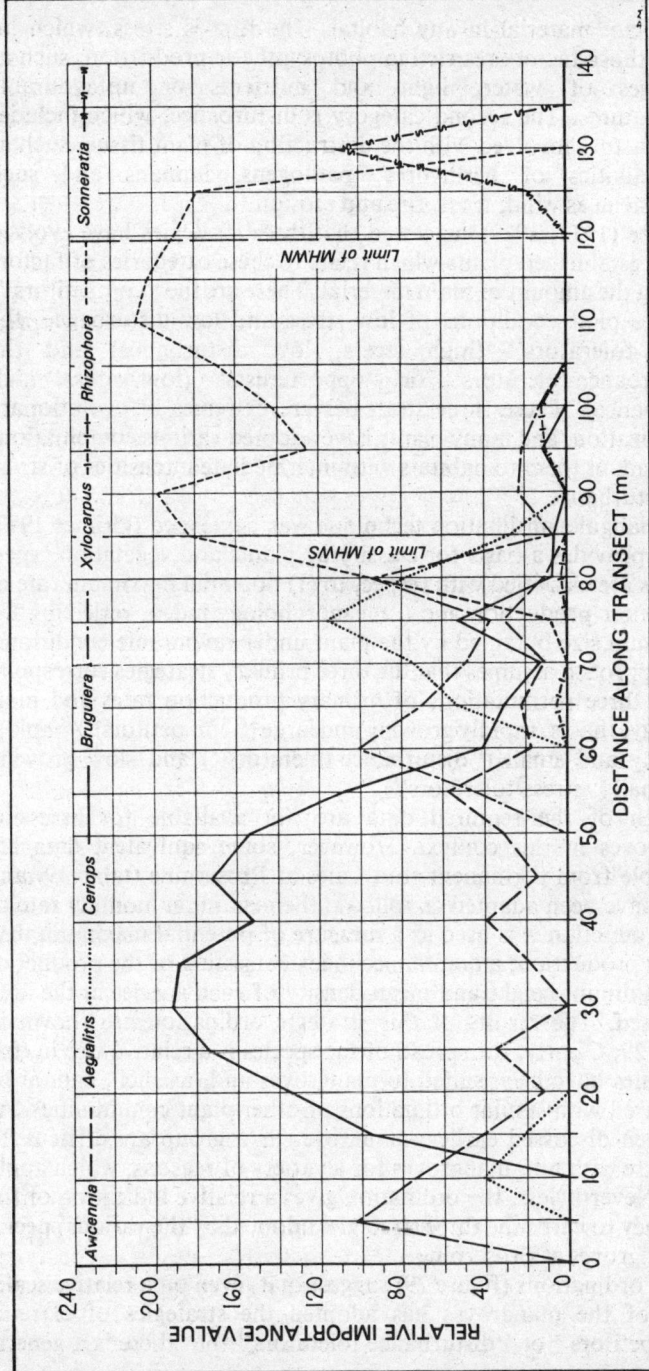

Figure 28 Open shoreline zonation at Princess Charlotte Bay using the distribution of relative importance values of the various species across a transect at right angles to the shoreline (from Elsol and Saenger 1983).

dead plant material in any habitat. The first is stress, which includes those factors restricting photosynthetic production, such as shortages of water, light and nutrients or unfavourable temperatures. The second category is disturbance, which includes those factors involved with the destruction of plant tissue, such as the activities of herbivores, pathogens, humans and such phenomena as wind, frost, fire and erosion.

Grime (1973, 1979) suggested that three strategies have evolved among established plants which relate to these categories of factors limiting the amount of plant material. These are the "competitors", which exploit conditions of low stress and low disturbance, the "stress-tolerators" (high stress, low disturbance) and the "disturbance-tolerators" or "opportunists" (low stress, high disturbance). These three strategies are extremes of evolutionary specialization, and many plants have adopted various combinations which adapt them to habitats with intermediate intensities of stress and disturbance.

A triangular ordination technique was developed (Grimes 1977) which provides a basis for classifying plants and vegetation types. Species are classified with respect to (1) potential maximum rate of dry-matter production and (2) a morphology index, reflecting the maximum size obtained by the plant under favourable conditions. This approach assumes that the three primary strategies correspond to the three permutations of primary production rates and morphology, that is, rapidly growing and large ("competitors"), rapidly growing and small ("disturbance-tolerators") and slow growing and small ("stress-tolerators").

Much of the required data are not available for assessing mangroves in this context. However, some equivalent data are available from permanent study sites at Proserpine (table 19) and these have been adapted as follows: the maximum monthly rate of leaf production was used as a measure of potential maximum dry-matter production; a dominance index consisting of the product of the maximum height and mean density of each species in the area was used. The results of this strategic ordination are shown in figure 29. Clearly, the spread of the species is a relative one in that the scales have been suited to mangroves and, as such, cannot be compared with similar ordinations of other plant communities. As has been discussed earlier, mangroves as a group are difficult to compare with non-mangroves for a variety of reasons, which apply here. Nevertheless, the ordination gives a relative indication of the tendency towards the three strategies adopted by the various species of mangroves at Proserpine.

The ordinations (figure 29) suggest that, even on a relative scale, none of the mangroves has adopted the strategies of extreme "competitors" or "disturbance-tolerators", but there is a general

Table 19 Data used to derive growth and dominance indices for the strategic analysis of Proserpine mangroves

Species	Maximum monthly leafing rate (lvs./1000 day^{-1})	Maximum height (m) H	Density (No./1,000 m^2) D	Dominance index $H \times D$
Cynometra iripa	0.5	3.5	4	14
Bruguiera parviflora	2.9	10.	2	20
Bruguiera gymnorhiza	3.4	12	2	24
Heritiera littoralis	3.5	15	6	90
Bruguiera exaristata	4.3	10	18	180
Ceriops tagal	6.1	10	102	1020
Avicennia marina	6.2	15	62	930
Rhizophora stylosa	6.7	15	208	3120
Aegiceras corniculatum	7.2	3.5	183	640
Acanthus ilicifolius	7.9	1.5	582	873
Osbornia octodonta	16.4	5	2	10
Xylocarpus granatum	28.4	12	8	96
Xylocarpus australasicus	46.6	12	14	168
Lumnitzera racemosa	48.1	8	44	352
Excoecaria agallocha	131.6	10	136	1360

Source: Saenger (unpubl. data).

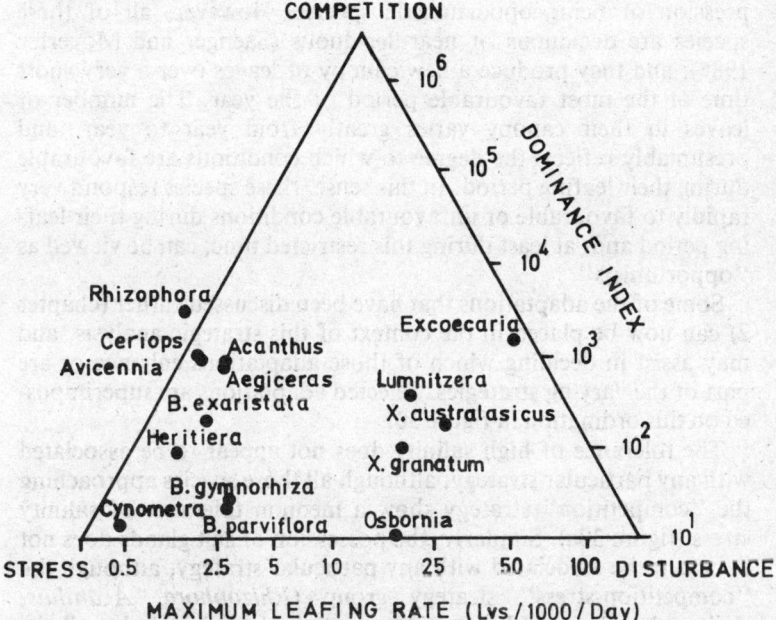

Figure 29 Strategic ordination of mangrove species at Repulse Bay, Queensland. Note that the dominance index was calculated using the maximum height of each species in the study area multiplied by that species' density in number m^{-2}.

distribution of these species towards the "stress-tolerator" strategy. The numerically most abundant and widespread species — *Avicennia marina*, *Rhizophora stylosa*, *Aegiceras corniculatum* and *Ceriops tagal* — appear to have a combined "competitor/stress-tolerator" strategy which would enable them to persist during unfavourable periods on the one hand and to exploit favourable periods reasonably efficiently on the other. *Avicennia* is probably the most-studied member of this group; its ability to grow in a wide range of habitats appears to be due to its response to increasing stress by reducing its growth rate and adjusting its growth habit. It is also worth noting, in relation to the previous comments concerning the competitive interactions between *Avicennia* and *Ceriops*, that these two species appear to be almost identical in their adopted strategy which, together with their similar tolerances to certain environmental conditions, suggests that they are indeed competing with each other. *Acanthus* and *Aegiceras* is another possible competitive pair that should be investigated where their distributions overlap.

It may seem surprising to find species such as *Excoecaria agallocha*, *Xylocarpus* spp., *Lumnitzera racemosa* and *Osbornia octodonta* tending towards the "opportunist" strategy. As slow-growing members of the landward fringes, they do not give the impression of being opportunistic species. However, all of these species are deciduous or near-deciduous (Saenger and Moverley 1985), and they produce a new canopy of leaves over a very short time at the most favourable period of the year. The number of leaves in their canopy varies greatly from year to year, and presumably reflects the degree to which conditions are favourable during their leafing period. In this sense, these species respond very rapidly to favourable or unfavourable conditions during their leafing period and, at least during this restricted time, can be viewed as "opportunists".

Some of the adaptations that have been discussed earlier (chapter 2) can now be placed in the context of this strategic analysis, and may assist in deciding which of those adaptations enhance or are part of the varying strategies. Selected adaptations are superimposed on this ordination in figure 30.

The tolerance of high salinity does not appear to be associated with any particular strategy, although all those species approaching the "competition" strategy show a medium tolerance of salinity stress (figure 30a). Similarly, the possession of salt glands does not appear to be associated with any particular strategy, although the "competition-stress" strategy group (*Rhizophora*, *Acanthus*, *Avicennia*, *Ceriops*, *B. exaristata* and *Aegiceras*) includes all the species with salt glands. Figure 30 also shows that there is little relationship between the ability to tolerate high soil salinities and the

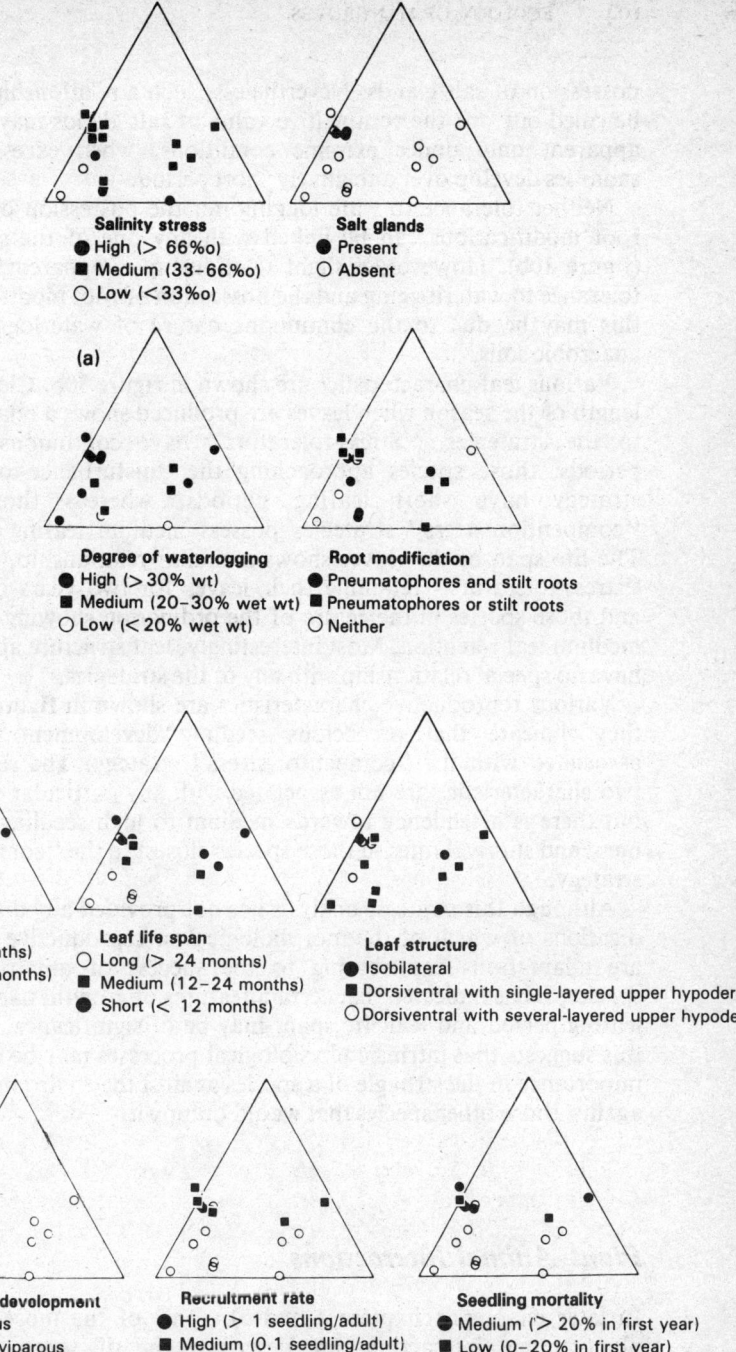

Figure 30 Superimposition of various characteristics on the ordination shown in figure 29: (a) salinity of their environment, presence of salt glands, degree of waterlogging and root modifications; (b) length of leafing period, leaf life span and leaf structure; and (c) type of seedling development, recruitment rate and seedling survival.

possession of salt glands. Nevertheless, such a relationship cannot be ruled out, for the competitive value of salt glands may become apparent only under extreme conditions, when excessive soil salinities develop over a relatively short period.

Neither tolerance to waterlogging nor the possession of certain root modifications can be linked with any one of the strategies (figure 30b). However, a slight relationship is apparent between tolerance to waterlogging and the possession of root modifications; this may be due to the continuous nature of waterlogging and anaerobic soils.

Various leaf characteristics are shown in figure 30b. Clearly, the length of the season when leaves are produced shows a relationship to the strategies. "Stress-tolerators" have continuous leafing periods, those species approaching the "disturbance-tolerator" strategy have short leafing periods, whereas those with "competition-stress" strategies possess medium leafing periods. The life span of the leaves shows a similar relationship, with the "stress-tolerators" retaining their leaves for two years or more, and those species in the centre of the ordination showing short to medium leaf retention. Most interestingly, leaf structure appears to have no special relationship with any of the strategies.

Various reproductive characteristics are shown in figure 30 and they indicate that precocious seedling development may be associated with the "competitor-stress" strategy. The remaining two characteristics are not associated with any particular strategy, but there is a tendency towards medium to high seedling recruitment and survival rates in those species closest to the "competitor" strategy.

Although this strategic analysis has not provided any distinct indications of which of the morphological or reproductive features are adaptations contributing to the success of any particular species, it has indicated that certain features of growth, particularly leafing period and leaf life span, may be of significance. In turn, this suggests that intrinsic physiological processes may be of prime importance in the struggle of a species against the environment and against those other species that would occupy it.

Plant–Animal Interactions

In later chapters (chapters 4 and 5), some of the more specific plant–animal interactions are described and discussed. The intimate link between certain animals and mangrove pollination is examined; the mutualistic cohabitation of mangroves with such animals as ants and butterflies is discussed; the occurrence of various termite species and their effects on mangroves are noted; and the various species of boring organisms and their roles in living and decomposing wood are summarized.

The plant-animal interactions discussed in this chapter are those that are of widespread significance or which directly or indirectly affect the physical environment in which mangroves grow.

Sediment Turnover

Probably the best example of this kind of interaction is the activities of crabs and mud lobsters in reworking the sediments among the mangroves (Macnae 1966; Bennett 1968). Mud lobsters (*Thalassina anomala*) build large tunnelling burrows which are generally recognized by the mound of fresh mud up to 75 centimetres high around their entrances. The burrows are U-shaped and extend up to 1.5 metres below the surface. Their entrances generally are blocked by a mud plug during the day, but at night they are opened when the mud lobster emerges to feed on surface muds. These burrowing activities have various effects, but the enormous amounts of mud these animals bring to the surface help to mix the soils and to change their surface characteristics. Often, the soil brought to the surface is rich in organic matter and pyrite and may be characterized by strong sulphate reduction and the presence of FeS (Andriesse, van Breeman and Blokhills 1973). This fresh mud oxidizes on the surface and often forms localized patches of highly acidic muds (acid-sulphate soils). Gradually, however, as the mud mounds age, the sulphur content decreases as a result of leaching, and these slightly raised areas then are suitable for mangrove colonization. In *Rhizophora* or *Bruguiera* forests, such elevations are colonized by the mangrove fern (*Acrostichum speciosum*) in small discrete patches.

The mud lobster provides an example of a species which can markedly alter the mangrove environment: the burrows allow drainage of and interchange between surface water and subsoil water; the mud is turned over, with subsurface muds placed on the surface where they can be oxidized which, in turn, leads to their acidification. Once the sulphides are oxidized to sulphates, they can be leached from the mounds, allowing the mounds to be colonized by the mangrove fern which is not able to grow at the general level of the surrounding mud surface.

Other burrowing organisms have similar effects, although generally on a smaller scale. The burrows of fiddler crabs, mudskippers and even the mud crab (*Scylla serrata*) allow drainage, mixing and a degree of aeration of subsurface waters in the mangroves, and in this way enhance the growth of mangroves.

Grazing and Trampling

Another important example of plant–animal interaction in the mangroves is that of grazing and trampling. The importance of grazing in the mangrove ecosystem is not well documented, but probably has the effect of maintaining the mangrove community at a lower level of plant biomass that would occur in its absence. In this sense, grazing and trampling are not unlike other regularly continuing disturbances.

Mangrove foliage contains significant quantities of minerals, vitamins, amino acids, proteins, fat and crude fibre, and is thus a nutritious food source for herbivores (Kehar and Negi 1953; Sokoloff, Redd and Dutcher 1950; Sundararaj 1954). In fact, Morton (1965) found that when used as cattle feed the leaves of *Rhizophora mangle* increased the yield of milk. Consequently, it is not surprising that mangrove foliage is grazed by cattle, sheep, goats and camels.

In Australia, grazing of mangroves by coastal species of wallabies and kangaroos is important locally, and in the Northern Territory the naturalized buffalo (*Bubalus bubalis*) grazes mangroves in substantial numbers. Probably of far greater significance, however, is the widespread grazing by insects (Johnstone 1981). Leaves of *Avicennia*, *Bruguiera*, *Rhizophora*, *Ceriops* and *Heritiera* often are found with serrated edges owing to damage by insects (particularly tettigonids) and by crabs of the family Grapsidae. Heald (1969) estimated a mean grazing effect on Florida mangrove leaves of 5.1 per cent of the total leaf area, ranging from 0 per cent to 18 per cent on a leaf-by-leaf basis. Beever, Simberloff and King (1979) found grazing by an arboreal grapsid crab (*Aratus pisonii*) on *Rhizophora mangle* leaves to range from 0.4 per cent to 7.1 per cent of total leaf area. The significance of these activities in litter breakdown (Malley 1978) and in the export and recycling of organic matter is being investigated currently; Johnstone (1981) has suggested that approximately one-fifth of all mangrove leaf material is diverted to herbivorous rather than detrital food chains.

Wilson and Simberloff (1969) and Simberloff and Wilson (1969, 1970) found over one hundred insect species on small mangrove islands in Florida Bay. They also found that the insect population re-established quickly after fumigating these islands with methyl bromide, and that it reached pre-fumigation levels within one year. After that, total species numbers remained more or less constant although there was considerable species turnover from year to year.

The size and diversity of the insect populations suggest that insect grazing is a significant factor in the structure and function of mangrove communities. Onuf, Teal and Valiela (1977) tested this experimentally by comparing two mangrove islands in Florida (see page 83): one island (high-nutrient area) had breeding colonies of pelicans and egrets, and the other (low-nutrient area) did not. The effect of the input of nutrients by birds on plant growth and reproduction has been noted already. More striking, however, was the significant stimulation of herbivory by insects in response to nutrient enrichment. Larvae of five lepidopteran species that fed on leaves and/or buds were either more abundant or present in the high-nutrient area only, as was the mangrove borer (*Poecilips rhizophorae*) that infested propagules before they dropped from the parent tree. This resulted in a fourfold greater loss to herbivores (26 per cent of total leaf area lost to grazing) and more than offset the increased leaf production owing to high nutrient input.

The observed difference in grazing in the two areas disappeared when the birds seasonally migrated from their nesting area at the high-nutrient island. This relationship between birds, nutrient enrichment and insect damage illustrates the complex interactions that occur in mangroves, as in other vegetation types. However, the effects of large nesting or roosting aggregations in mangroves such as those of white ibis (*Threskiornis molucca*) in northern Australia, lesser noddies (*Anous tenuirostris*) on the Albrolhos Islands in Western Australia and fruit bats (*Pteropus alecto* and *P. conspicillatus*) in eastern Australia undoubtedly enhance the local nutrient status, and may simultaneously be increasing the general level of herbivory.

At present, it is not known whether any of the mangrove species have developed specialized defence mechanisms against grazing. Many mangrove leaves are extremely tough and high in tannins (chapter 2); this feature may reduce their palatability. *Acanthus ilicifolius* has leaves with spinous margins which may discourage grazing, and it contains saponins (Minocha and Tiwari 1981) which may render it unpalatable. Several species of *Melaleuca* occurring near swampy margins of mangroves contain various oils (Jones and Harvey 1936), of which nerolidol recently has been demonstrated to have anti-feeding properties effective against larvae of the gypsy moth, *Lymantia dispar* (Doskotch et al. 1980). The mangrove *Excoecaria agallocha*, which contains a milky sap, rarely shows evidence of grazing, and it seems likely that the latex discourages grazing either by its taste or by its toxic properties.

Other animal–plant interactions include the damage done to the foliage by nesting birds and leaf-weaver ants. Leaf-weaver ants (*Oecophylla smaragdina*) weave leaves into nests within the mangrove canopy and, in turn, provide some protection to the mangroves by preying on herbivorous insects. Nevertheless, for mangroves of the family Rhizophoraceae which have strictly terminal growing points, these ant nests effectively inhibit the further development of the affected shoot, and thus can impair the full development of the tree. Similar results have been noted where large aggregations of nesting birds occur. Nesting birds may use twigs pruned from the top of the mangrove canopy to build massive nests (for example white ibis), and they commonly peck at and damage the young growing tips within reach while attending their young.

Interactions Expressed as Structure

All of the interactions discussed in the previous sections produce a range of physico-chemical settings and of mangrove communities, which differ in their function and structure.

These various attributes can be used to classify mangrove communities (see page 119), a process which through its data reduction can provide some overview of the types of mangrove communities and how dominant interactions shape them.

There are, however, a number of situations where the mangroves are zoned, and these provide a particularly good opportunity to investigate and perhaps answer many ecological questions, for these zoned communities can be treated as the outcome of a natural experiment (Pielou 1977). Especially where uniformly sloping environmental gradients are involved, the zonation pattern can be interpreted and used to study the plant–environment interactions of the constituent species.

Two types of zonation are discussed below: the parallel zonation along open shorelines and the longitudinal zonation along rivers. As pointed out by Elsol and Saenger (1983), both types are superimposed in the lower reaches of rivers and their deltas and may produce diverse floristic assemblages with highly complex patterns that cannot be interpreted even with sophisticated techniques of pattern analysis (Bunt and Williams 1980, 1981). Consequently, both types of zonation are discussed separately, although a preliminary attempt at their integration is then made.

Parallel Shoreline Zonation

The phenomenon of rather predictable, often monospecific zones of mangroves parallel to shorelines is well documented (table 20), and there is general agreement on the sequence of zones (Macnae 1966, 1967; Saenger et al. 1977). However, the underlying causes as to why mangroves so frequently appear in zones are far from clear. Snedaker (1982) critically reviewed suggested causes and found that they fell into the following general categories: plant succession, geomorphology, physiological ecology and population dynamics.

Table 20 References dealing with zonation of mangrove communities in Australasia

AUSTRALIA
General
Macnae 1966; Saenger et al. 1977; Lear and Turner 1977

Queensland
Macnae (1966, 1967, Townsville and Cairns); Spenceley (1983, Townsville); Saenger and Hopkins (1975, Gulf of Carpentaria); Saenger and Robson (1977, Port Curtis); Graham et al. (1975, Trinity Inlet); Elsol and Saenger (1983, Princess Charlotte Bay); Shanco and Timmins (1975, Bustard Bay); Thom (1975, low wooded islands on Great Barrier Reef)

Northern Territory
Specht (1958, Arnhem Land); Hegerl et al. (1979, Alligator Rivers region)

Western Australia
Semeniuk, Kenneally and Wilson (1978, general); Thom, Wright and Coleman (1975, Cambridge Gulf); Sauer (1965, Port Hedland); Semeniuk (1980, King Sound); Congdon (1981, Blackwood River salt marshes)

South Australia
Osborn and Wood (1923, Port Wakefield)

Victoria
Bird (1971, Westernport Bay); Bridgewater (1975, Westernport Bay)

Tasmania
Kirkpatrick and Glasby (1981, general salt marshes); Curtis and Sommerville (1947, Boomer Marsh); Guiler (1951, Pipe Clay Lagoon)

New South Wales
Kratochvil, Hannon and Clarke (1973, Sydney); Hutchings and Recher (1974, Careel Bay)

New Zealand
Chapman (1977, general); Chapman and Ronaldson (1958, Auckland Isthmus)

Papua New Guinea
Chapman (1977, general); Womersley (1983, general); Percival and Womersley (1975, general); Johnstone (1983, Hood lagoon)

Oceania
Baltzer (1969, Dumbea River, New Caledonia); Chapman (1977, general)

Zonation as the spatial expression of plant succession was the earliest view, going back to Curtiss (1888). This view interprets zoned mangrove communities as a sequence of seral communities from seawards to landwards. It hinges on the apparent ability of *Rhizophora* to build and colonize new land (primary succession) by trapping sediments among its root system into which fall the viviparous propagules that colonize the newly won land (Davis 1940; Richards 1964). Further build-up of the substrate allows other mangrove species to invade and eventually replace the *Rhizophora* (secondary succession), until build-up exceeds the level of tidal inundation. At this point, terrestrial species that are not salt-tolerant invade and replace (out-compete) the mangroves.

The view that plant succession is the basis for mangrove zonation is logical and appealing (Snedaker 1982). However, with more detailed study of succession in ecosystems, it is becoming increasingly clear that mangrove zonation does not conform to the general characteristics of secondary succession (Snedaker 1982; Johnstone 1983).

Because of the undoubted ability of mangroves to trap sediment and thus build land, mangrove zonation has been interpreted as a response to geomorphic change. Snedaker (1982) and Woodroffe (1983) summarized the evidence of geomorphic control over vegetational patterns and species assemblages, particularly landform patterns and vegetation. Thom (1967, 1975) and Thom, Wright and Coleman (1975) investigated vegetation in Mexico and Australia. They were able to relate species assemblages, distributions and overall spatial organization to the depositional and erosional histories and to subsidence, compaction, freshwater discharge and sea-level rise.

As discussed earlier, mangrove development and zonation are historically bound to the geomorphic process of a region through the particular soils and soil conditions that these processes have produced. Consequently, more direct information on mangrove growth, development and zonation clearly can be obtained from more direct (physiological and ecological) studies of the soil–mangrove relationship.

Zonation as a physiological response to tidally maintained gradients has received considerable attention since the classical work of Watson (1928). The interactions between surface hydrology and salinity on the one hand and mangrove zonation on the other has been reviewed extensively (Macnae 1967, 1968; Clarke and Hannon 1967, 1969, 1970, 1971; Walsh 1974; Lugo and Snedaker 1974; Chapman 1976). However, as Snedaker (1982) pointed out, although good correlations may exist between salinity, tidal inundation and mangrove zonation, the physiological response of the mangroves to many of these features is so poorly known for most species that it is premature to conclude such correlations imply causality.

Clarke and Hannon (1967, 1969, 1970, 1971) have shown experimentally and in the field that species of salt-tolerant plants near Sydney do have definable tolerances and optima under specific conditions, and that these can be used to explain landward and seaward boundaries for each species (Clarke and Hannon 1971).

Other authors also have established experimental optima for several mangroves, especially in relation to salinity (McMillan 1971, 1974; Connor 1969; Kylin and Gee 1970; Cintron et al. 1978; Teas 1979; Downton 1982); some long-term field data on the soil salinities and degrees of waterlogging under which different species occur are given in tables 13 and 14.

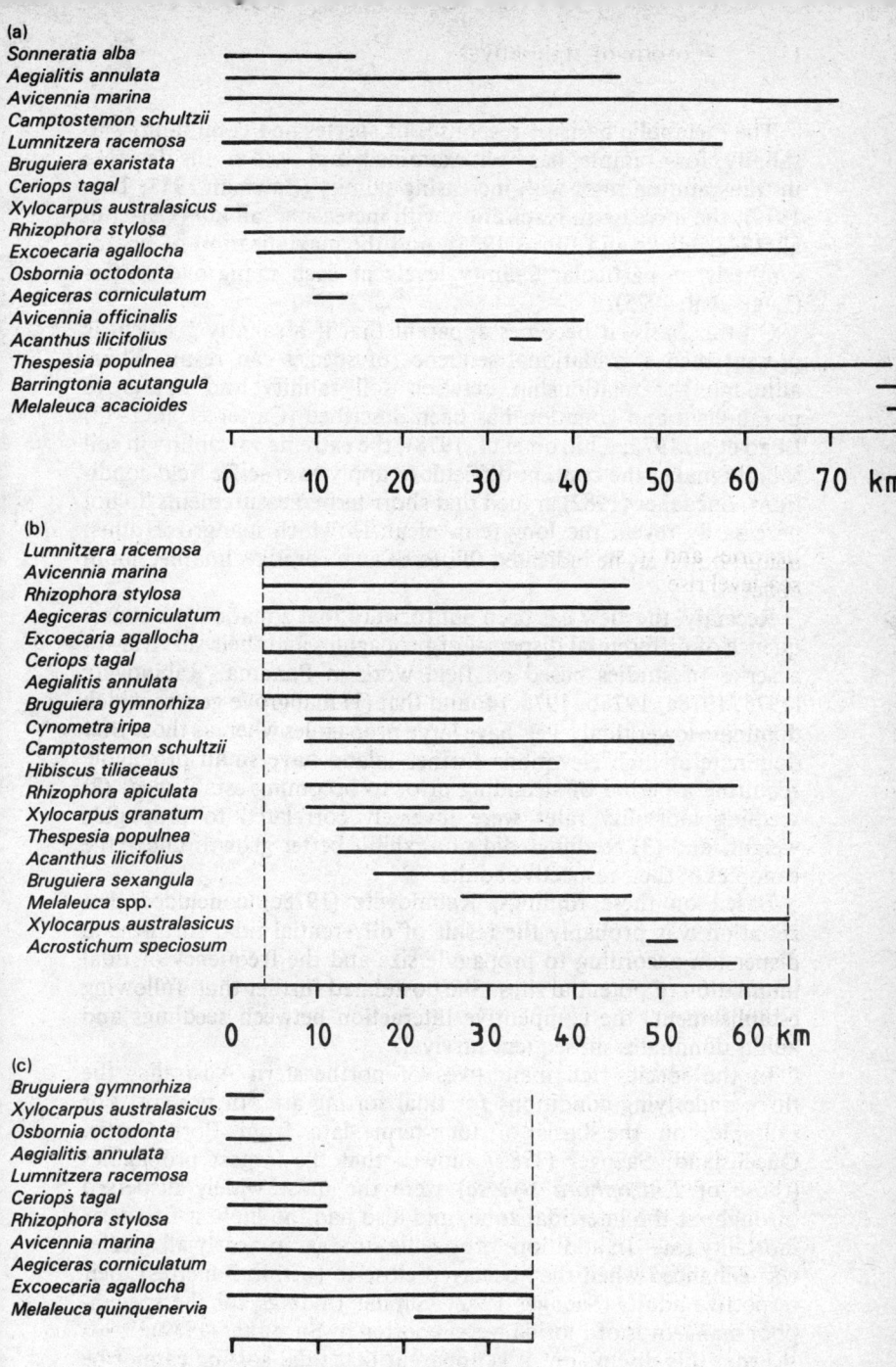

Figure 31 Upriver distribution patterns of mangroves in (a) East Alligator River, Northern Territory (after Hegerl et al. 1979, and from Wells and Saenger, unpubl. data); (b) Watson River, Cape York (Saenger, unpubl. data); and (c) Calliope River, central Queensland (Saenger, unpubl. data).

The metabolic basis of responses of species and communities to salinity, for example, has been examined, and rests on the decrease in transpiration rates with increasing salinity (Bowman 1917; Teas 1979), the increase in respiration with increasing salinity (Carter et al. 1973; Hicks and Burns 1975), and the maximization of photosynthesis at particular salinity levels in each mangrove species (Lugo et al. 1975).

On this basis, it becomes apparent that if a salinity gradient is present then a gradational sequence of species can result. Thus, although the relationship between soil salinity and mangrove metabolism and zonation has been described (Carter et al. 1973; Lugo et al. 1975; Cintron et al. 1978), the extreme variability in soil salinity makes the concept difficult to apply to specific field conditions. Snedaker (1982) argued that short-term measurements do not necessarily reveal the long-term mean to which mangroves must adapt; however, he indicated this to be an error in technique, not in concept.

Recently, the view has been put forward that zonation is a consequence of differential dispersal of propagules and their survival. In a series of studies based on field work in Panama, Rabinowitz (1975, 1978a, 1978b, 1978c) found that (1) mangrove genera which dominate lower tidal levels have large propagules whereas those that dominate at high elevations further inland have small propagule requiring a period of stranding prior to becoming established, (2) seedling mortality rates were inversely correlated to propagule weight, and (3) seedlings did not exhibit better growth under the canopies of their respective adults.

Based on these findings, Rabinowitz (1978c) concluded that zonation was probably the result of differential tidal sorting and dispersion according to propagule size and the frequency of tidal inundation of potential sites. She postulated further that, following establishment, the competitive interaction between seedlings and adults dominates subsequent survival.

In the species-rich mangroves of northeastern Australia, the three underlying conditions for tidal sorting are not present. For example, on the basis of long-term data from Port Curtis, Queensland, Saenger (1982) showed that the largest propagules (those of *Rhizophora stylosa*) were the most widely dispersed throughout the intertidal zone, and also had the highest first-year mortality rate. In addition, propagule survival in nearly all species was enhanced when they occurred close to (within 2 metres) their respective adults (Saenger 1982). Similar findings for the species-poor mangroves of Florida were reported by Snedaker (1982).

From this discussion, it is apparent that tidal sorting cannot be invoked as a universal mechanism, and that in at least some mangrove communities it has a minor role, if any at all.

Snedaker (1982) concluded that of the four views mentioned above, those advocating geomorphology and environmental physiology appeared to be most relevant in furthering an understanding of zonation and plant succession in the intertidal zone. He argued that there is a temporal tendency for each species to assume competitive dominance in its preferred zone. Whether this occupation is guided by physical forces or results from interspecific competition, the species which can maximize its photosynthetic output with greatest metabolic efficiency dominates in competition with other species. The concept of a zone or environmental preference implies that each mangrove species does have a preferred optimum and a limit of tolerance related to the metabolic cost of existence along an environmental gradient. Variations in that gradient may either last long enough to result in competitive exclusion or domination by a previously subordinate competitor or fluctuate around a long-term mean which enhances the likelihood of survival of the existing dominant and, thus, the zone.

Longitudinal Upriver Zonation

Although sharing some features with the parallel zonation of shorelines, upriver zonation has been recognized as a distinct phenomenon since the descriptive accounts of Myers (1935) on the riverine vegetation of South America.

Myers defined upriver zonation as the definite sequence of plant communities along the course of a stream, determined not by edaphic factors of the area through which the river flows, but by factors dependent on the stream itself (for example, its width in a given place or the distance from the sea) and thus recurring in an essentially similar sequence in all the streams of the region where modification by humans has not obscured it.

As shown by the species distributions in figure 31, sequences of mangrove communities also occur in Australian river systems, and show some similarities despite the great geographical and geological differences among the three river systems considered. For example, both *Avicennia marina* and *Excoecaria agallocha* have wide upriver distributions, whereas *Ceriops tagal* and *Lumnitzera racemosa* have limited, downriver distributions. On the other hand, *Aegiceras corniculatum* and *Xylocarpus australasicus* show very different upriver distributional patterns in the three river systems (Bunt, Williams and Clay 1982; Elsol and Saenger 1983; Hegerl et al. 1979).

Bunt, Williams and Clay (1982) related upriver distributions of mangroves to upriver distance and salinity gradients (see fuller discussion on p. 70) in rivers of northeastern Australia. They found that *Rhizophora stylosa*, *R. apiculata*, *Sonneratia alba* and *Ceriops tagal* were mainly from downstream, high-salinity areas, whereas *Heritiera littoralis*, *Excoecaria agallocha*, *Acrostichum* sp., *Aegiceras corniculatum* and *Rhizophora mucronata* occurred principally in upstream, low-salinity areas. Three other species, including *Avicennia marina*, showed no correlations with site and were found over almost the entire salinity range of the river systems.

Correlations with distance from river mouth were always better than those with salinity, and in the Lockhart River, where virtually no salinity gradient was found, Bunt, Williams and Clay (1982) concluded that certain of the mangroves were responding to some aspect of distance other than salinity. Furthermore, their study indicated that distance does not simply act as an integrated measure of salinity, or at least not universally.

Myers (1935) tentatively identified three factors which influence the upriver zonation, including (1) width of the river, (2) character of the water and (3) distance from the sea. According to him, the width of the river was of importance because it determined whether the waterway acted as a light gap, exposing the vegetation on the bank to full sunlight. The character of the water largely depends on catchment characteristics, whereas the distance from the sea is important because salinity is a function of that distance and of the size of the river.

Two additional factors can be suggested which appear to correlate with upriver distance: upstream gradients of decreasing salinity fluctuation (or range) and increasing turbulent flow. Clearly, the salinity range to which a mangrove species is exposed may be just as important as the absolute levels of salinity, and may influence upriver distributions of individual species. Leaf thickness in *Rhizophora mangle*, for example, is related more to salinity fluctuations than to absolute levels of salinity (Camilleri and Ribi 1983).

There is little evidence to support any direct effects of turbulent flow on mangroves, although flow characteristics can affect the meanders in a river, which in turn may affect species absence or presence. Erosion of concave banks and accretion on convex lobes are largely associated with the intermittent seasonal flow of floodwaters. The species composition of actively accreting convex lobes are generally strikingly different from nearby, non-accreting river banks (Elsol and Saenger 1983).

Clearly, the phenomenon of upriver zonation is still poorly understood. Gradients of absolute salinity or of degree of salinity fluctuation are clearly involved (Bunt, Williams and Clay 1982), but other factors such as the width of the stream and its geomorphological characteristics may also have an effect. As has been discussed already under parallel shoreline zonation, if a salinity gradient is present, then a gradational species sequence can result, although in river systems the sequence may be secondarily modified by river width and sedimentary characteristics. The fact that Bunt, Williams and Clay (1982) found better correlations with distance than with salinity may be referable to shortcomings of measuring salinity. As previously discussed, short-term measurements in a river system are unlikely to reveal long-term means or ranges to which mangroves must adapt. As in the case of soil salinities in shoreline zonation, this can be viewed as an error in technique rather than concept (Snedaker 1982). Consequently, it seems appropriate that, until detailed long-term salinity studies can be related to upriver mangrove distributions, the view that salinity gradients are important should not be discarded despite some of its presently known imperfections.

Unifying Both Zonation Types

In figures 19 and 22, the gradients of soil salinity and waterlogging are shown for four permanent studies sites at Gladstone, Queensland, monitored over nine years. To these tidally maintained gradients, those for percentage submergence and tidal frequency can be added. In this way, four intertidal gradients are established, each of which is related to tidal levels in a way specific for this particular locality.

Figure 32 shows these gradients together with the topographic distributions of the intertidal communities in the study areas. Two approaches can now be adopted to describe the upper and lower limits of these communities and to compare these with other localities.

The traditional approach has been to relate the limits of communities to the various tidal levels, and because these are well established for most coastal sites, comparisons can be made easily. This approach assumes that the various communities have a constant relationship to specific tide levels, an assumption that is difficult to justify on the one hand, and tautological on the other.

The alternative approach is to relate the upper and lower limits to each of the four gradients, and to use these to define boundary conditions for each of the plant communities. This provides ecologically meaningful conditions which may limit the distributions of the plant communities. To compare the distributions at different localities, the level at which the boundary conditions are identical must be determined. In other words, although the gradients are all tidally maintained, tide levels as such are not used but rather certain cut-off points along each of the four gradients.

The relationships of these cut-off points to tidal levels will vary

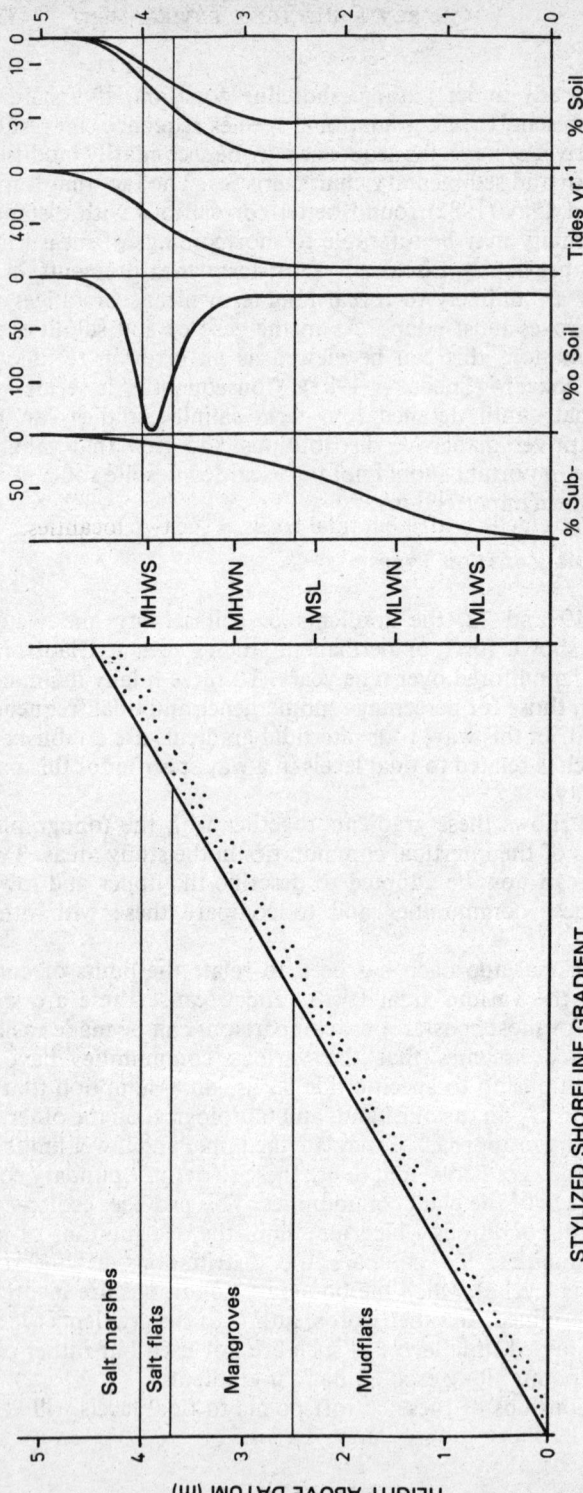

Figure 32 Integration of vegetational boundaries with gradient-related and tidally induced boundary conditions based on data collected from study areas in Gladstone, Queensland, 1975-1983.

from one locality to another, but the ecologically significant cut-off points can be expected to have a constant relationship with each of the plant communities under consideration. By a detailed comparison of boundary conditions at several sites, the more important boundary conditions can be identified as these will remain relatively constant in relation to each plant community from site to site.

Agreement is apparent in the boundary conditions for Port Curtis and Repulse Bay (table 21) with the exception of the number of tides per year for the saltflat and saltmarsh boundaries. This is significant in that the three boundaries involved are clearly at different tidal levels at the two localities, yet, ecologically, there is little to separate them. For example, critical per cent submergence time is 1 and 10 at Repulse Bay and Port Curtis respectively, and both soil salinity and soil water levels are similar at the two sites. The same conditions in terms of soil salinity and waterlogging obviously can be attained at different tidal levels at the two localities.

Table 21 Boundary conditions for the various intertidal plant communities at Port Curtis and Repulse Bay (Saenger, unpubl. data).

Community	Boundary	% Submergence	Tides/ year	% Soil water	$^o/_{oo}$ Soil chlorinity
Mudflats	upper	< 30	< 720	> 40	~ 20
Mangroves	lower	> 30		> 40	
	upper			< 5	> 70
Salt flats	lower	> 15	> 5/ > 420*		< 70
	upper	< 1 < 10*	< 2/ < 360*		< 70
Salt marshes	lower	> 1 > 10*	> 2/ > 360*	> 15–20	> 70
	upper				
Fringing vegetation	lower		> 0		> 5

* Given for Repulse Bay and Port Curtis respectively.
Source: Saenger (unpubl. data).

For each of the plant communities, several boundary conditions are given. This has been done because, on the present data, no single critical boundary condition can be identified. On the other hand, no tidally related boundary condition is provided for the upper limit of saltmarshes and this is consistent with their extensive non-coastal occurrence. The lower boundary conditions for saltmarshes are given, and when these are compared with those determined in other areas (table 22), good agreement is found, particularly with soil water and soil salinity levels. In turn, this suggests that soil salinity and the degree of waterlogging are the major factors in determining the lower limits of these plant communities.

A similar approach could be used to determine boundary conditions for each species individually. At present, the available data

Table 22 Comparative boundary conditions for the lower limit of salt marshes

Location	% Submergence	Tides/ year	% Soil water	°/oo Soil chlorinity
Port Curtis	10	360	15–20	< 70
Repulse Bay	1	2	15–20	< 70
NSW[1]		420		< 55
South Aust.[2]			21.8	
Tasmania[3]	0–30			

[1] Clarke and Hannon (1971)
[2] Osborn and Wood (1923)
[3] Guiler (1951)

are insufficient to do this, although the parallel zonation along open shorelines provides a hint of which species are most tolerant of waterlogging and high salinities.

Mangroves, saltmarsh plants and other vegetation can be broadly grouped by their tolerance with respect to salinity and waterlogging. This has been attempted in table 23, and the results may be interpreted as follows. On the open shoreline where salinities range from medium to high, the plant sequence is likely to follow the waterlogging gradient, that is, high at the seaward margin and low at the landward margin. On the other hand, in the upriver situation where waterlogging ranges from medium to high, the plant sequence is likely to follow the salinity gradient, that is, high salinities at the mouth and low salinities in the upper reaches. Both of these likely sequences are indicated by arrows in table 23.

Table 23 Tolerance of mangrove, saltmarsh and fringing plants to soil salinity and waterlogging

SALT TOLERANCE

WATERLOGGING TOLERANCE	High	Medium	Low
High	Most mangroves	Heritiera, Pemphis, Hibiscus, Cynometra, Aegiceras, Xylocarpus granatum, Acrostichum	Melaleuca spp. Casuarina glauca Most freshwater aquatics
Medium	Clerodendron inerme Lumnitzera racemosa Sporobolus virginicus	Camptostemon Excoecaria Most strand plants	Some swamp margin plants
Low	Most saltmarsh species	Triglochin procera	Most upland plants

UPRIVER SEQUENCE →

↓ OPEN SHORELINE SEQUENCE

On two identically shaped coastlines where one is fully open shoreline and the other is a river mouth discharging water of reduced salinity, the vegetation sequence is likely to be as shown in figure 33. Salt flats are absent upstream because very high salinity levels do not occur there. In addition, the tidal influence decreases upstream with the result that the tidal zone becomes narrower and the terrestrial fringing vegetation approaches the river bank. Those plants adapted to medium-low salinities become increasingly common and are able to out-compete most of the species tolerant of high salinities. This, in turn, would result in the loss of the most seaward mangrove zones somewhere in the lower reaches of the river at the same time that the driest landward zones (salt flats and saltmarshes) are lost. The middle and landward mangrove zones, because of their medium tolerance of both salinity and waterlogging, would extend furthest upriver. Gradually, the mangrove zone would become dominated by those species able to optimize growth in medium to low salinity conditions (such as *Aegiceras*, *Heritiera*, *Hibiscus*, *Cynometra*, *X. granatum* and *Acrostichum*).

Irregular coastlines would tend to disrupt the idealized pattern shown in figure 33, and may introduce various site-specific anomalies. Nevertheless, the general patterns described can be recognized with sufficient frequency to suggest that gradients of salinity and waterlogging are the interconnecting features of upriver and shoreline zonation.

Classification of Mangrove Communities

Any of the various attributes (physico-chemical, functional, structural) of mangrove communities may be useful in classifying them, and the appropriate selection of attributes depends upon the purpose of the classification. However, three classificatory schemes appear to be of some universal value in comparing mangrove communities, two at the medium to local scale and the other at the medium to regional scale. Brief summaries and discussions of these schemes are given below.

Classification Using Structural Attributes

Specht (1970) developed a structural classification of evergreen plant communities which uses those properties reflecting the amount of photosynthetic tissue (contributing to energy input) and the biomass of respiring aerial plant tissue (involved in energy output). The properties used are (1) the height and life form of the tallest stratum (which provides an estimate of the biomass) and (2) the "foliage projective cover" (FPC) of the tallest stratum. The FPC is the proportion of photosynthetic tissue vertically above the landscape. Ideally, it should be measured using some crosswire

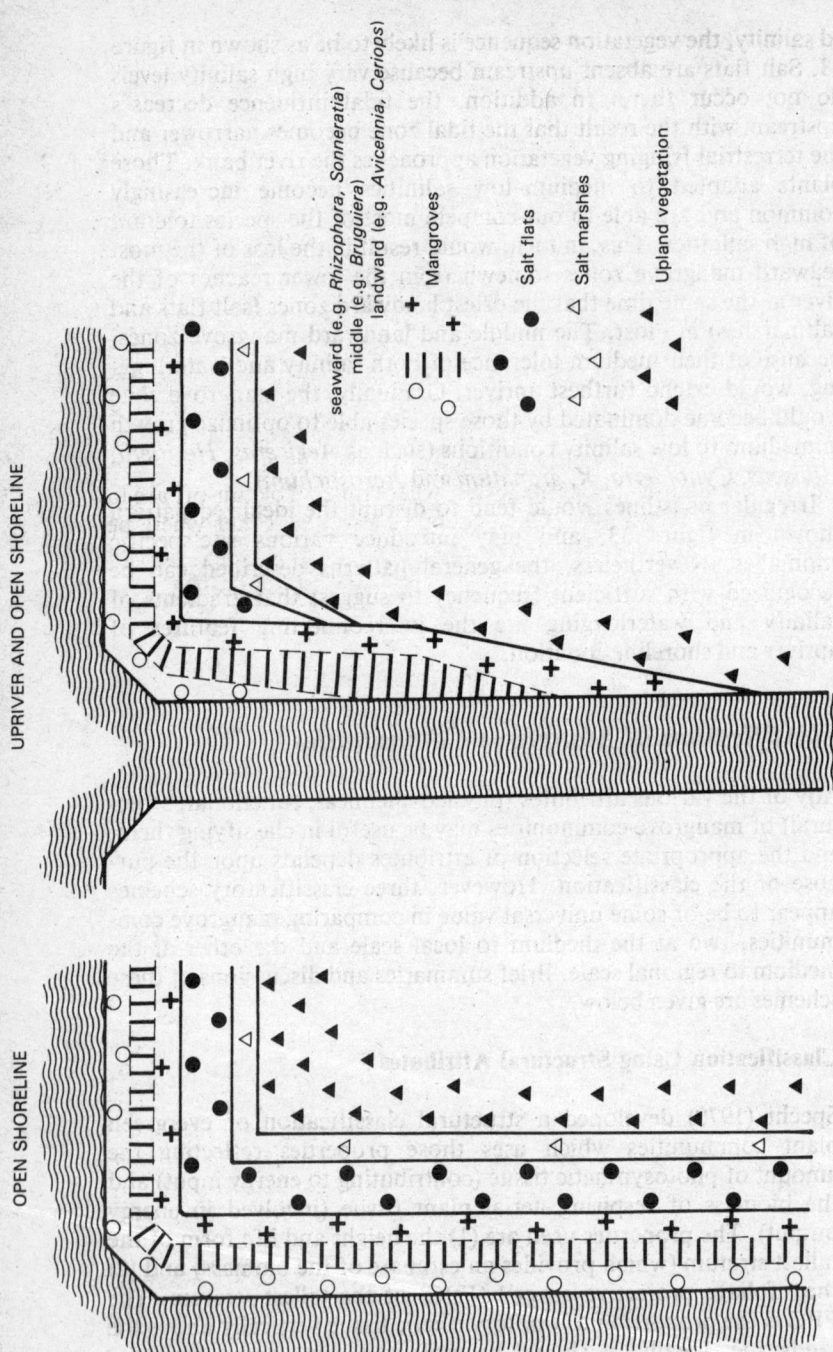

Figure 33 Stylized zonational sequences along open shorelines and into adjacent river mouths.

device to determine the presence or absence of foliage vertically above a large number of randomly selected points in the community.
Using these two properties, the identification of structural formations can be achieved. These formations (such as forest or woodland) can be defined further by including the name of the dominant genus or species (such as *Avicennia* woodland). The general absence of well-developed understorey and shrub strata in mangrove communities and their marked tendency towards dominance by one species of canopy tree mean that it is rarely necessary to seek further precision. Using this classification, the range of mangrove structural formations that have been encountered are shown in table 24. The most common of these are closed communities. Only rarely do trees exceed 30 metres in height. Open canopies are associated with high salinity sites, often at or near high-water spring levels, where rainfall or run-off are low or moderately seasonal. Open canopies also may occur where persistent waterlogging is a feature of the environment and, in some other instances, in dwarfed communities whose structure is not readily explained as yet.

Table 24 Structural formations of Australian mangrove communities

Life form and height of tallest stratum		Foliage projective cover of tallest stratum		
		Dense 70–100%	Mid dense 30–70%	Sparse 10–30%
Trees*	30 m	Tall closed forest	—	—
Trees	10–30 m	Closed forest	—	—
Trees	5–10 m	Low closed forest	Low open forest	Low woodland
Shrubs**	2–8 m	Closed scrub	Open scrub	Tall shrublands

* A tree is defined as a woody plant more than 5 m tall usually with a single stem.
** A shrub is defined as a woody plant less than 8 m tall, usually with many stems at or near the base.

The assumption is made in this classification that the communities are mature, that is, fully reflecting the constraining effects of water balance, soil fertility, temperature and light. In practice, this assumption is not difficult to meet as successional response by FPC is rapid (Specht and Morgan 1981), minimizing any error arising from this parameter; the age/size structure of the population in relation to neighbouring sites permits a reasonable assessment of the developmental (successional) phase of the ecosystem. This scheme can be used validly also where the community is not mature because of disturbance if that disturbance is regular. In this instance, regular disturbance can be viewed as an integral part of the environment.

In terms of understanding the ecological biogeography of mangrove plant communities, this classification is extremely useful. The selected parameters (FPC and height) relate to community growth and, consequently, allow some functional interpretations of structural variation to be made. An assessment of mangrove plant communities in these terms conveys clues about the environment and will enable some approximate predictions to be made about the direction of any change in the community following environmental manipulations.

Classification Using Physiographic and Structural Attributes

Lugo and Snedaker (1974) suggested that, as mangrove communities exhibit a tremendous range of form, a convenient system of classification can be based on the geomorphic and hydrological processes that induce that form. From their work in Florida, they subsequently identified six major community types based largely on their physiographic setting. Correlation between community physiography and community structure was found to be high, and led Lugo and Snedaker (1974) to recognize, on structure alone, one community type whose physiography was extremely variable and poorly understood.

Each of the six community types (figure 34) has its own characteristic set of environmental variables, such as soil type and depth, soil salinity range and tidal flushing rates. In addition, each community type has characteristic ranges of primary production, litter decomposition and carbon export along with differences in nutrient recycling rates and community components. The types are:

1. *Overwash mangrove forests*: These occur on the smaller low islands and finger-like projections of large land masses in shallow bays and estuaries. Their positions and alignments obstruct tidal flow, and thus they are overwashed frequently by tides and much of the organic matter is washed away. In Florida, all local mangrove species may be present but *Rhizophora mangle* usually dominates. Maximum height is about 7 metres.
2. *Fringe mangrove forests*: These form thin fringes along protected shorelines and islands, being best developed along shorelines whose elevations are higher than mean high tide. This community type generally shows characteristic zonation. The low velocities of the incoming and retreating tides and the dense, well-developed stilt root systems entrap all but the smallest organic debris. Because of the relatively open exposure along shorelines, the fringe forest is occasionally affected by strong winds, causing breakage and resulting in the accumulation of relatively large amounts of debris among the stilt roots.

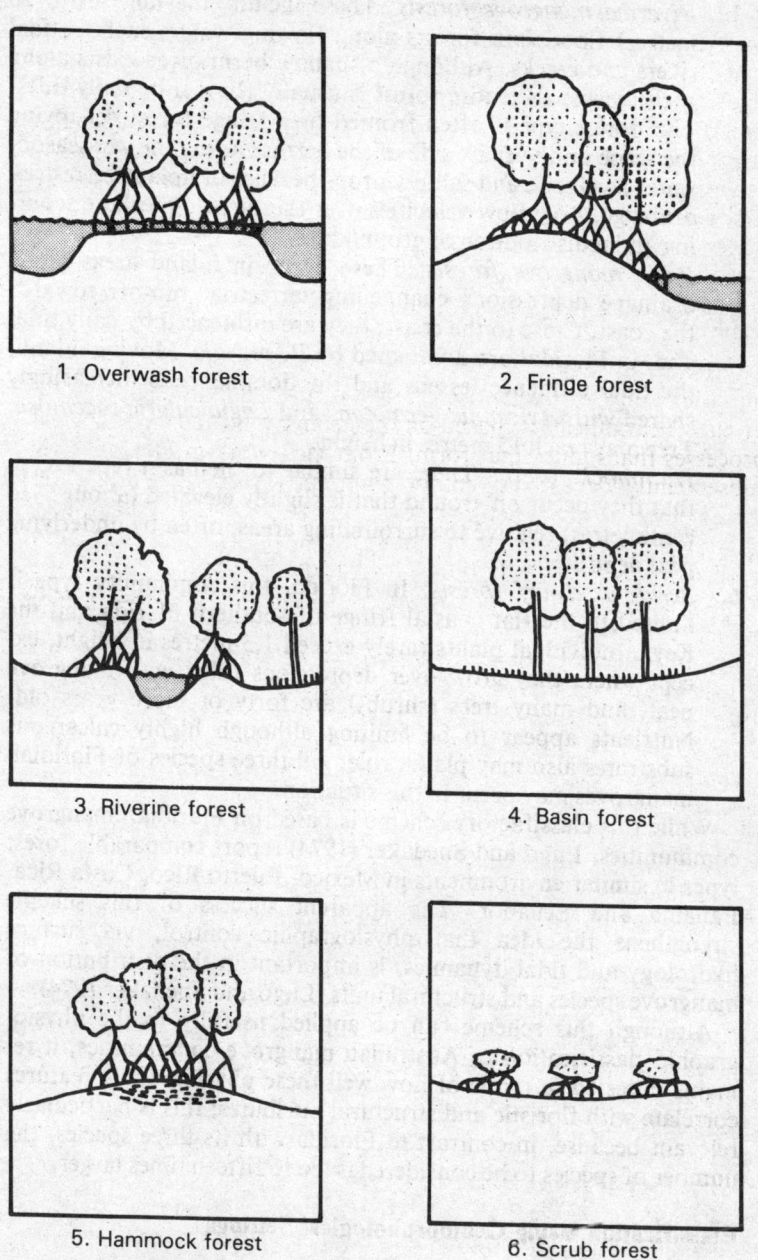

Figure 34 Classification of mangrove environments using physiographic characteristics (adapted from Lugo and Snedaker 1974).

3. *Riverine mangrove forests*: These include the tall (up to 20 metres) floodplain forests along flowing waters such as tidal rivers and creeks. Although a shallow berm often exists along such creeks, the entire forest is usually flushed by daily tides. This forest type is often fronted by a fringe forest occupying the slope on the creek side of the berm. During the wet season, water levels rise and salinity drops because of upland terrestrial drainage. Low flow velocities over the surface preclude scouring and redistribution of ground litter.
4. *Basin mangrove forests*: These occur in inland areas along drainage depressions channelling terrestrial run-off towards the coast. Close to the coast, they are influenced by daily tides and, in Florida, are dominated by *R. mangle*. Moving inland, the tidal influence lessens and the dominance is increasingly shared with *Avicennia germinans* and *Laguncularia racemosa*. Tree may reach 15 metres in height.
5. *Hummock forests*: These are similar to the basin type except that they occur on ground that is slightly elevated (about 5-10 centimetres) relative to surrounding areas, often by underlying peat deposits.
6. *Scrub or dwarf forests*: In Florida, this community type is limited to the flat coastal fringe of southern Florida and the Keys. Individual plants rarely exceed 1.5 metres in height, except where they grow over depressions filled with mangrove peat, and many trees (shrubs) are forty or more years old. Nutrients appear to be limiting although highly calcareous substrates also may play a role. All three species of Floridian mangroves may occur in this situation.

While this classifactory scheme is based on Floridian mangrove communities, Lugo and Snedaker (1974) report comparable forest types in similar environments in Mexico, Puerto Rico, Costa Rica, Panama and Ecuador. The apparent success of this scheme strengthens the idea that physiographic control, via surface hydrology and tidal dynamics, is important in the distribution of mangrove species and structural units (Lugo and Snedaker 1974).

Although this scheme can be applied usefully to the physiographic classification of Australian mangrove communities, it remains untested in terms of how well these physiographic features correlate with floristic and structural attributes; this is particularly relevant because, in contrast to Florida with its three species, the number of species to be considered is ten to fifteen times larger.

Classification Using Geomorphological Settings

Comparisons of thirty-four major river systems by Wright, Coleman and Erickson (1974) in terms of particular sets of physico-chemical variables (such as river discharge, wave energy regimes, river-mouth morphology and delta-plain landform suites) revealed that deltas tend to cluster together into a relatively few categories. Further generalization (Coleman and Wright 1975; Wright 1978) resulted in the classification of a number of general delta types, with each reflecting a particular combination of processes and physico-chemical controls.

Thom (1982) used this classification as the basis for a broader classification of coastal settings in which mangroves grow, and has described five types of terrigenous sedimentary coasts (figure 35). Although relatively unimportant on a global scale, the carbonate (coral coast) setting, where sediment accumulation is either from *in situ* growth of coral reefs or from deposition of carbonate particulates, should be added to this broad scheme; consequently six classes of settings now are recognized.

Galloway (1982) used this scheme to interpret patterns in aerial photographs of mangroves around the Australian coastline, recognizing twenty-six regional patterns within the six classes of settings. The six classes are described below together with some of the variations described by Galloway (1982).

Alluvial Plains

This setting is characteristic of coasts with a low tidal range and where the discharge of fresh water and sediment leads to rapid deposition of terrigenous sands, silts and clays to form deltas. These deltas build seawards over flat offshore slopes composed of fine-grained sediments. Such slopes help dampen wave energy and any tendency for longshore drift. The delta consists of multiple branching distributaries forming elongate, finger-like protrusions, resulting in a highly crenulate coastline with shallow bays and lagoons between and adjacent to the distributaries.

The active distributary region is predominantly an area of high freshwater discharge, so that salt-tolerant plants are not common. However, where abandoned distributaries occur into which saline waters penetrate seasonally or more frequently, salt-tolerant vegetation will develop. The area around these distributaries is also relevant to this setting as longshore drift of muds and reworking of sands and shells by waves influence plant establishment and regeneration, a phenomenon particularly striking on chenier plains. Thus, parts of the alluvial plain may contain an array of habitats where mangroves can establish or be maintained. Such plains are subject to rapid rates of subsidence and changes in freshwater discharge point and deposition, and are consequently characterized by a high degree of physico-chemical diversity and rapid habitat change.

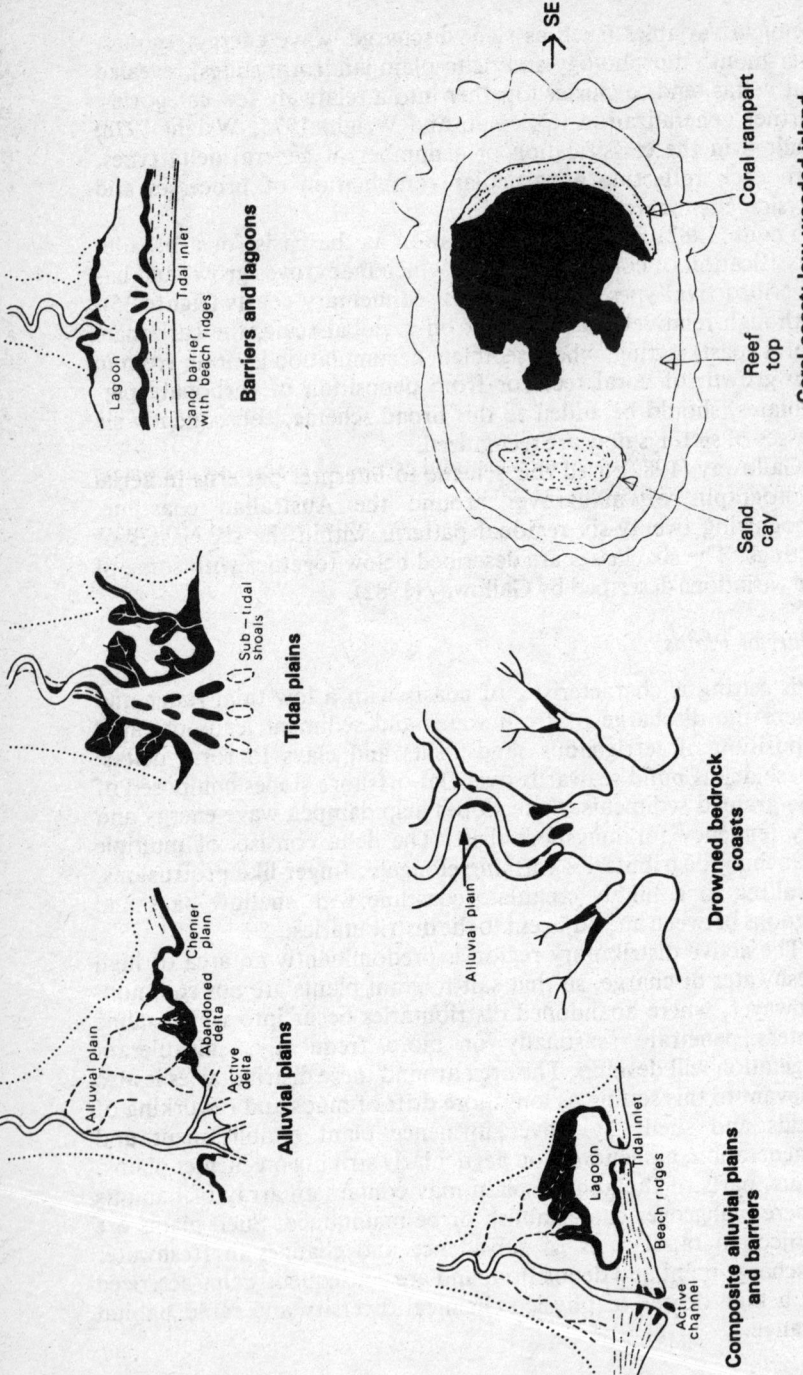

Figure 35 Classification of mangrove environments using geomorphological characteristics (after Thom 1982).

Examples of this setting include the eastern side of Exmouth Gulf, Western Australia; the southern shores of Van Diemens Gulf in the Northern Territory; the southeastern Gulf of Carpentaria (Rhodes 1982; Saenger and Hopkins 1975) and Princess Charlotte Bay in Queensland (Elsol and Saenger 1983); and the shores of the South Australian gulfs (Butler et al. 1977a).

Tidal Plains

This setting occurs on coasts where high tidal ranges and associated strong bidirectional tidal currents predominate. These currents are responsible for the dispersion of sediments brought to the coast by rivers, and in the offshore zone they form elongate sand bodies. Wave power is often low because of frictional damping over broad intertidal shoals. The main river channels are typically funnel-shaped and are fed by numerous tidal creeks; these creeks are often separated by extensive tidal flats.

Examples include the Ord River (Thom, Wright and Coleman 1975), Fitzroy River (Jennings and Coventry 1973) and King Sound (Semeniuk 1980) in Western Australia; Fitzroy River and the mud islands and shores of Moreton Bay, Queensland (Flood 1980); and the shores of Westernport Bay, Victoria (Enright 1973).

Barriers and Lagoons

This setting is characterized by much higher wave energy than previous ones, and by relatively low amounts of river discharge. Offshore barrier islands, barrier spits and bay barriers are typical of this setting. Small finger-like deltas prograde into these water bodies without significant opposition from marine forces. Considerable tidal modification may occur within the barrier system. Where the barriers project from the coast or link islands to the mainland, sheltered water in their lee provides sites for extensive mangroves if a sediment supply is available. Salt-tolerant plants occur around the margins of the lagoon in a variety of habitats.

Examples include the western shores of the Gulf of Carpentaria, in the Northern Territory; Eyre Peninsula, South Australia; and Port Curtis and Cape Bowling Green, Queensland.

Composite Alluvial Plains and Barriers

This setting represents a combination of high wave energy and high river discharge. Sand carried to the sea by the river is rapidly redistributed by waves along shore to form extensive sand sheets. Much of the sand deposited on the inner continental shelf during lower sea-levels is reworked landward during periods of rising or stable sea-levels. The result is a coastal plain dominated by sand beach ridges and narrow discontinuous lagoons with an alluvial plain to landward.

Salt-tolerant plants such as mangroves are concentrated along abandoned distributaries and in areas near river mouths and adjacent lagoons. Where the tidal range is large and the climate dry, as in the case of the Burdekin delta, there is a spread of saline habitats to interdistributary areas which are periodically inundated by high spring tides.

Examples include the Burdekin Delta and Repulse Bay, Queensland; and the Purari Delta of Papua New Guinea.

Drowned Bedrock Coasts

This setting can be described as a drowned river valley complex. The depth of deposition is confined by a bedrock valley system which has been drowned by rising sea-levels. Neither marine nor river deposition has been sufficient to infill what is an open estuarine system. However, the heads of the valleys may contain relatively small river deltas which are little modified by waves, and often maintained by self-scouring (Bunt and Wolanski 1980). At the mouth of the drowned valley bordering the open sea, a tidal delta may occur, composed of marine mud and sand reworked landward during rising sea-levels.

Mangroves flourish in the fine sediments at the heads of the drowned tributary valleys, and in lagoons behind bay barriers near the mouth of the estuary.

Examples include Broken Bay, New South Wales; Port Darwin, Northern Territory; Kimberley Coast, Western Australia; Hinchinbrook Channel, Newcastle Bay, Shoalwater Bay, and Trinity Inlet, Queensland (Bird 1970).

Coral Coasts

Two expressions of this setting occur: mangroves may grow on terrestrial sediments which have accumulated behind fringing reefs, or they may occur on coral sediments (sand) on platform reefs. Stoddart (1980) provided a detailed account of the mangrove communities occurring on the coral islands of the northern Great Barrier Reef, while Thom (1975) described their general response to varying substrate and energy conditions.

As Galloway (1982) has shown, this scheme has considerable practical value when used on a regional scale, as it is based on both structural and dynamic characteristics of a particular section of coastline. For these reasons, this scheme also may have considerable value when applied on a more local scale, although to date it has not been used in such a context.

4. Associated Flora

Apart from the mangroves themselves, a number of characteristic species of other plants occur in the mangrove communities throughout Australia. Many of these species occur in other communities also, but there are some which are confined to mangroves. The association of these species with mangroves may be an intimate one; conversely, the association may be strictly casual or fortuitous, but these associated species increase the diversity of the mangrove communities.

In this chapter, a basic taxonomic and ecological outline is given for the various associated plant groups including bacteria, fungi, algae, lichens, epiphytes, mistletoes and saltmarsh plants, as well as other species of vascular plants which occur at the margins of mangroves.

Bacteria

As discussed elsewhere (chapter 8), bacteria, together with the fungi, form an important component of the mangrove community, and as decomposers play a central role in the functioning of the ecosystem.

Bacterial numbers in mangroves and related communities are high (table 25); sediments carry populations of heterotrophic bacteria two or three orders of magnitude higher than the waters above them, and the populations on clays and muds are usually several orders of magnitude greater than those on sandy substrates (Millis 1981).

The estuarine marine environment contains bacteria belonging to the same range of genera as found in fresh water, although an absolute requirement for sodium ions has been shown in most marine species. Marine bacteria are generally smaller than non-marine forms, and a large proportion consists of gram-negative rods; most of these are actively motile, flagellated forms. Cocci (spherical bacteria) are less common than the rod forms. Many marine bacteria are sedentary, attaching themselves to solid surfaces by means of a mucilaginous holdfast. In this way, bacteria may form a surface film on muds and make them more susceptible to subse-

Table 25 Bacterial numbers in water and sediments of mangrove and associated communities

Locality	Habitat	Bacterial Numbers per ml Water	Sediment	Reference
Westernport Bay, Vic.	Open mudflats		10^5-10^7	Bavor and Millis 1976
	Mangrove muds		10^6-10^9	Bavor and Millis 1976
	Seagrasses	$4 \times 10^3 - 6 \times 10^5$	10^7-10^9	Rudov 1977
Botany Bay, NSW	Estuarine mud		$8 \times 10^4 - 2.5 \times 10^7$	Wood 1953
Lake Macquarie, NSW	Estuary	$50-10^4$	$3 \times 10^5 - 6.5 \times 10^6$	Wood 1959
Moreton Bay, Qld	Seagrasses		$0.6-2 \times 10^9$	Moriarty 1981
	Seagrasses		$1.4-5.7 \times 10^9$	Moriarty 1980
	Sand flats		$0.9 \times 10^{-8} 1 \times 10^9$	Moriarty 1980
Low Isles, Qld	Mangroves	10^5	2.4×10^9	Bavor 1978
Delaware Inlet, NZ	Sand flats		$0.3-1.2 \times 10^9$	Gillespie and Mackenzie 1981
	Seagrass		$0.55-1.5 \times 10^9$	Gillespie and Mackenzie 1981
	Salt marsh		6.5×10^9	Gillespie and Mackenzie 1981
	Open mudflats		1.5×10^9	Juniper 1981

quent algal, seagrass or mangrove growth (Marshall, Stout and Mitchell 1971; Corpe 1974).

Taxonomic data on bacteria are limited for mangrove communities, but the few that there are suggest that many species are common; for example, Chou et al. (1980) recorded only ten species of bacteria from Singaporean mangroves. No bacterial species lists are available for Australian mangroves, but an indication of species likely to occur can be gleaned from the list of bacteria found in the Lake Macquarie estuarine system in central New South Wales (table 26); twenty species were recorded including autotrophic sulphur and iron bacteria.

Fungi

The fungal flora of mangroves is inconspicuous but nevertheless diverse and abundant; the majority of fungi are microscopic and few macroscopic species are common. Nevertheless, fungi play an important role in mangrove communities, particularly in conjunction with bacteria in the rapid breakdown of leaf litter (Fell et al. 1975).

Table 26 Bacteria recorded from the Lake Macquarie estuarine system in New South Wales

Autotrophs
  Sulphur bacteria
    Beggiatoa mirabilis
    Chlorobium sp.
    Chromatium sp.
    Desulphovibrio desulphuricans
    Thiobacillus denitrificans
    T. thioparus
    Thiothrix sp.
    Thiovulum sp.
  Iron bacteria
    Sphaerotilus sp.

Heterotrophs
  Actinomyces sp.
  Bacillus megatherium
  B. sphaericus
  B. subtilis
  Cornynebacterim flavum
  C. globiforme
  C. miltinum
  Mycoplana citrea
  M. dimorpha
  Sarcina lutea
  Staphylococcus candidus
  S. roseus

Source: Wood (1959).

There are few studies (Cribb and Cribb 1955; Leightley 1980; Chandrashekar and Ball 1980; Pegg and Alcorn 1982) on the fungi of Australian mangroves communities, and even the taxonomic data are rudimentary. However, since many genera of fungi are ubiquitous, overseas studies can be drawn upon to provide some general information.

Fungi may be divided according to the microhabitat they occupy within the mangroves; certain species may occur in more than one microhabitat. The main microhabitats are (1) the mangrove canopy, (2) the trunks and aerial roots and (3) the soil.

A number of parasitic and saprophytic fungi, including ascomycetes (Vizioli 1923; Batista, ca Silva Maia and Vital 1955), basidiomycetes (Creager 1962; McMillan 1964) and deuteromycetes (Chandrasekar and Ball 1980), have been recorded from leaves in the mangrove canopy. Fell et al. (1975) showed that as the leaves were shed into the water, some species of fungi disappeared whereas others, such as *Pestalotia*, the most prevalent genus, were able to persist throughout the decay process. Newell (1976) listed and described the parasitic and saprophytic fungi that occurred on the *Rhizophora mangle* seedlings that were still on the parent trees.

Six genera were recorded, including *Cladosporium*, *Pestalotia*, *Alternaria*, *Zygosporium*, *Penicillium* and *Aureobasidium*.

It should be noted that the genera of parasitic and saprophytic fungi recorded from mangrove canopies are generally large, widespread genera, often with species pathogenic to a number of hosts. For example, the genera *Pestalotia*, *Phyllosticta*, *Cladosporium*, *Nigrospora* and *Cercospora* all have species affecting terrestrial plants. In some instances, the same species affects both mangroves and terrestrial plants; for example, *Nigrospora sphaerica* causes leaf rot in *Rhizophora mangle* and squirter's disease in bananas.

Kohlmeyer (1969) examined the fungi associated with mangroves and compiled a list of thirty-one species of marine fungi and forty-four species of terrestrial fungi. Whereas most of the terrestrial species were associated with living leaves, the marine species were associated either with the roots or with rotting mangrove wood.

While the mangrove is alive, fungi can grow on the trunk and the proproots. Of these, the polyporous fungi (bracket or shelf fungi) are the most conspicuous because of their size and colour. Nineteen species of polypore are presently known from Australian mangroves (table 27) but this undoubtedly represents only a small proportion of those actually present.

Table 27 Basidiomycetes from Australian Mangroves

Species	Habitat
POLYPORACEAE	
Ganoderma ochralaccatum	—
Ganoderma williamsianum	—
Gleophyllum abietinum	dead *Rhizophora stylosa* wood
Lenzites sp.	dead *Camptostemon schultzii* wood
Osmoporus carteri	—
Osmoporus floccosus	—
Osmoporus sp.	—
Phellinus badius	—
Phellinus caliginosus	—
Phellinus gilvus	dead *Ceriops tagal* wood
Phellinus lloydii	*Rhizophora stylosa* trunk and proproots
Phellinus ramosus	*Avicennia marina* trunk and dead *Xylocarpus granatum* wood
Phellinus spadiceus	—
Pseudofavolus polygrammus	dead *Rhizophora stylosa* wood
Pseudofavolus tenuis	dead *Ceriops tagal* wood
Trametes hirsuta	—
Trametes mulleri	—
Tyromyces grammocephalus	—
THELEPHORACEAE	
Stereum lamellatum	—

Source: Hegerl et al. (1979), Hegerl and Davie (1977), Saenger (unpubl. data).

Olexa and Freeman (1975) recorded a gall-forming fungus, *Cylindrocarpon didymum*, from the stilt roots and stems of *Rhizophora mangle* in Florida. Similar galls have been noted on Australian *Rhizophora* species but the causative fungus has not been determined.

Once the tree has died, the wood comprises a suitable substrate for wood-decaying fungi. Leightley (1980) used discs from five mangrove species and found that a single ascomycete (*Lulworthia grandispora*) and two imperfect fungi (*Phialophora and Phoma*) were the primary decay fungi in the wood. Foot (1980) also studied the deterioration of Australian mangrove woods and found that the rate of decay varied according to genus and occurred in the following order: *Excoecaria* > *Avicennia* > *Ceriops* > *Bruguiera* = *Rhizophora*

It was found (Foot 1980) that above extreme high tide there was little decay of the timber, and the main species involved were basidiomycetes (mostly polypores) and microfungi of the genera *Alternaria, Cladosporium, Penicillum* and *Trichoderma*. However, those samples submerged in sea water suffered greater decay and were colonized by typically marine species such as *Lulworthia grandispora, Leptosphaeria australiensis* and *Didymosphaeria rhizophorae*.

The soil fungi of mangroves can be subdivided conveniently into those fungi occurring in the soil itself and those which are associated with the breakdown of mangrove leaves on the soil surface. Clearly, however, both of these groups depend on organic material on or in the soil.

Fungi are the primary decomposers of mangrove leaves because of their ability to degrade cellulose and lignin, which together form the major component of cell walls in the leaves. Fell and Masters (1973) studied the process of leaf degradation of mangroves. They isolated sixty-six genera of fungi and observed a sequence of infestation. Within the first week of leaf fall, the most abundant initial invaders were phycomycetes including *Thraustochytrium, Schizochtytrium, Phytophthora vesicula, P. bahamensis, P. epistomium, P. mycoparasitica* and *P. spinosa*. Other primary invaders included *Aspergillus, Penicillium, Trichoderma, Fusarium, Curvularia* and *Drechslera*. After the second week, the cellulytic marine ascomycete *Lulworthia* invaded the decaying leaves, and after three weeks *Zalerion varium* was found on the majority of them.

Newell (1976) reported a similar sequence of infestation in fallen *Rhizophora* seedlings (figure 36); the infestation sequences of seedlings, leaves and wood were also compared and all showed qualitatively and quantitatively diffcient fungal floras (Newell 1976).

Date	Nov. 1970 Dec. Jan. Feb. Mar. Apr. May June July Aug. Oct. Nov. Dec. 1971 1972 ----→
Seedling location	pre-abscission ----→ post-abscission, estuarine subsurface environment
Seedling condition	mature, healthy →all viable --→viable-senescent-dead --→senescent-dead--→dead
Mycoseral species	*Cladosporium cladosporioides* *Pestalotia* sp. *Cladosporium* sp. *Alternaria alternata* *Zygosporium masonii* *Aureobasidium pullulans* *Pestalotia* sp. *C. cladosporioides* *Septonema* sp. *Penicillium steckii* *A. alternata* *Aspergillus repens* *Thraustochytrium* sp. *Lulworthia grandispora* → *Zalerion varium* → *Flagellospora* sp. → *L. medusa* var. *biscaynia* *Pestalotia* sp. *Labyrinthula* sp. → *Thraustochytrium* sp. → } in senescent-dead seedlings *Keissleriella blepharospora* *Cytosporina* sp. *Cytospora rhizophorae* *Pestalotia* sp. *Thraustochytrium* sp. } in viable-senescent seedlings *Trichoderma viride* *Penicillium roseopurpureum* *Papulospora halima* *Phytophthora vesicula*
Mycoseral stages	pre-abscission superficial invaders post-abscission superficial invaders subepidermal invaders, in dead and living tissue decayed tissue invaders
Analogous stages	weak parasites + common and restricted primary saprophytes secondary saprophytes stage I: Ascomycetes and Deuteromycetes secondary saprophytes stage II: "soil fungi"

Figure 36 Successional stages in the fungal breakdown of seedlings of *Rhizophora mangle* in Florida, USA (from Newell 1976).

The majority of genera isolated from decaying leaves and seedlings are ubiquitous saprophytes that generally are associated with the breakdown of plant material (Fell et al. 1975), and have been reported from seagrasses (Meyers et al. 1965), mangrove muds (Rai, Tewari and Mukerji 1969; Swart 1958, 1963) and saltmarshes (Dickinson 1965), as well as terrestrial and freshwater communities (Hering 1965; Kaushik and Hynes 1968). It would appear that these species are able to tolerate the saline conditions of mangroves but do not require them. Other mangrove saprophytes appear to have a positive sodium chloride requirement for growth; an example is *Phlyctochytrium mangrovis* (Ulken 1975).

In recent studies of mangrove death in Queensland, Pegg and Foresberg (1981) and Pegg and Alcorn (1982) isolated at least five different species of *Phytophthora* from mangrove soils and mangrove litter; most, apparently, are undescribed species. It appears that these species constitute the normal primary decomposers of mangrove leaf litter in Australian mangroves. However, one of these *Phytophthora* species seems to have some parasitic ability as well (Pegg, Gillespie and Foresberg 1980), and it can infect the roots of *Avicennia marina* (and any wounds on its stem), causing the absorbing rootlets to become black and decayed, and ultimately causing the death of the tree. In this way, widespread mortality of *Avicennia marina* has occurred in parts of the Queensland coastline from Princess Charlotte Bay (Weste, Cahill and Stamps 1982) to Moreton Bay (Pegg and Foresberg 1981). A similar dieback of *Avicennia marina* has been reported from New South Wales, but it appears that, in this instance, it is not related to any species of *Phytophthora* (West, Thorogood and Williams 1983).

Soil fungi from mangroves are poorly known, particularly those from Australia. Swart (1958, 1963) investigated the mangrove soil fungi in eastern Africa by direct inoculation of soil samples on to culture medium. It was found that phycomycetes were absent except in some landward soil samples, that ascomycetes were rare except for the genera *Aspergillus* and *Penicillium*, and that basidiomycetes were absent altogether. Many species of deuteromycetes (particularly Moniliales) were found throughout the area.

Lee and Baker (1972a, 1972b, 1973) surveyed the soil fungi of Hawaiian mangroves and commented on the absence of phycomycetes. They suggested that this may be an artefact of the isolation techniques employed. Other investigators have reported a number of phycomycetes from mangrove soils (Rai, Tewari and Mukerji 1969; Volz and Jerger 1972; Ulken 1970, 1983), and in view of their predominance in the surface litter, it seems likely that phycomycetes are present in most mangrove soils.

The large fruiting bodies of various mushrooms also have been found on Australian mangrove soils; these include species belonging to *Agaricus*, *Pholiota* and *Inocybe*.

Algae

There are relatively few studies of the algae of mangroves since most of the species are small, are not as aesthetically pleasing as are some of the larger seaweeds, and are taxonomically a rather difficult group. Mangrove algae form a reddish, brownish or greenish fur on mud, pneumatophores, stilt roots and the bark of the lower parts of tree trunks. Unlike the larger algae of rocky shores and coral reefs, they do not appear to have an obvious zonation dependent on tidal levels. However, the rise and fall of the tide does operate as a major factor in the growth of these algae since it affects the prevailing sailinity, the temperature of the soil and the water loss from the algae thallus during low tide.

In most parts of the tropics and subtropics, mangroves support a *Catanella-Bostrychia-Caloglossa* community, with different species of these genera dominating in different areas (Post 1963). In some areas, blue-green algae have been reported on pneumatophores with siphonous algae, particularly *Vaucheria*, being prevalent on mud. In temperate mangroves, *Caloglossa* often forms pure stands (Post 1963; Davey and Woelkerling 1980; Beanland and Woelkerling 1982; King 1981a, 1981b, 1981c).

Mangrove algae may be divided according to their microhabitat within the community but, like the fungi, certain species may occur in more than one microhabitat. The main microhabitats are (1) the mud surface, (2) the surface of tree trunks and roots and (3) the upper branches and canopies.

The algae on the mud surface can be subdivided into unicellular algae (mostly diatoms) and multicellular algae, dominated by blue-green species. Butler et al. (1977) recorded forty-one species of diatoms from the mud surface of South Australian mangrove communities, whereas McLeod (1969) found no diatoms on the mud surfaces in southeastern Queensland.

Womersley and Edmonds (1958) recorded four species of multicellular algae from the mangrove/saltmarsh mud surface in South Australia and McLeod (1969), Cribb (1979) and Atherton and Dyne (1975) reported fifty-eight species from southeastern Queensland mud surfaces, of which thirty-four species were blue-green algae. From central Queensland, one diatom and fourteen multicellular algae have been recorded (Saenger et al. 1977, unpubl. data).

Blue-green algae often form characteristic mats on the salt flats of tropical and subtropical coasts. The most common species here are *Microcoleus lyngbyaceus, Schizothrix arenaria, S. calciola* and *S. tenerrima* (Cribb 1979). These filamentous species form a tough leathery skin over the surface, which, in the absence of rain and between periods of spring tides, can crack with the mat curling up to expose the dried soil underneath. On inundation, an unbroken skin is re-formed and within 15-30 minutes the formation of numerous small bubbles indicates that photosynthesis has resumed.

Mudflats seaward of mangroves are dominated by green algae and even in the tropics and subtropics their appearance is highly seasonal. The monthly biomass production of these algae — mainly *Monostroma crepidinum* and species of *Rhizoclonium* — from mudflats adjacent to central Queensland mangroves is shown in figure 37. Growth occurs in the winter months.

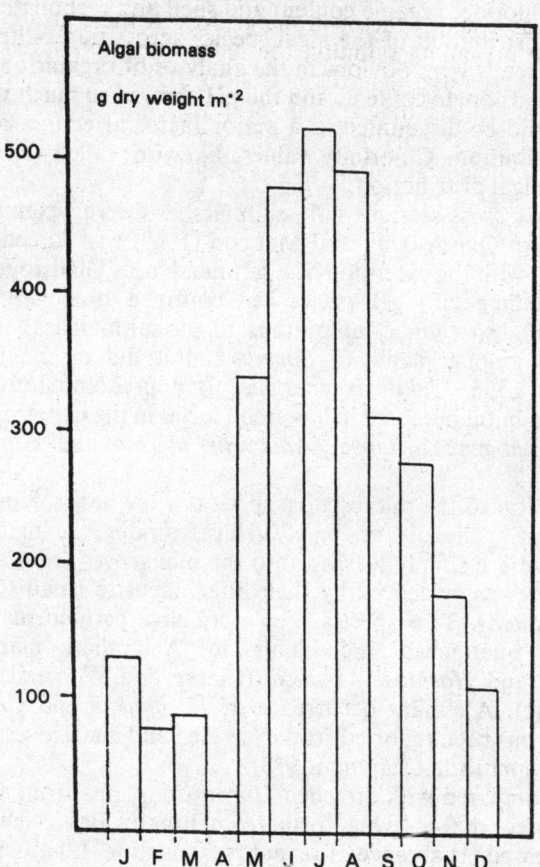

Figure 37 Algal standing crop on mudflats at Gladstone, Queensland, showing seasonal variation in abundance.

The microscopic algae of mud are important in the soil economy for (1) increasing the organic content of the soil through photosynthesis, (2) promoting the disintegration of soil particles thereby increasing the water-holding capacity of the soil, (3) secretion of mucilaginous substances which bind particles together to form a matrix, and (4) in the case of heterocystous blue-green algae, increasing the nitrogen content of the soil by atmospheric fixation of nitrogen (Jones 1974; Whitney, Woodwell and Howarth 1975). Owing to the already high organic content of most mangrove soils, it seems unlikely that three of these functions have any special significance, but the secretion of mucilaginous substances may indeed be important in consolidating the mud surface at least in some areas.

McLeod (1969) attempted to correlate soil characters such as pH, per cent chloride, organic content and shell and carbonate content with the distribution of the algal species across tidal saltmarshes. No clear trends were obvious in the analyses of organic content or shell and carbonate content, and the pH showed so much variation that it could be discounted as a major factor affecting saltmarsh algal distribution. Chlorinity values, likewise, failed to correlate well with algal distribution.

Thirteen cross-sections of saltmarshes were examined in southeastern Queensland, and McLeod (1969) was forced to conclude that, with the exception of a saltmarsh near Gladstone, only a small percentage of algal species were confined to one level. *Phormidium angustissimum*, ubiquitous in all saltmarshes, was best developed among plants of *Suaeda* but it did extend into the mangrove zone. *Calothrix crustacea* is a predominantly lower-marsh alga but it persisted in low frequencies in the mangrove zone, as did *Anabaena torulosa*, *Anacystis marina* and *Gloeocapsa alpicola*.

In addition to the microscopic species, a few macroscopic mud algae occur, generally on bare surfaces among the mangroves. Often these are simply washed into the mangroves from adjacent rocky shores as evidenced by their attachment to small stones or shell fragments. Two species, however, have permanent populations of unattached individuals in Australian mangroves: *Gracilaria* and *Hormosira banksii* (Clarke and Womersley 1981; King 1981c). A similar occurrence of *H. banksii* and *Gracilaria secundata* has been reported from New Zealand mangroves (Moore 1950; Chapman and Chapman 1973).

When compared with attached *Hormosira* plants from adjacent rocky shores, a free-living form from Botany Bay, New South Wales, showed (1) absence of sexual reproduction, (2) absence of a holdfast and (3) a generally smaller, more compact thallus (King 1981c).

The epiphytic algal flora of mangrove trunks, pneumatophores and stilt roots is rather poor in species compared with adjacent rocky shores (Saenger et al. 1977; Davey and Woelkerling 1980; King 1981b; Beanland and Woelkerling 1982). Nevertheless, ninety-four species of algae have been recorded from trunks, pneumatophores, stilt roots and canopies of mangroves in Australia (table 28). Species of red algae are the most frequent, and the genera *Bostrychia*, *Caloglossa* and *Catanella* are represented by eleven, four and two species respectively. Although various geographical comparisons have been made (Davey and Woelkerling 1980; Beanland and Woelkerling 1982; King 1981b; King and Wheeler 1985), the data are too incomplete for any reliance to be placed on them.

Table 28 Numbers of algal species recorded from various substrates in Australian mangrove communities

Phylum	Number of algal species			
	On pneumatophores and stilt roots	On lower trunks	On upper trunks and leaves	Total
Chlorophyta	19	6	4	24
Phaeophyta	7	—	—	7
Rhodophyta	55	7	—	55
Cyanophyta	7	1	—	7
Chrysophyta	1	—	—	1
Total	89	14	4	94

Source: Saenger et al. (1977); Atherton and Dyne (1975); Cribb (1979); Davey and Woelkerling (1980); King (1981); Beanland and Woelkerling (1982).

The algal growth on pneumatophores and stilt roots appears to be more luxurious in deep shade than in open, fully illuminated areas. Most of the algal species show a preference for shade, although *Enteromorpha clathrata* appears to be exceptional in its light requirement, growing in brightly illuminated open spaces (Atherton and Dyne 1975). Beanland and Woelkerling (1983) similarly found changes in frequency distributions of algal species growing on shaded or exposed pneumatophores in South Australia, and they concluded that the tree canopy influences the species frequency of algae but not necessarily the total cover, biomass or diversity.

Whether the preference for shade in most species is a direct response to light or whether shade merely reduces the degree of desiccation during low tides is not known.

In overseas studies, it has been shown that *Bostrychia tenella* can grow above high tide level provided that it remains continuously shaded (Almodovar and Biebl 1962); Gayral (1966) showed that species of *Bostrychia* are capable of withstanding extreme desiccation.

Few other factors have been investigated in relation to these epiphytic algae; Davey and Woelkerling (1980) showed that estuarine influences reduced the pneumatophore flora on *Avicennia marina* in southern Australia. The tolerance of these algae to hypersaline conditions is not known, but based on their broad geographical distributions most species appear to be tolerant of wide ranges in water and air temperature, Yarish, Edwards and Casey (1979) showed that the American species of *Bostrychia*, *Catanella* and *Caloglossa* tolerated widely varying salinities, but the responses of the Australian species are not known.

As far as such biotic factors as competition and grazing are concerned, even fewer data are available. Cribb (1979) recorded traces of *Cladophorella calcicola* in the faecal pellets of the gastropod *Bembecium auratum, and Trentepohlia* cells in those of the gastropods belonging to the *Littoraria* complex and *Cerithidea largillierti*.

A number of algal species have been recorded from the leaves and upper branches of Australian mangroves, near the upper tidal levels or where they are subjected to salt spray (Cribb 1979). These algae include *Apatococcus lobatus*, which forms a green, mealy layer on the upper branches of *Avicennia marina*, orange patches of *Trentepohlia rigidula* on the trunks and branches of *Rhizophora*, *Bruguiera* and *Ceriops*, small crusts of the green alga *Pseudendoclonium submarinum* which grows particularly well in the salt glands of *Avicennia marina* on the upper leaf surface and among the grey felt of hairs on its lower surface, and *Trentepohlia odorata* which grows between and sometimes within the leaf-hairs of *Avicennia marina* in some cases imparting a light orange-pink tinge to parts of the undersurfaces.

Lichens

The factors influencing growth and development of mangroves were discussed in chapter 3. These same factors are also important in determining the occurrences of lichen species in the mangrove vegetation. Lichens consist of a fungus which has formed a permanent union with a blue-green, green or yellow-green alga; the algal partner in this relationship is termed a phycobiont, and in the majority of lichens is one of three algal genera, namely *Trebouxia*, *Trentepohlia* or *Nostoc* (Ahmadjian 1967).

Those lichen species that occur in mangroves are almost entirely species also common on nearby terrestrial vegetation. In the monsoonal parts of Australia, lichens appear to be rare in all terrestrial vegetation (Specht 1958; Specht, Salt and Reynolds 1977; Saenger et al. 1977; Rogers and Stevens 1981), but more species are being found in rainforests and mangroves than in other vegetation types (Roger 1977; Stevens 1979; Stevens and Rogers 1979; Stevens 1981).

Stevens (1979) surveyed the lichens on mangrove bark along the eastern Australian coast and recorded 56 species in the tropical region (north of lat. 23°30'), 77 in the subtropics (between lat. 23°30' and 30°S) and 45 species in the warm temperate zone (between lat. 30° and 38°S). There was a total of 105 species, with some occurring in more than one zone. The distribution of these species along the coastline indicates that a replacement of species takes place (to a greater or lesser extent within different genera) with change of latitude. At any one locality in the tropical and temperate latitudes there is a rather constant number of species; the number of species at a given locality in the subtropical region was high (figure 38). The richness of the lichen flora in the subtropics appears to be caused by an overlap of the southern limit of tropical species with the northern limit of temperate ones. Replacement of species within

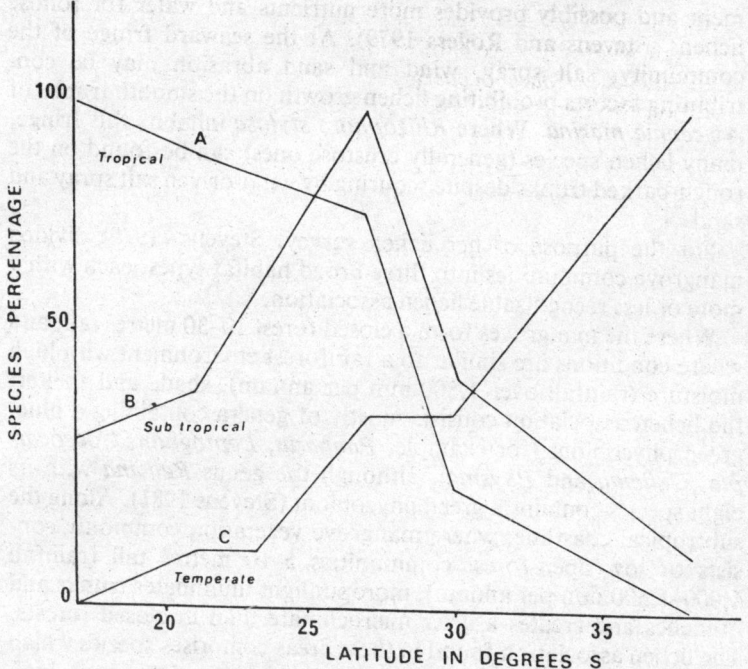

Figure 38 The percentage of lichens growing on mangroves at varying latitudes along the eastern Australian coastline: A = the tropics; B = the subtropics; C = the temperate zone (from Rogers and Stevens 1981).

each genus with change in latitude is best illustrated by those genera with a large number of species (Stevens 1979); data for the genera *Relicina*, *Parmotrema*, *Ramalina* and *Teloschistes* are shown in figure 39.

The presence or absence of lichens at any locality can be related to the influence of various climatological and substrate factors. Conditions of light, moisture, humidity, wind, exposure, salt and bark texture, and chemistry will prove favourable or restrictive for the establishment and growth of lichens (Stevens and Rogers 1979). For example, Stevens (1979) found that although complete substrate specificity did not occur, there was a preference by some species of lichens for certain mangrove barks. This preference was related to bark texture (rough or smooth), the pH of the bark, or the particular habitat of the mangrove species concerned. Consequently, bark characteristic may have a major influence on the types of lichens present. For example, in South Australian mangroves comprised solely of *Avicennia marina* with relatively smooth trunks, no macrolichens occur despite apparently suitable climatological conditions. Rough bark, on the other hand, more readily traps lichen diaspores, offers greater protection in establishment and possibly provides more nutrients and water for foliose lichens (Stevens and Rogers 1979). At the seaward fringe of the community, salt spray, wind and sand abrasion may be contributing factors prohibiting lichen growth on the smooth trunks of *Avicennia marina*. Where *Rhizophora stylosa* inhabits this fringe, many lichen species (generally crustose ones) can be found on the rough-barked trunks despite scouring by wind-driven salt spray and sand.

For the purpose of her lichen survey, Stevens (1979) divided mangrove communities into three broad habitat types, each with a more or less recognizable lichen association.

Where the mangroves form a closed forest 10-30 metres tall, and where conditions are similar to a rainforest environment with high moisture (rainfall over 1,500 mm per annum), shade and shelter, the lichen association consists mostly of genera containing a blue-green phycobiont (for example, *Pannaria*, *Leptogium*, *Coccocarpia*, *Collema* and *Physma*), although the genus *Relicina* with its eight species contains a green phycobiont (Stevens 1981). Along the subtropical coastline, where mangrove vegetation commonly consists of low open-forest communities 5-10 metres tall (rainfall 1,000-1,500 mm per annum), more sunlight illuminates trunks and branches and creates a drier microclimate than in closed forests. The lichen association found in these areas comprises species which are tolerant of full sunlight, varying humidity and high temperatures. The genera found include *Ramalina*, *Usnea*, *Teloschistes*, *Dirinaria*, *Pyxine*, *Heterodermia* and *Parmotrema*.

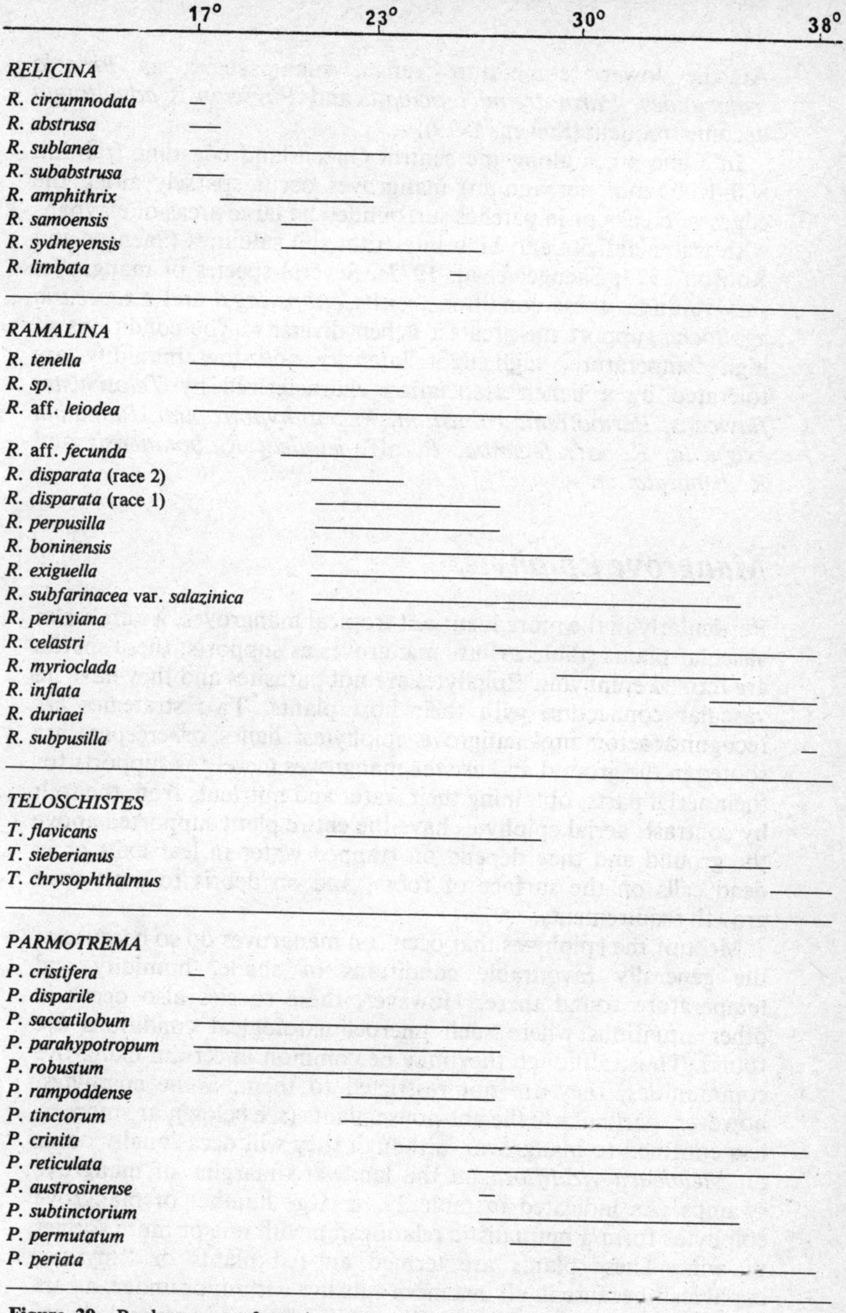

17°   23°   30°   38°

RELICINA
R. circumnodata
R. abstrusa
R. sublanea
R. subabstrusa
R. amphithrix
R. samoensis
R. sydneyensis
R. limbata

RAMALINA
R. tenella
R. sp.
R. aff. leiodea

R. aff. fecunda
R. disparata (race 2)
R. disparata (race 1)
R. perpusilla
R. boninensis
R. exiguella
R. subfarinacea var. salazinica
R. peruviana
R. celastri
R. myrioclada
R. inflata
R. duriaei
R. subpusilla

TELOSCHISTES
T. flavicans
T. sieberianus
T. chrysophthalmus

PARMOTREMA
P. cristifera
P. disparile
P. saccatilobum
P. parahypotropum
P. robustum
P. rampoddense
P. tinctorum
P. crinita
P. reticulata
P. austrosinense
P. subtinctorium
P. permutatum
P. periata

Figure 39 Replacement of species within several lichen genera with change in latitude along the eastern Australian coast (redrawn from Stevens 1979).

At the lower temperature range, such species as *Physcia tribacioides*, *Parmotrema reticulata* and *Physciopsis adglutinata* become frequent (Stevens 1979).

In some areas along the central Queensland coastline (rainfall 800-1,000 mm per annum) mangroves occur sparsely along the edges of creeks or in patches surrounded by large areas of claypans with water deficits and high interstital soil salinities (Saenger and Robson 1977; Saenger et al. 1977). Several species of mangroves survive under these conditions, but *Ceriops tagal* and *Excoecaria agallocha* support the greatest lichen diversity. The conditions of high temperature, high light intensity and low humidity are tolerated by a lichen association characterized by *Teloschistes flavicans*, *Parmotrema robustum*, *P. parahypotropum*, *Ramalina exiguella*, *R.* aff. *fecunda*, *R.* aff. *leiodea*, *R. boninensis* and *R. disparata*.

Mangrove Epiphytes

Particularly in the more luxuriant tropical mangroves, a number of vascular plants (table 29) use mangroves as supports; these species are termed epiphytes. Epiphytes are not parasites and they have no vascular connection with their host plants. Two strategies are recognizable among mangrove epiphytes: lianes or creepers are rooted in the ground and use the mangroves merely as supports for their aerial parts, obtaining their water and nutrients from the soil; by contrast, aerial epiphytes have the entire plant supported above the ground and they depend on trapped water in leaf axils or in dead cells on the surface of roots, and on debris to meet their growth requirements.

Most of the epiphytes that occur on mangroves do so because of the generally favourable conditions of shade, humidity and temperature found there. However, these species also occur in other situations where such microclimatological conditions are found. Thus, although they may be common in certain mangrove communities, they are not restricted to them. Some epiphytes, however, particularly the ant-house plants (see below), are more or less confined to mangroves, although they will occasionally occur on *Melaleuca viridiflora* on the landward margins of mangrove swamps. As indicated in table 29, a large number of mangrove epiphytes form a mutualistic relationship with one or more species of ants. These plants are termed ant-fed plants or "myrmecophytes" because in all instances ants live within, or under, a part of the plant that appears specialized for harbouring ants and absorbing nutrients from the ants' piles of debris (Thompson 1981). Plants of *Myrmecodia beccarii* are common in the more northerly

Table 29 Epiphytes recorded from Australian mangroves

Family/Species	Type of epiphyte	Common hosts
Asclepiadaceae		
Ischnostemma carnosum	liane	Several
Gymnanthera nitida	liane	Several
Dischidia nummularia (= D. ovata)	aerial	Xylocarpus granatum and Heritiera littoralis
Dischidia major (= D. rafflesiana)	aerial	Xylocarpus granatum, and Heritiera littoralis, Excoecaria agallocha
Hoya australis	liane	Several
Fabaceae		
Derris trifoliata	liane	Lumnitzera racemosa, Heriteria littoralis
Capparadaceae		
Capparis lucida	liane	Ceriops tagal, Excoecaria agallocha, Lumnitzera racemosa
Orchidaceae		
Bulbophyllum minutissimum	aerial	Bruguiera gymnorhiza
Cymbidium canaliculatum	aerial	Xylocarpus granatum, Heritiera littoralis
Cymbidium maididium	aerial	Avicennia marina, Bruguiera gymnorhiza, Xylocarpus granatum, Heritiera littoralis
Dendrobium discolor	aerial	Bruguiera spp., Heritiera littoralis
Dendrobium rigidum	aerial	Bruguiera spp.
Dendrobium speciosum	aerial	Rhizophora stylosa
Dendrobium teretifolium	aerial	Bruguiera gymnorhiza
Polypodiaceae		
Platycerium bifurcatum	aerial	Heritiera littoralis
Polypodium elongata	aerial	Xylocarpus granatum
Polypodium confluens (= Drynaria rigidula) (= Polypodium rigidulum)	aerial	Xylocarpus granatum
Asplenium australasicum	aerial	Heritiera littoralis, Xylocarpus granatum
Rubiaceae		
Hydnophytum formicarum	aerial	Several
Myrmecodia antoinii	aerial	Several
Myrmecodia beccarii	aerial	Several

mangroves and a plant may weigh up to several kilograms. Their swollen bulb-like stems are honeycombed with tunnel-like galleries in which an ant species — *Iridomyrmex cordatus* — and the larvae of a butterfly — *Hypochrysops apollo* — live in an apparently symbiotic relationship.

The ant-plant relationship has been established in *Dischidia* (Janzen 1974; Rinz 1980), *Hoya* (Merrill 1945), *Hydnophytum* (Huxley 1978; Rickson 1979) and *Myrmecodia* (Janzen 1974;

Huxley 1978). It has been suggested that the relationship has evolved in response to a nutrient-poor (low nitrogen) habitat (Janzen 1974; Huxley 1978; Thompson 1981), and this has been demonstrated experimentally in *Hydnophytum formicarium* (Rickson 1979; Janzen 1974) and *Myrmecodia* cf. *tuberosa* (Huxley 1978).

In *Hydnophytum formicarium*, two types of cavities occur in the swollen base — rough-walled, water-absorbing chambers and smooth-walled non-absorbing cavities. The ant *Iridomyrmex cordatus* places its debris in the rough-walled cavities instead of tossing it out of the nest (Janzen 1974). Rickson (1979) showed that when radioactively labelled *Drosophila* larvae were carried into the plants by the ants, radioactive compounds moved into the plant and were translocated up the stem. Huxley (1978) showed that when ants were given food tagged with radioactive tracers, the tracers were preferentially deposited in the rough-walled cavities of *Myrmecodia*, most probably through defaecation. She also found that survival of *Myrmecodia* seedlings in the field was significantly higher when the plants were associated with *I. cordatus* than when the plants either had no ants or were associated with ant species other than *I. cordatus*.

Thompson (1981) argued that the phenomenon of ant-fed plants is associated with the following ecological conditions: nutrient-poor environments, epiphytic habit generally on open-canopy trees, and low moisture availability. These conditions are typical of those prevailing in the tops of mangrove canopies and possibly explain the disproportionately large number of ant-fed species among mangrove epiphytes.

Mistletoes

A number of mistletoes (family Loranthaceae) occur on tropical and subtropical mangroves (Barlow 1966). Mistletoes are parasitic plants which have normal leaves and are capable of photosynthesis, but which have become modified to tap into the xylem tissue of their host plant and to obtain their water from the host's vascular system. Apart from a penetrating structure termed an "haustorium", the mistletoes have virtually eliminated their root system (Barlow 1967) and they have no contact with soil. Being xylem sap parasites, the mistletoes must adapt to the salt concentration in the host's vascular system. Different strategies of coping with salt have evolved in their mangrove hosts (chapter 2), but the means by which the mistletoes cope with this problem remain to be investigated.

Of the seventy species of mistletoes recorded from Australia, many have become specialized for life on a particular host species or group of host species (Barlow 1981, 1984); evolution of increased efficiency as a parasite often involves reduction in the degree of disturbance caused to the host plant, since the survival and vigour of the parasite usually depend on the survival and vigour of the host. Among those species to be seen on mangroves (table 30) are some which are more or less exclusive to mangrove hosts (such as *Amyema mackayense, A. thalassium*), and even among mangroves they may be restricted to a few species. Other species (table 30) are

Table 30 Mangrove mistletoes with mangrove and non-mangrove host species

Mistletoe	Mangrove host species	Other host species
Amyema thalassium	*Avicennia marina* *Excoecaria agallocha* *Bruguiera* spp.	None
Amyema mackayense	*Rhizophora* spp. *Avicennia marina* *Ceriops tagal* *Lumnitzera racemosa* *Osbornia octodonta*	One record on *Flindersia* sp.
Lysiana subfalcata spp. *maritima*	*Ceriops tagal* *Rhizophora* spp.	*Casuarina glauca* *Myoporum acuminatum*
Amyema glabrum	*Bruguiera* spp. *Rhizophora* spp. *Avicennia marina* *Xylocarpus* spp.	Occasionally on rainforest species, particularly *Achronychia muelleri*, also on *Commersonia bartramia*
Amyema conspicuum ssp. *conspicuum*	*Xylocarpus australasicus* *Avicennia marina* *Ceriops decandra*	*Alphitonia* spp., *Casuarina* spp., *Alyxia spicata*, *Torraea brownii*, *Callistemon* spp. and *Acacia crassicarpa*
Amyema congener ssp. *congener*	*Avicennia marina* *Lumnitzera littoralis* *Ceriops* spp. *Excoecaria agallocha*	Many species including *Casuarina* spp. *Banksia integrifolia, Eucalyptus* spp., *Acacia* spp. *Callistemon viminalis* and *Callistemon petrei*
Dendrophthoe glabrescens	*Excoecaria agallocha* *Rhizophora* spp.	Many species including *Eucalyptus* spp., *Acacia* spp. and *Casuarina* spp.

common on both mangrove and non-mangrove hosts (such as *Amyema glabrum*), and a third group generally prefers non-mangrove hosts but may occasionally be found in mangroves.

The distributions of mangrove mistletoes are shown in figure 40; for the non-exclusive species, only their distribution on mangrove hosts is indicated. The geographic range of the various mangrove mistletoes seems to correlate with the species richness of the mangrove host species — the northern and northeastern mangroves appear to have the most species. No single mistletoe species occurs throughout the entire range of mangrove distribution, and significant areas of mangroves (for example, western Gulf of Carpentaria, southern Western Australia, Victoria and South Australia) apparently possess no mistletoes.

Mistletoe seeds are distributed by birds, and some birds such as the mistletoe bird (*Dicaeum hirundinaceum*) live in close association with the mistletoe plant, feeding on its fruit. This particular bird defaecates in such a way that the undigested seeds fall not to the ground but rather on to the branch on which it is perched. A sticky layer on the seed cements it to the branch, where it germinates spontaneously. The emerging root is highly modified in the form of a sucker pad, which becomes attached to the host and develops an haustorium, which quickly penetrates the host's xylem tissue. Parasitism is thus established in the early seedling stage; the mistletoe eventually becomes a woody shrub (Barlow 1967).

Primack, Duke and Tomlinson (1981) showed that the flowers of the mangrove mistletoe *Lysiana maritima* are adapted to pollination by birds, and various nectar-feeding birds have been recorded as visiting the flowers — such visitors include the white-throated honeyeater (*Melithreptus albogularis*), the yellow-breasted sunbird (*Nectarinia jugularis*) and the mistletoe bird (*Dicaeum hirundinaceum*). The flowers of *Lysiana maritima* have a red calyx as an attractant and sticky pollen which facilitates attachment to the bird's bill. Birds fly actively between plants, probing all flowers. Many flowers have slits in the calyx tube where the birds have apparently torn the flowers to get at the nectar. However, because the ovary of these flowers is inferior, it is protected from damage by the birds.

Five species of butterflies are known to feed on the leaves and flowers of mangrove mistletoes (Common and Waterhouse 1972), including *Delias nigrina*, *Ogyris amaryllis*, *Canadalides margarita*, *Hypochrysops digglesii* and *H. narcissus*. One subspecies, *O. amaryllis hewitsoni*, feeds exclusively on *Amyema mackayense*, a more or less mangrove-exclusive mistletoe. The close association between these butterflies and the mangrove mistletoes suggests they also may play a role in pollination.

Although the distribution gaps (figure 40) in southern Australia can be partly explained by the mistletoe bird not being associated with mangroves in these regions (although it occurs there; Saenger et al. 1977), the gaps in the Gulf of Carpentaria are more difficult

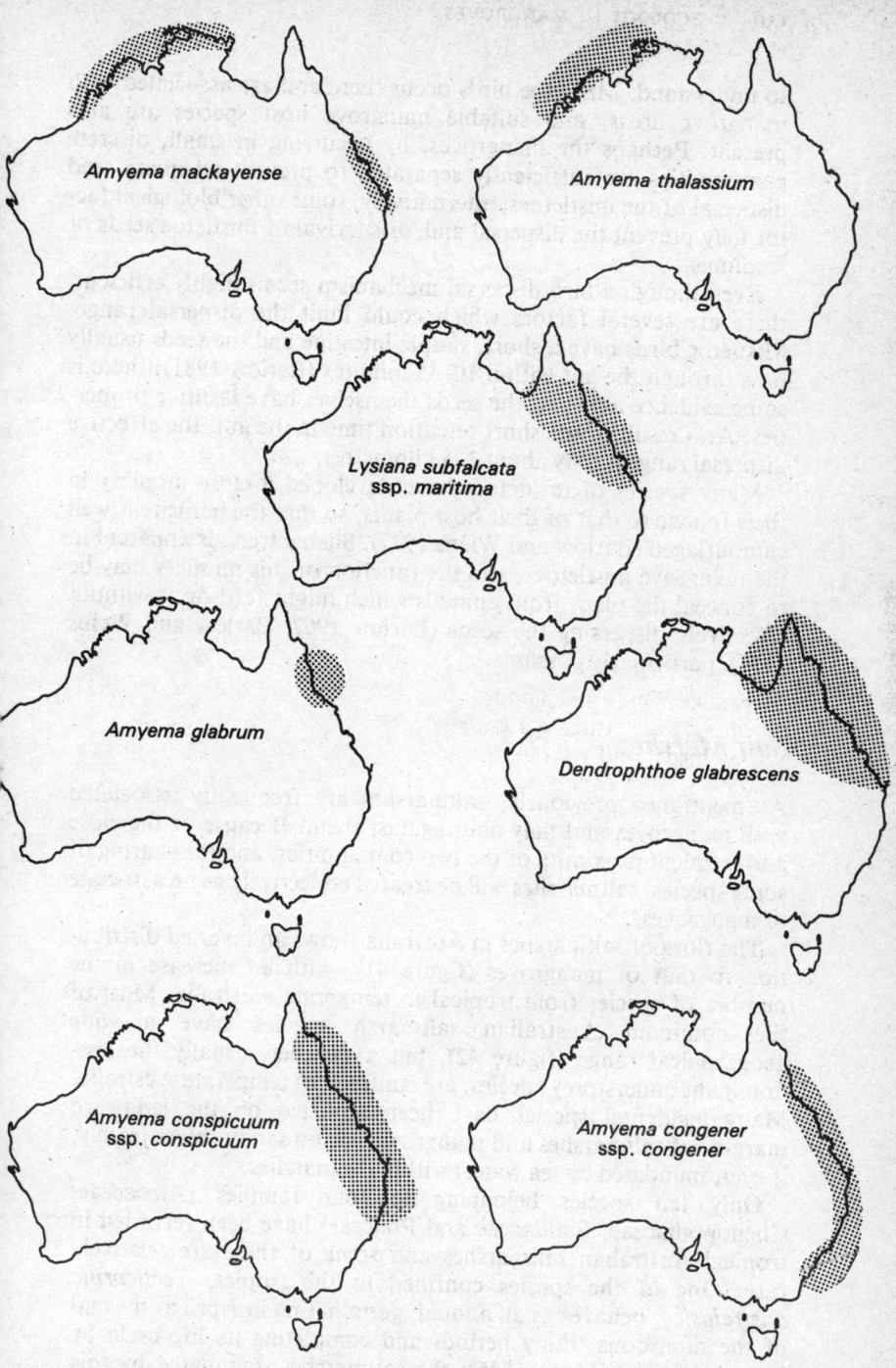

Figure 40 Distribution in Australia of mistletoes confined to or commonly growing on mangrove hosts.

to understand. Mistletoe birds occur there and are associated with mangrove areas, and suitable mangrove host species are also present. Perhaps the mangroves, by occurring in small, discrete communities are sufficiently separated to prevent adequate seed dispersal of the mistletoes; alternatively, some other biological factor may prevent the dispersal and/or survival of mistletoe seeds or seedlings.

Even though a bird dispersal mechanism seems highly efficient, there are several factors which could limit the dispersal range. Mistletoe birds have a short, simple intestine and the seeds usually pass through the gut within 10-45 minutes (Barlow 1981); there is some evidence also that the seeds themselves have laxative properties. As a result of the short retention time in the gut, the effective dispersal range is only about 3-5 kilometres.

Many species of mistletoes have developed a close mimicry in their foliage to that of their host plants, so that the parasite is well camouflaged (Barlow and Wiens 1977). Such a trend is apparent in the mangrove mistletoes, and the function of this mimicry may be to conceal the plant from animals which might feed on it without effectively dispersing the seeds (Barlow 1967; Barlow and Weins 1977), particularly possums.

Salt Marshes

As mentioned previously, saltmarshes are frequently associated with mangroves and may abut against them. Because of the close and frequent proximity of the two communities, and the sharing of some species, saltmarshes will be treated collectively as an associate of mangroves.

The flora of saltmarshes in Australia shows an inverted distribution to that of mangroves (figure 41), with an increase in the number of species from tropical to temperate Australia. Most of the dominant Australian saltmarsh species have a wide geographical range (figure 42), but a number, usually the less-dominant understorey species, are confined to temperate Australia. Many incidental species have been reported on the landward margins of saltmarshes and mangroves, or on sandy ridges (rarely, if ever, inundated by sea water) within the marshes.

Only ten species belonging to four families (Aizoaceae, Chenopodiaceae, Batidaceae and Poaceae) have been recorded in tropical Australian saltmarshes and some of these are relatively rare. One of the species confined to the tropics, *Tecticornia australasica*, behaves as an annual, germinating in April at the end of the monsoonal rainy period, and completing its life cycle by November (van Royen 1956); the saltmarshes dominated by this

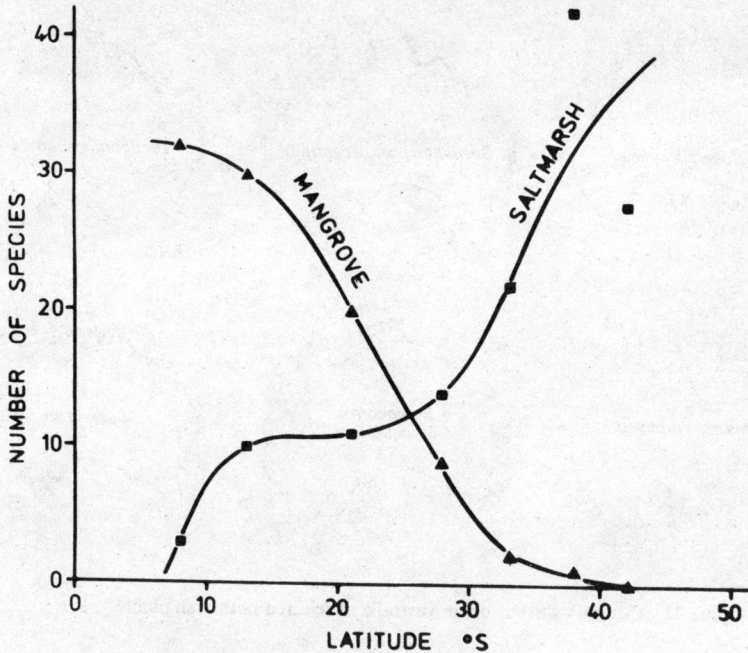

Figure 41 Relationship between the number of species of mangroves and saltmarsh plants and latitude along eastern Australia.

species may be devoid of macrophytes during the rest of the year. The other typically tropical saltmarsh species is the perennial *Batis argillicola*. A few additional species such as *Sesuvium portulacastrum*, *Xerochloa barbata* and (in New Guinea) *Eriachne pallescens*, which are found at the landward edge of the saltmarsh, also may be considered as typically tropical.

Subtropical saltmarshes have a greater species richness than tropical ones, and on the eastern Australian coast contain the endemic *Suaeda arbusculoides* (Smith 1969).

Towards temperate Australia, the saltmarsh flora increases markedly in species richness. In southern Australia, fifty-two species and nine subspecies belonging to sixteen families have been recorded (Saenger et al. 1977; Specht 1981; Kirkpatrick and Glasby 1981). Some of these species are semi-succulent shrubs 0.5–2.0 metres tall, comprising low shrubland (foliage projective cover 10–30 per cent) to low open shrubland (foliage projective cover 0–10 per cent) formations.

Salt-tolerant grasses (*Distichlis*, *Pucciniella* and *Sporobolus*) and herbs (nineteen genera) may occur with the succulent,

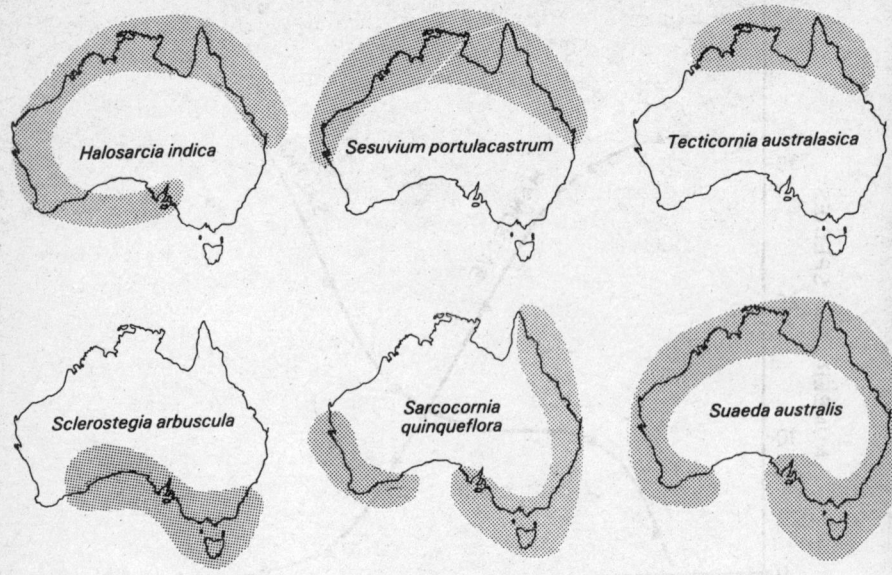

Figure 42 Coastal distribution in Australia of selected saltmarsh plants.

rhizomatous chenopod *Sarcocornia quinqueflora* to form either an understorey to the low shrubland formation, or alone as a herbland or closed herbland formation.

The Chenopodiaceae is the dominant and generally characteristic family of Australian saltmarshes; its fourteen genera contain thirty species of saltmarsh shrubs distributed as follows: *Atriplex* — 4; *Bassia* — 1; *Enchylaena* — 1; *Halosarcia* (Syn. *Arthrocnemum* in part) — 9; *Hemichroa* — 2; *Maireana* (Syn. *Kochia*) — 2; *Rhagodia* — 1; *Salsola* — 1; *Sarcocornia* (Syn. *Salicornia*) — 2; *Sclerostegia* (Syn. *Arthrocnemum* in part) — 2; *Suaeda* — 2; *Tecticornia* — 1; *Theleophyton* — 1; and *Threlkeldia* — 1.

In addition to the Chenopodiaceae, the following families contain genera with all or most species confined to the saltmarsh habitat: Aizoaceae (2 genera), Batidaceae (1 genus), Convolvulaceae (1 genus), Plumbaginaceae (4 genera), Poaceae (6 genera), Primulaceae (1 genus) and Scrophulariaceae (2 genera). It is of note that, with the exception of the Batidaceae, all of the above families possess other genera which grow only in non-saline, terrestrial environments; the only family having both saltmarsh and mangrove representatives is the Plumbaginaceae, although different genera are involved in the two habitats.

Like the mangrove flora, many of the saltmarsh plants have adaptations which aid survival in their generally drier, saline habitat. Succulence is probably one of the most obvious features and, as discussed in chapter 2, apparently is related to the physiological dryness of the habitat; this condition is found in nearly all of the chenopodiaceous saltmarsh genera and also occurs in *Sesuvium* (Aizoaceae) and *Batis* (Batidaceae).

Water-storing aqueous tissue is found in many species also, particularly those showing succulence (De Fraine 1912; Fahn 1963; Wilson 1980), and "passage cells" which may be involved in internal water movement form a network in the palisade tissue of many species of *Halosarcia* (Tolken 1967; Wilson 1980).

Salt glands have been reported from *Limonium* (Ziegler and Luttge 1966), of which there are two native and two introduced species in Australian saltmarshes, and from *Frankenia* (Campbell and Thomson 1976), *Spartina* (Klugh 1909), *Sporobolus* (Liphschitz and Waisel 1974) and *Samolus* (Adam and Wiecek 1983).

In contrast to dicotyledonous glands, those of the grasses are small, two-celled structures. Although these are the only saltmarsh plants known to have salt glands, many species have a physiological adaption to the high salinity concentrations in their environment. Malcolm (1964) showed that *Halosarcia pterygosperma* and *H. pergranulata* were able to germinate successfully in saline solutions with only 50 per cent reduction at 8 g/l NaCl and 20 g/l NaCl respectively. Similarly, Smith-White (1981) and Gallagher (1979) showed that the saltmarsh grass *Sporobolus virginicus* was able to survive and grow at 100 per cent sea water, although it showed some reduction when compared with growth at 10 per cent sea water.

As with the mangroves, the mechanism of salt tolerance is not known in saltmarsh plants at present, but it has generated interest in its potential economic value, particularly with regard to stabilizing dunes and reclaimed saline areas in coastal and inland localities. Smith and Malcolm (1959) discussed the use of *Maireana brevifolia* and *Atriplex* spp. for growing on salt-affected soils in Western Australian agricultural areas, but these species were not suited to waterlogged conditions. The germination studies on *Halosarcia* by Malcolm (1964) were part of a programme for using these species as fodder plants on highly saline, winter-waterlogged soils.

Although the root systems of saltmarsh plants may have some physiological or biochemical adaptations, they do not show the morphological diversity of mangroves. On this basis, it can be assumed that most saltmarsh plants cannot survive or remain vigorous in waterlogged, anaerobic soils, and field studies on these plants indicate that anaerobic waterlogging is a major factor limiting their distribution, particularly their seaward boundary (Clarke and Hannon 1971).

Fringing Species

A large number of species of trees and shrubs occur at the landward margins of mangroves and saltmarshes, and will occasionally grow within these communities. Some of these fringing species occur in a number of habitats whereas others appear to grow only in the ecotonal community. Saenger et al. (1977) listed the more important species.

Species which are characteristic of the fringing community and which never or only rarely occur in other situations include *Hibiscus tiliaceus, Thespesia* spp., *Cerbera manghas* and *Clerodendron inerme. Diospyros littorea* and *D. ferrea* var. *geminata* also appear to belong to this group (Duke, Birch and Williams 1981; Elsol and Saenger 1983). In Papua New Guinea and southeastern Asia, the leguminous *Intsia bijuga* forms a distinct ecotone between mangroves and the abutting rainforests (Taylor 1959).

Those species which commonly occur in the fringing community but also occur in other (but related) situations, such as sandy beaches or coral shingle, inland salt flats or fresh to brackish swamps, include *Pemphis acidula, Myoporum acuminata* and *M. insulare, Melaleuca quinquenervia* and *Casuarina glauca*.

The number of species which sometimes occur in the fringing community but which occur widely elsewhere, is high, and the species vary in different areas of Australia. For example, in Princess Charlotte Bay in northern Queensland, Elsol and Saenger (1983) found the following non-mangrove species growing within tidal limits: *Canthium latifolium, Flagellaria indica, Maytenus emarginata, Melaleuca acacioides, M. leucadendron, M. stenostachya, M. viridiflora, Premna acuminata, Strychnos lucida* and *Terminalia subacropta*.

5. The Fauna of Mangroves

The fauna of mangroves reflects the convergence of two types of environment: (1) a sheltered marine muddy intertidal habitat and (2) a terrestrial forest. In temperate areas, mangroves are usually layered woodlands less than 10 metres in height; in tropical areas they may be open or closed woodlands or forests, often multi-layered and generally with distinct zonation. As the terrestrial fauna is restricted largely to the forest canopy, and the marine fauna is found mainly on the substrate and in the lower levels of the forest, vertical partitioning of the two dominant kinds of fauna occurs (figures 43(a) and (b)). A third but minor component, a freshwater fauna, may be present along the margins of mangroves in their upper regions.

Within the two major faunal components, subdivisions occur. For example, the terrestrial fauna may be restricted to the canopy

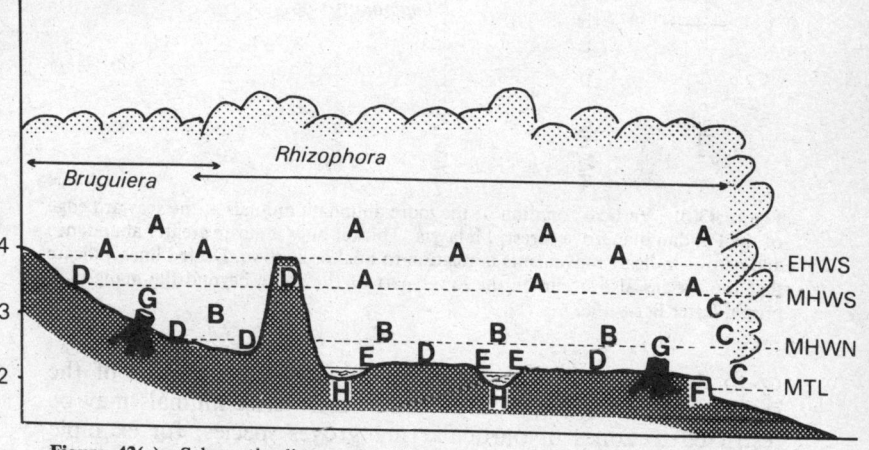

Figure 43(a) Schematic diagram to illustrate the partitioning of the mangrove habitat, as it affects the fauna. Horizontal dimensions are very much shortened compared with the vertical ones. A = higher stems and leaves of mangrove trees; B = lower stems and roots of mangrove trees; C = seaward fringe of mangrove vegetation; D = mud surface; E = edges of small creeks flowing into the mangroves; F = coast mud bank at mangrove edge; G = dead wood; H = small creek (after Berry 1963).

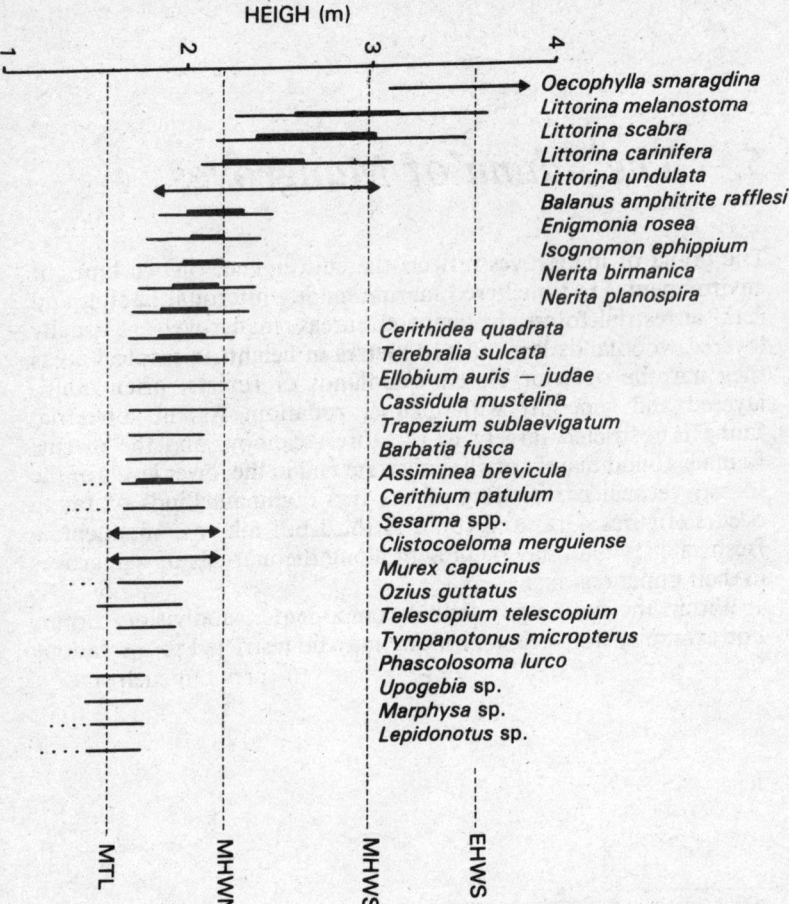

Figure 43(b) Vertical zonation of the more abundant animals at the seaward edge of the Pandan mangrove forest, Malaysia. Thicker lines indicate greater abundance and arrows indicate movements in relation to tidal oscillation. Dotted lines indicate that the species also occur on the lower part of the shore beyond the mangroves proper (after Berry 1963).

or to the trunks of trees or be found only on the floor of the mangroves during low tide. In some cases, these animals may be restricted to zones of particular mangroves species; for example insects may prefer the rough bark of *Rhizophora* to the smooth bark of *Avicennia*, some possums feed selectively on the pollen produced by certain mangrove species and some bats feed mainly on the nectar of *Sonneratia* flowers. Similar subdivisions occur within the marine fauna, some animals being restricted to the wetter frontal zones of the mangroves among the *Rhizophora*.

In many cases mangroves are contiguous with adjacent terrestrial forests and some of the terrestrial fauna occurs in both habitats. Often these animals are using the mangroves simply as an extension of the terrestrial habitat but in some cases the mangroves provide an essential seasonal source of food or a suitable site for breeding or laying eggs (for example, for some insects). Particular species may vary in the number of habitats they occupy in a given area, and their occurrence may vary geographically. Ford (1982) postulated that in some areas certain species of birds are restricted to mangroves, whereas in other geographical regions the same species do not occur in mangroves at all, but rather in other types of habitats with closed canopies. He suggested that the degree to which a population is restricted to mangroves appears to be a function of three main factors: (1) the proximity of structurally similar habitats, (2) the presence of competitors in these similar habitats and (3) the selective pressure that operated on geographical isolates during climatic cycles in the Pleistocene (when several closed-canopy inhabitants underwent ecological shifts) (figure 44; table 31). Ford also suggested that specialization to mangroves has occurred in response to particular kinds of food not found elsewhere in closed-canopy habitats and through association with the structure and microclimate of mangroves (a warm mesic habitat with good overhead cover for concealment and protection). Birds are probably the easiest group on which to carry out such studies. Are there other taxa which extend their ranges by using mangroves in areas where no other suitable terrestrial forests exist? For much of the terrestrial fauna living in mangroves, either permanently or as visitors, little detailed information on habitat requirements is available. Terrestrial biologists seem reluctant to work in mangroves!

The terrestrial fauna of mangroves can be influenced by the adjacent vegetation type. A given type of mangrove may have very different kinds of animals depending on whether the adjacent habitat is tropical lowland forest, eucalypt forest, *Juncus* sedgelands or *Melaleuca* swamp (Ford 1982). In some cases, mangroves may provide a refuge for terrestrial species from adjacent habitats. Animals feeding on the muddy substrate or moving through the mangrove forest on the ground will be restricted by tidal cycles. In areas where mangroves abut on nearby forests, animals may move extensively between the two habitats, depending on changes in the relative availability of food or water in the two habitats at different times. Similar movements may occur between mangroves and nearby marine habitats. Birds such as herons may breed or roost in the mangroves, yet feed on nearby mudflats and seagrass beds or on adjacent coral reefs.

The terrestrial fauna can be affected by the presence of perma-

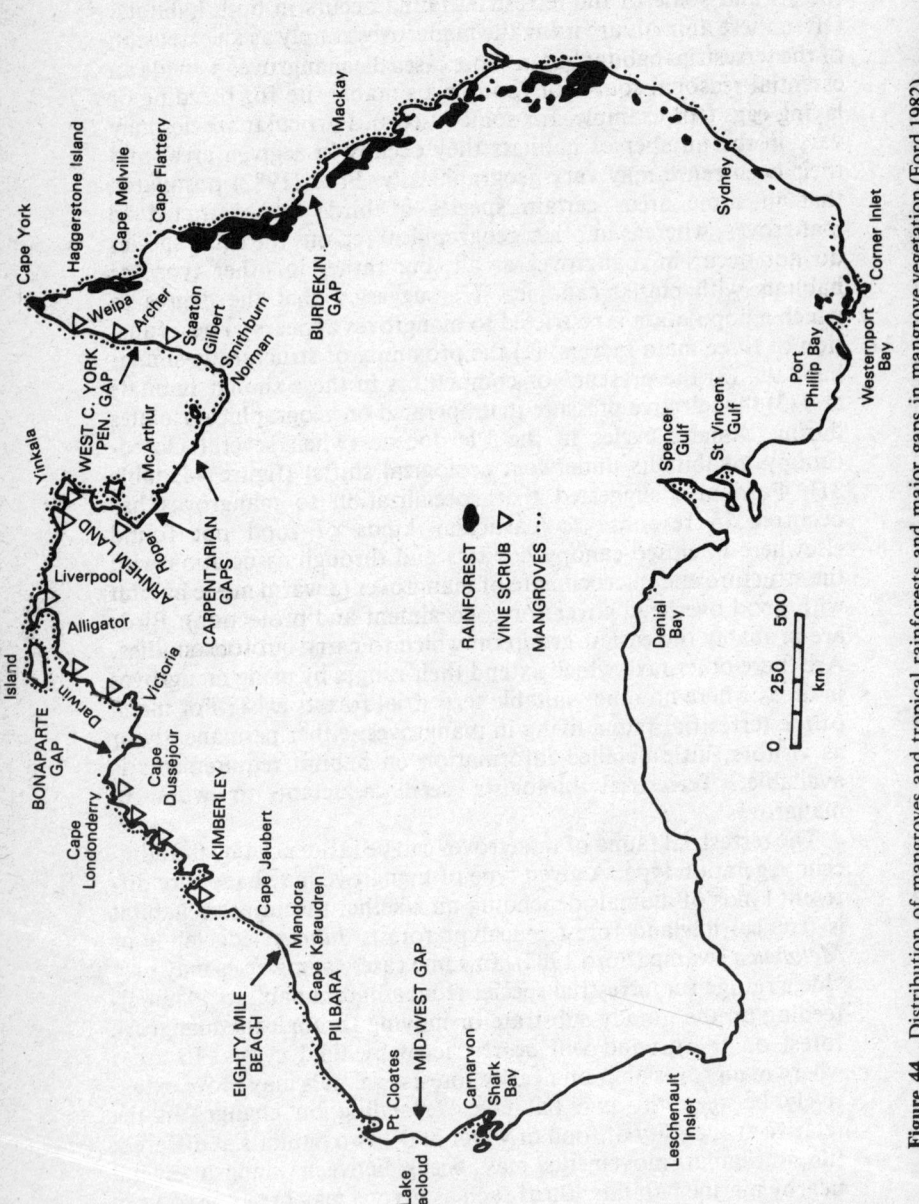

Figure 44 Distribution of mangroves and tropical rainforests and major gaps in mangrove vegetation (Ford 1982).

nent fresh water, the frequency of flooding (McCormick 1978), climatic variation in rainfall, and evaporation and tidal patterns, all of which determine water availability and its temporal distribution at critical times of the year for both sedentary and migratory species.

For the marine fauna, vertical zonation, both across the substrate from low-water level to high-water level and up the tree trunks (Wells and Slack-Smith 1981), occurs in response to periodicity of tidal inundation and drainage patterns (Macnae 1967; Yates 1978). This pattern may be modified by tidal creeks within the mangroves. Behavioural, physiological and morphological adaptations to water loss will largely determine the tidal levels at which particular species can live.

The distribution of animals may vary during a tidal cycle or during their life cycle (Bell et al. 1984; Cockcroft and Forbes 1981; Milward 1982; Staples 1980a, 1980b; Subrahmangam and Coutlas 1980). Like the terrestrial fauna, the distribution of the marine fauna may be determined by the zonation of the mangroves themselves. For example, the prop roots of *Rhizophora* and buttress roots of *Bruguiera* may provide a larger surface area or more crevices for marine animals in comparison with *Avicennia* trunks. The density of pneumatophores and the amount of debris and fallen logs on the forest floor may be important for cryptic and wood-inhabiting species by providing cover and protection from desiccation. Fine sediment tends to be trapped around the bases of trees with well-developed buttress roots like *Bruguiera*, and polychaetes are often in considerable densities in such sediment (Hutchings, unpubl. data). Similarly, the fine sediment forming the mounds of *Thalassina* turrets are often populated by polychaetes. Thus any zonation patterns of *Thalassina* will be imposed on the polychaete species living in the burrow mounds. A detailed study of sediments in mangroves in South Australia has been carried out by Butler et al. (1977a). Only *Avicennia marina* occurs in that region and the sediment it traps appears to be an important factor in determining which polychaetes occur in the mangrove stands.

The density of seedlings and the canopy cover which provides shade can be important for species living on the surface of the mud. Herbivorous species may exhibit feeding preferences for particular mangrove species. The marine fauna should be compared with adjacent intertidal muddy habitats in order to ascertain which species are common to several habitats and which are restricted to mangroves. This may vary over the geographical range of the species, or during the life cycle. Juvenile soldier crabs (*Mictyris*) occur in mangroves and seagrass beds, whereas adults occur on nearby mudflats.

Salinity gradients along a river will determine which species

Table 31 Distribution of birds adapted to mangroves and their origins

	Western Australia (25°-14°S)	Pilbara (20°-23°S)	West Kimberley (14°-18°S)	Cambridge Gulf (±15°S)	Arnhem Land (11°-14°S)	Gulf of Carpentaria (15°-18°S)	Northwest Cape York Peninsula (S.E. to 19°S)	Central east coast (19°-29°S)	Southeast coast south of 29°S	Habitat of ancestor if known
Ardea sumatrana	A		A	A	A/B	A				M/WV
*Butorides striatus		A	A	A	A	A⊞	A	A		M
*Eulabeornis castaneoventris			A		A	A				LRF
Chrysococcyx minutillus			B	B	B	B		C⊞		LRF
Alcedo azurea			?							WV
Alcedo pusilla					B		C			RF/SI
*Halcyon chloris			A	A	A	A	A	A?		OF
*Eopsaltria pulverulenta			A⊞		A⊞					LRF
*Microeca tormenti			A⊞							LRF
*Pachycephala melanura		A	A	A	A	A	A			LRF
*Pachycephala lanioides	A	A	A	A	A	A				LRF
Colluricincla megarhyncha			C	?	C		A?	A⊞		LRF
*Myiagra ruficollis			A	A	A	A	A⊞	C⊞		LRF

Species	A	B	?	B	B	A	B/C	B	A	Habitat
Myiagra alecto	A							B	A	LRF
*Rhipidura rufifrons dryas							⊞			RF/SE
*Rhipidura phasiana	A	A	A	A	A	A	A			LRF
Gerygone magnirostris		A				⊞	B		B	M
*Gerygone tenebrosa	A	A	A							LRF
*Gerygone laevigaster		⊞	A	A	A	A	A ?	A	A	LRF
Lichenostomus versicolor					A		A			SW
*Myzomela erythrocephala						A				LRF
Zosterops lutea	A	A	A	A	A	A	?	A		LRF
Cracticus quoyi							⊞	A	A	LRF
No. of species	6	9	20	19	20	16	17	14	3	

A = entirely or mainly in mangroves
B = frequently in mangroves
C = equally in all sorts of closed canopy vegetation
M = mangroves
LRF = lowland rainforest
WV = waterside vegetation
OF = open forest
RF = rain forest
SF = schlerophyll forest
SW = savannah woodland

Solid lines indicate more or less continuous range; breaks in lines indicate major breaks in range (> 50 km)
⊞ indicate subspecific change, i.e., populations on either side belong to different subspecies
Species* are obligate mangroves species, i.e., virtually confined to it
Remainder are primary faculative species, i.e., restricted to it for at least part of their range
Source: After Ford (1982) and Schodde, Mason and Gill (1982).

penetrate upstream (providing suitable habitats are otherwise present) as salinity affects the zonation of mangroves (chapter 3).

Superimposed upon all these factors are latitudinal gradients along the coast. Variations in the hinterland will affect rainfall, drainage patterns and run-off.

Rainfall is distinctly seasonal in the tropical parts of Australia; in these areas soil salinity may fluctuate from almost zero during prolonged flooding to 37 0⁰/oo (Saenger 1979). In temperate mangroves, rainfall is far less seasonal, although there may be prolonged periods of drought. However, in some areas such as southern Australia and northwestern Western Australia very little freshwater run-off occurs. In these areas, high salinities may be experienced throughout the year. Thomas (pers. comm.) has recorded salinities of 45.4–63.3⁰/oo at 120 cm depth in the sediments at Port Augusta; sea water in this area is 44.3–48.5⁰/oo. At Torrens Island near Adelaide, sediment salinity at 120 cm ranges from 25.9–41.8⁰/oo, with sea water varying between 36.5 and 38.3⁰/oo. Thomas suggested that the increased sediment salinities at the top of the gulf occur because the salinity of sea water is higher there and because of saline ground water. There is virtually no freshwater run-off.

Such low run-off will severely restrict terrestrial sources of nutrients and detritus and high salinities may restrict infaunal species. Flood conditions may wash away infaunal and epifaunal species or restrict the feeding of these marine animals (see p. 237). Other latitudinal gradients are those of temperature, incidence of frosts and day length. Virtually nothing is known about the effect of day length on the marine mangrove fauna. Whether breeding in the fauna is related to changes in temperature or changes in day length is also unknown. With decreasing latitude, the number of marine species increases as do species restricted to mangroves (figure 45). This may be a function of temperature or, more likely, it may be related to a combination of factors.

Seasonal and yearly fluctuations in the fauna of mangroves may occur, especially in species which recruit from pelagic larvae and die shortly after breeding. Species of nudibranch molluscs appear seasonally in temperate mangroves (Rudman, unpubl. data) but detailed population studies on the mangrove fauna have rarely been carried out (Robinson et al. 1983a, 1983b). In other nearby habitats (coral reefs and seagrass beds) extensive fluctuations occur (Sale 1983; Hutchings 1981; Hutchings and Murray 1982; Hutchings and Recher 1974).

As already mentioned, catastrophic events, such as flooding or severe storms, lead to extensive tree death and sediment disturbance and may cause local extinctions among the fauna. Again, little work has been done to document these changes. Alterations to drainage patterns through flood mitigation works (Briggs 1977b;

Pressey and Middleton 1982) or reclamation (Moss 1983) may lead to permanent changes in the marine fauna.

Finally, local geographical conditions affect the mangrove fauna. For example, small islands of mangroves may have different faunas from extensive coastal areas of mangroves. No data are available on the minimum area of mangroves needed to maintain a viable community of animals. Larger stands of mangroves may provide sufficient refuges for some of the fauna to survive during adverse conditions such as floods. Reseeding of smaller, more adversely affected areas may take place from such refuges.

Composition of the Fauna

Like the flora (chapter 4), the fauna of mangroves is composed of a characteristic assemblage of species (both vertebrates and invertebrates) derived from those in terrestrial, marine and freshwater environments. Macnae (1968) pointed out that a mangrove, like any other intertidal area, represents a transition from the sea to the land and hence its fauna may be derived from either. The majority of the aquatic and substrate fauna have been derived from the sea, and the mangrove swamp clearly provides one route from sea to land. As a consequence of these origins, many species occurring in mangrove communities may have distributions extending over wide areas beyond them, and into a variety of different habitats and environments (Milward 1982). However, some species are mangrove specialists, that is, they occur primarily in mangroves. In temperate regions few, if any, mangrove specialists occur among the marine invertebrates, but with decreasing latitude the percentage of mangrove specialists increases (Hutchings and Recher 1982) (figure 45). The following factors may be involved. Mangrove occupy far larger areas in the tropics which are often contiguous, whereas in temperate regions they often occur in small isolated stands. Also, tropical mangroves are more diverse in terms of the species and the habitats they provide in comparison with temperate regions where only one or two species of trees occur. Such extensive diverse stands of tropical mangroves provide suitable living conditions and protection from predators and desiccation, and lead to faunal speciation and formation of mangrove endemics. In contrast, temperate patches of mangroves are too small and ephemeral for long-term adaptation and evolution of mangrove endemics, and the mangrove fauna is similar to that living in adjacent muddy intertidal areas. The past geological history of temperate areas also may have played a role. Recently, George and Jones (1984) suggested that the high diversity of crab species found in Mangrove Bay, North West Cape, is a result of isolation from previously more continuous coastal mangrove systems.

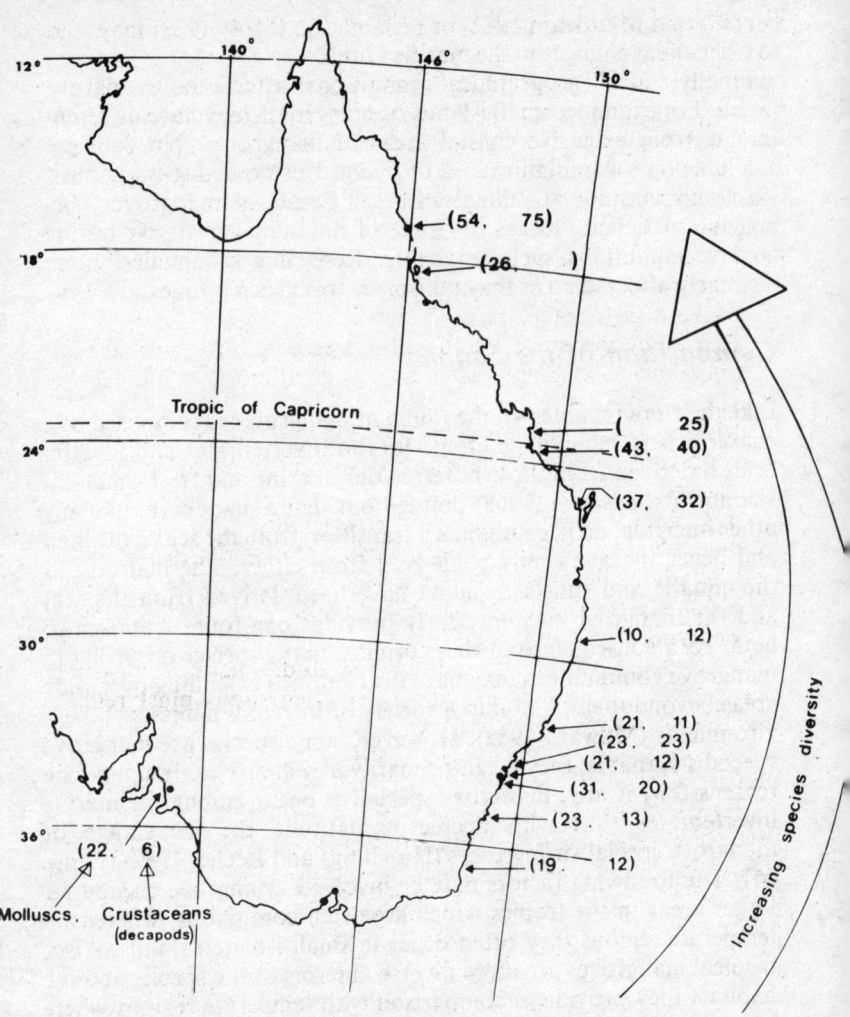

Figure 45 Several groups of marine invertebrates appear to be more abundant and richer in species in the tropics than in the temperate zones. Data on molluscs and crustaceans (mainly decapods) are most complete and are used here to illustrate this point for eastern Australian mangrove forests (after Hutchings and Recher 1982).

The mangrove fauna can be divided into permanent and temporary residents. The latter spend only part of their lives in the mangroves, such as birds which come into the mangroves to roost or feed, or prawns and juvenile fish which feed or seek protection there only during high tide (Bell et al. 1984; Schodde, Mason and Gill 1982; Staples 1980a, 1980b).

Terrestrial Fauna

For the terrestrial fauna, mangroves provide additional habitat or form a bridge or corridor between other types of habitat, constitute an island refuge, provide an isolated breeding site, or may be used as a feeding ground during a migratory passage. Although not usually primary habitat for the terrestrial fauna, mangroves may be important for the following reasons:
1. Fewer predators or competitors of terrestrial animals live in mangroves than in adjacent forests; this may be important during roosting or breeding.
2. Dense monospecific stands of mangroves may provide an abundant food supply (such as nectar) at critical times of the year.
3. Marine invertebrates may be an abundant source of potential food for some terrestrial species (such as crab-eating varanid lizards).
4. The flora comprises species that generally have fleshy, succulent leaves (but which may be high in tannin and salt).
5. Abundant detritus on the forest floor, inundated daily, perhaps is important for some species of insects associated with decaying vegetation.

However, it must be stressed that few studies have been carried out on the terrestrial fauna and these suggestions must remain tentative pending further study.

Vertebrates

Few vertebrates are restricted to mangroves. The majority are visitors including mammals, birds and reptiles.

Mammals

One of Australia's rarest mammals, a small, grey, white-bellied rat (*Xeromys myoides*), recently has been recorded as occurring in the mangroves of various localities in the Northern Territory and Queensland (Redhead and McKean 1975; Magnusson, Webb and Taylor 1976), where it forages at low tide among *Avicennia marina* and *Rhizophora stylosa* (Van Dyck, Baker and Gillette 1979). In the Northern Territory, *Xeromys* builds a nest of leaves and mud interlocked within the buttress trunks of *Bruguiera parviflora* (Magnusson, Webb and Taylor 1976). Breeding probably is restricted to neap tides, as high spring tides would inundate the nests. *Xeromys* is very agile, climbs trees, and in the mangroves appears to feed mainly on crabs. Mangroves are obviously an important habitat for this rodent, although it has been recorded from other habitats such as freshwater swamps and wet heath (Troughton 1943; Redhead and McKean 1975; Dwyer, Hockings and Willmer 1979).

Mammals periodically seen in mangroves are water-rats (*Hydromys chrysogaster*), canefield rats (*Rattus sordidus*), house mice (*Mus musculus*), mosaic-tailed rats (*Melomys* sp.), tree-rats (*Mesembriomys* sp. and *Conilurus* sp.), long-nosed bandicoots (*Perameles* sp.), short-nosed bandicoots (*Isoodon* sp.), northern brush-tailed possums (*Trichosurus arnhemensis*), swamp wallabies (*Wallabia bicolor*), feral cattle (*Bos javanicus*), buffalo (*Bubalus bubalis*) and feral pig (*Sus scrofa*) in the Northern Territory and central and northern Queensland (Hegerl et al. 1979). *Hydromys* is one of the more common inhabitants of mangroves. All of these species occur in other habitats as well as mangroves.

Two species of flying foxes, *Pteropus poliocephalus* and *P. alecto*, commonly come into mangroves to feed on blossoms or to camp. *Pteropus poliocephalus* occurs along the entire eastern Australian coast, whereas *P. alecto* is restricted to tropical mangrove areas. Another species, *P. conspicillatus*, occasionally camps in mangroves. Some species of tree-roosting bats, such as the flat-headed mastiff-bat (*Tadarida planiceps*) and the little northern mastiff-bat (*T. loriae*), are also found occasionally in mangroves (Hall and Richards 1979). The northern blossom bat, *Macroglossus lagochilus*, occurs in stands of *Sonneratia alba* in Western Australia (Straughan 1983).

Birds

Birds are a conspicuous component of all mangrove forests, although they are not present in large numbers. Schodde, Mason and Gill (1982) attempted to explain the low density of birds by the small variety of foraging surfaces resulting from the comparative uniformity of structure within the canopy of mangrove forests and by the limited area of many mangrove stands. Feeding time for ground foragers is restricted by diurnal flooding, and feeding in soft sediment also may pose problems. Perhaps suitable nest sites are limited. These reasons may explain why there are seldom large populations in mangroves, and why many species occurring in mangroves also occur in other habitats.

Over two hundred species have been recorded from Australian mangroves (Saenger et al. 1977). Of these, fourteen species are virtually restricted to mangroves, twelve species utilize mangroves as primary habitat in at least part of their range, and sixty species use it regularly throughout the year or in particular seasons (Schodde, Mason and Gill 1982; Ford 1982) (figure 46; tables 31, 32). This is a rich, mangrove bird fauna compared with other parts of the world

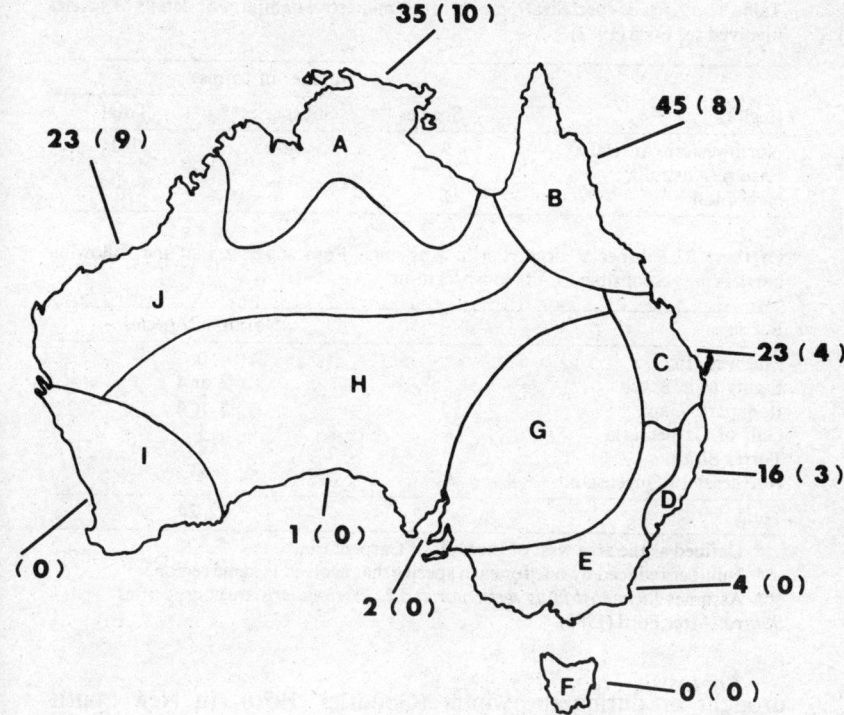

Figure 46 Kikkawa and Pearse (1969) divided Australia into regions based on the similarity of the bird fauna. These regions are convenient to illustrate the number of species of birds which occur regularly in mangroves. The number of bird species endemic to mangroves in each region is shown in parentheses (after Hutchings and Recher 1982).

(Macnae 1968). For example, New Guinea has only eleven mangrove specialists (Ford 1982).

For other species, such as some bitterns, crakes and rails, mangroves are important but not primary habitat. Many honeyeaters and lorikeets visit mangroves seasonally when mangroves are flowering and attracting insects. Other species such as the Torresian imperial pigeon (*Ducula spilorrhoa*) and pied cormorant (*Phalacrocorax varius*) breed in mangroves, benefiting from the isolation of mangrove stands. Some utilize mangroves during migration, such as southeastern Australian populations of the rufous and grey fantails (*Rhipidura rufifrons*, *R. fuliginosa*). Mangrove habitats may become very important refuges during

Table 32 Area of specialization of birds to mangrove habitat. For details of species involved see Ford (1982)

Region	Number of forms		
	Species	Subspecies**	Total
Northwestern Australia*	9	14–15	23–24
Eastern Australia	0***	8	8
New Guinea	1	9	10

Of these 22 subspecies occurring in Australia, Ford suggests that the following barriers are responsible for the subspeciation:

Barrier	No. of subspecies
Mid-western	0
Eighty Mile Beach	3 or 4
Bonaparte Gap	3 or 4
Gulf of Carpentaria	4
Torres Strait	8
Northeastern Queensland	4
Total	22–24

 * Defined as the area west of the Gulf of Carpentaria.
 ** Number reduced by one for each species that evolved in same region.
 *** Assumes *Lichenostomus versicolor* and *L. fasciogularis* are conspecific.
Source: After Ford (1982).

drought or during the winter (Goodrick 1970). In New South Wales, several species utilize mangroves and saline lagoons at these times, including the grey teal (*Anas gibberifrons*), black duck (*A. superciliosa*), blue-winged shoveler (*A. rhynchotis*), black swan (*Cygnus atratus*), musk duck (*Biziura lobata*), night heron (*Nycticorax caledonicus*) and the little pied cormorant (*Phalacrocorax melanoleucos*) (Miller 1972).

Southern temperate mangroves have few birds restricted to them. The lesser noddy (*Anous tenuirostris*) breeds on the Albrohos Islands, Western Australia. It does not occur in other habitats in that region. Most other temperate mangrove birds also occur in adjacent dry sclerophyll forests. At about 34°S, the latitude of Sydney, the first restricted species appears. With decreasing latitude both on the western and eastern Australian coasts, the total number of terrestrial mangrove birds, including those restricted to mangroves, increases (see figure 46), suggesting that as the diversity and complexity of mangrove forests increase there is a corresponding increase in the number of bird specialists. However, this simple relationship breaks down in northern Australia. In far northern Queensland, where the richest community, floristically (over thirty species), occurs there are fewer mangrove bird specialists (seven) than in northwestern Australia (thirteen) where only sixteen to twenty species of mangroves occur. Schodde, Mason and Gill (1982) and Ford (1982) explain these anomalies in terms of the origins of mangrove specialists. Detailed information on the exact ranges of each of the mangrove species, together with information on the numerous subspecies found in mangrove appears in Blakers, Davies and Reilly (1984).

Northwestern Australia is especially rich in mangrove-endemic bird species and both Ford (1982) and Schodde, Mason and Gill (1982) believes that this region played a very important role in the evolution of these endemics. Isolated populations of species in refuges in northwestern Australia became increasingly dependent upon mangroves as areas of rainforest and monsoonal forests shrank during arid conditions in the Pleistocene (Kemp 1978). Ford (1982) and Schodde, Mason and Gill (1982) suggested that the majority of endemic mangrove bird species evolved from rainforest-inhabiting species. These isolated stands of mangroves in northwestern Australia served as refuges for the dwindling stocks of formerly widespread rainforest species. As the mangrove areas became patchy, there were greater opportunities for isolation leading to speciation. In contrast, mangrove areas in northeastern Australia and perhaps in New Guinea remained contiguous with large tracts of rainforest throughout the Pleistocene. As there was continual interchange of birds between the two environments, speciation did not occur.

Ford (1982) suggested that the area in which species and subspecies became mangrove specialists has often become obscured by subsequent expansion in range. He suggested that the speciation of birds can be attributed directly to the gaps in mangroves at Eighty Mile Beach, the western side of Joseph Bonaparte Gulf, Gulf of Carpentaria, Torres Strait and possibly the Burdekin Gap. Two subspecies may have evolved in response to competitors rather than to physical barriers.

Ford (1982) discussed the patchy distribution of species. Several species have a rather continuous range across northern Australia, but have a patchy distribution in eastern Australia or are absent from most of the mangroves in the humid northeastern part of Queensland, although occurring in mangrove regions further south. It is unclear whether they are excluded by competition or find the mangrove vegetation in northeastern Queensland unsuitable. Ford suggested that competition is important as the contiguity of mangroves with rainforests in northeastern Australia allows many rainforest birds to occupy mangroves, whereas in northwestern Australia rainforest species are poorly represented.

A characteristic of mangrove forests is a patchy distribution, and Ford (1982) likened mangrove vegetation to a series of islands of habitat strung along the coastline. As a consequence, mangrove birds have a somewhat discontinuous distribution. Many species must be capable of moving between neighbouring mangrove patches, for differentiations between contiguous populations coincide mainly with wide geographical barriers. Diamond (1975, 1976) suggested that the mangrove golden whistler (*Pachycephela melanura dahli*) behaves as a supertramp, having spread from northern Australia across Torres Strait and east along southern New Guinea to the Bismarks. Ford (1982) listed which species are good dispersers (commonly found on offshore islands and in isolated patches of depauperate mangroves) and which are poor dispersers.

The theories of Ford (1982) and Schodde, Mason and Gill (1982) have been discussed in considerable detail, for these two papers are almost the only ones to consider in depth the origin of mangrove endemics or to explain current distributional patterns. It is hoped that other groups of mangrove endemics, be they terrestrial or marine, will be subjected to similar treatment in the near future.

The most southerly mangrove endemic, the mangrove heron (*Butorides striatus*), occurs around Sydney. The mangrove warbler (*Gerygone magnirostris*) occurs from Newcastle northwards. In northern New South Wales, four species restricted to mangroves occur, the two additional species being the mangrove honeyeater (*Meliphaga fasciogularis*) and the mangrove kingfisher (*Halcyon chloris*) (Miller 1972).

Tropical mangrove specialists include the mangrove kingfisher (*Halcyon chloris*), mangrove bittern (*Butorides striatus*), black butcher-bird (*Cracticus quoyi*), red-headed honeyeater (*Myzomela erythrocephela*), rufous-banded honeyeater (*Conopophila albogularis*), varied honeyeater (*Lichenostomus versicolor*), mangrove honeyeater (*L. fasciogularis*), shining flycatcher (*Myiagra alecto melvillensis*, mangrove robin (*Eopsaltria pulverulenta*), brown whistler (*Pachycephala simplex*), white-breasted whistler (*P. lanioides*), mangrove golden whistler (*P. melanura*), dusky warbler (*Gerygone tenebrosa*), large-billed warbler (*G. magnirostris*), mangrove warbler (*G. levigaster*), yellow silvereye (*Zosterops lutea*) and the broad-billed flycatcher (*Myiagra ruficollis*) (Frith 1973). Recently, Ford (1983) examined geographic variation in some of the mangrove birds.

Reptiles

Reptiles are common in tropical mangroves, but are rarely seen in temperate ones. Two species of pythons (*Liasis fuscus*, *L. olivaceus*) utilize mangroves as peripheral habitats, and are attracted into the mangroves by large camps of flying foxes on which they feed. The carpet snake (*Morelia spilotes*) also occurs in mangrove habitats. Several other species occur but they are aquatic

Plate 1 Dense mangrove communities of numerous species characterize the subtropical and tropical coastlines as shown here at Repulse Inlet near Proserpine, Queensland. (Photograph by P. Saenger)

Plate 2 The mangrove palm (*Nypa fruticans*) is common throughout south-east Asia, often forming extensive stands. In Australia, like here in Harmer Creek, Cape York, this species forms narrow riverine fringes only at a few scattered localities on Cape York. (Photograph by P. Saenger)

Plate 3 In areas of high freshwater seepage, the boundary between mangrove and freshwater communities may be extremely fine. Shown here on the west coast of Cape York, the landward mangrove margin grades into mangrove fern and sedges while a metre further on, dense stands of tea-tree (*Melaleuca* spp.) form almost pure stands. (Photograph by P. Saenger)

Plate 4 In temperate New South Wales, *Avicennia marina* often forms park-like communities with a low understorey of *Aegiceras corniculatum*, as shown here on the Shoalhaven River. (Photograph by P. Saenger)

Plate 5 Only *Avicennia marina* can grow in the low temperatures of the southern coastline of Australia forming monospecific and often dwarfed communities as shown here at Leschenault Inlet, Western Australia. (Photograph by P. Saenger)

Plate 6 Mangrove zonation on the foreshore of Admiralty Gulf, Western Australia. Typically, the zones are quite marked with a seaward *Avicennia-Sonneratia* zone, a *Rhizophora* zone, a mixed zone and finally, at the junction with the salt-flat, a *Ceriops* zone. (Photograph by T. Farrell)

Plate 7 Extensive salt-flat development in the south-eastern Gulf of Carpentaria, showing the mangroves confined to the foreshore and as narrow fringes along tidal creeks. (Photograph by P. Saenger)

Plate 8 Zonation on the openshore around the mouth of the Wildman River, showing the distinct zonation, the reduction of the zones on approaching the river mouth, and the partial loss of some of the seaward zones due to cyclone damage. (Photograph by P. Saenger)

Plate 9 Saltwater couch (*Sporobolous virginicus*) can form extensive salt-meadows to the landward of the mangroves, as here at Hayes Inlet, just north of Brisbane. These salt-meadows are seasonally important feeding areas for ibis, egrets and herons. (Photograph by P. Saenger)

Plate 10 Algal growth on the mudflats to the seaward of the mangroves, as seen here at Raglan Creek, Central Queensland, provides a rich source of food for many grazers, especially during the winter months. (Photograph by P. Saenger)

Plate 11 One of the more common epiphytes of the mangroves of Cape York, the button plant (*Dischidia nummularia*) hangs from the branches and trunks of *Xylocarpus granatum*. (Photograph by P. Saenger)

Plate 12 The ant plant (*Myrmecodia antoinii*), inhabited by ants of the genus *Pheidole*, a common epiphyte of the tropical mangroves of northern Australia. (Photograph by P. Saenger)

Plate 13 *Derris trifoliata*, a climber of the pea-family, is common in the mangroves of northern Australia. Crushed material of this species was used as a fish poison by some of the coastal Aborigines. (Photograph by P. Saenger)

Plate 14 *Tecticornia cinerea* is one of the few tropical saltmarsh species, occurring on the margins of salt-flats in the drier areas of northern Australia. (Photograph by P. Saenger)

Plate 15 *Phytophthora*-induced dieback in *Avicennia marina* at Port Curtis, Queensland. (Photograph by P. Saenger)

Plate 16 The coppicing ability of *Avicennia marina* is illustrated by this specimen which having been blown over, has developed a series of new trunks. (Photograph by P. Saenger)

Plate 17 The mud-lobster (*Thalassina anomala*) seen here in the mangroves of Gladstone, Queensland, plays a significant role in the turnover of mangrove muds and in the oxygenation of the subsurface layers. (Photograph by S. Parish)

Plate 18 *Cassidula angulifera* feeds on algae and organic matter on the surface of the mud and helps recycle some of this material through its faecal pellets which can be seen scattered over the mud surface. (Photograph by P. Saenger)

Plate 19 In high rainfall areas, as shown here near Cairns, sugar cane crops can be grown on mangrove lands with mixed success, resulting in a continual encroachment on these coastal wetlands. (Photograph by P. Saenger)

Plate 20 In many areas, mangroves are still seen as wastelands, needing reclamation; this attitude allows such ill-conceived developments as Raby Bay, near Brisbane, to destroy valuable and productive coastal wetlands. (Photograph by P. Saenger)

Plate 21 The desire of Australians to live on water-frontage land on navigable waterways, has resulted in the proliferation of cut-and-fill canal estates in coastal wetlands. (Photograph by P. Saenger)

Plate 22 Bank stabilization of the new floodway associated with the Brisbane International Airport was achieved using mangroves as the most cost-effective way. Seen here after 4 years, both the white mangrove (*Avicennia marina*) and the river mangrove (*Aegiceras corniculatum*) were flowering and fruiting. (Photograph by P. Saenger)

Plate 23 *Periophthalmus vulgaris* (mudskipper) with orobranchial chamber expanded perched on a twig, Three Mile Creek, Townsville, Queensland. (Photograph by R. Nursall)

Plate 24 *Periophthalmus vulgaris* on the mud showing well-positioned turret eyes, Three Mile Creek, Townsville, Queensland. (Photograph by R. Nursall)

Plate 25 The crab *Heloecius cordiformes* at entrance to its burrow, feeding, Careel Bay, New South Wales. (Photograph by P. Whalan)

Plate 26 Close up of *Heloecius cordiformes*, Careel Bay, New South Wales. (Photograph by P. Whalan)

Plate 27 The gastropod *Austrocochlea* sp. on the trunk of *Avicennia marina*. (Photograph by P. Whalan)

Plate 28 Crab hole at the landward margin of the mangroves, Fullerton Cove, Hunter River, New South Wales. (Photograph by P. Hutchings)

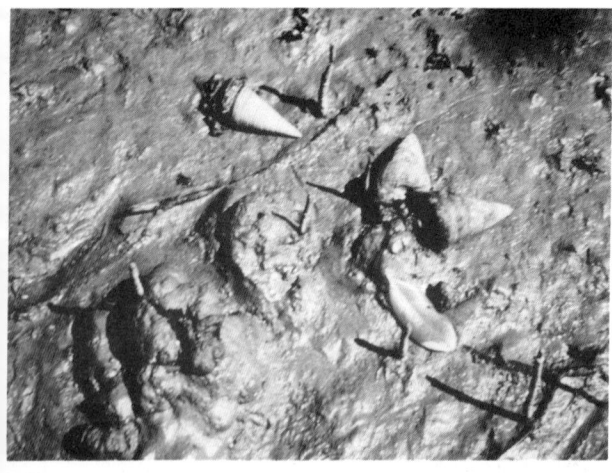

Plate 29 The grazing gastropod *Telescopium* on mudflats in front of mangroves, Arnhem Land. (Photograph by P. Hutchings)

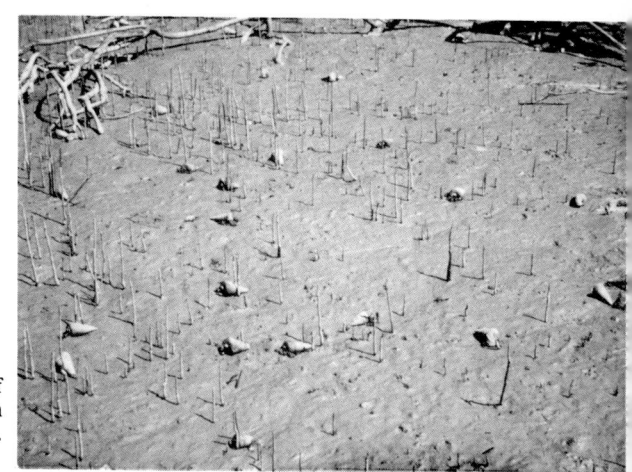

Plate 30 Dense aggregations of the gastropod *Telescopium* on mudflats in Arnhem Land. (Photograph by P. Hutchings)

Plate 31 Variety of gastropods grazing among the pneumatophores on mudflats in Arnhem Land. (Photograph by P. Hutchings)

Plate 32 Encrusted pneumatophores, mainly with oysters, on mudflats in Arnhem Land. (Photograph by P. Hutchings)

Plate 33 Base of *Avicennia marina* on seaward margin encrusted with oysters. (Photograph by P. Hutchings)

Plate 34 Undersurface of submerged logs, covered in encrusting organisms, mainly oysters, Careel Bay, New South Wales. (Photograph by P. Whalan)

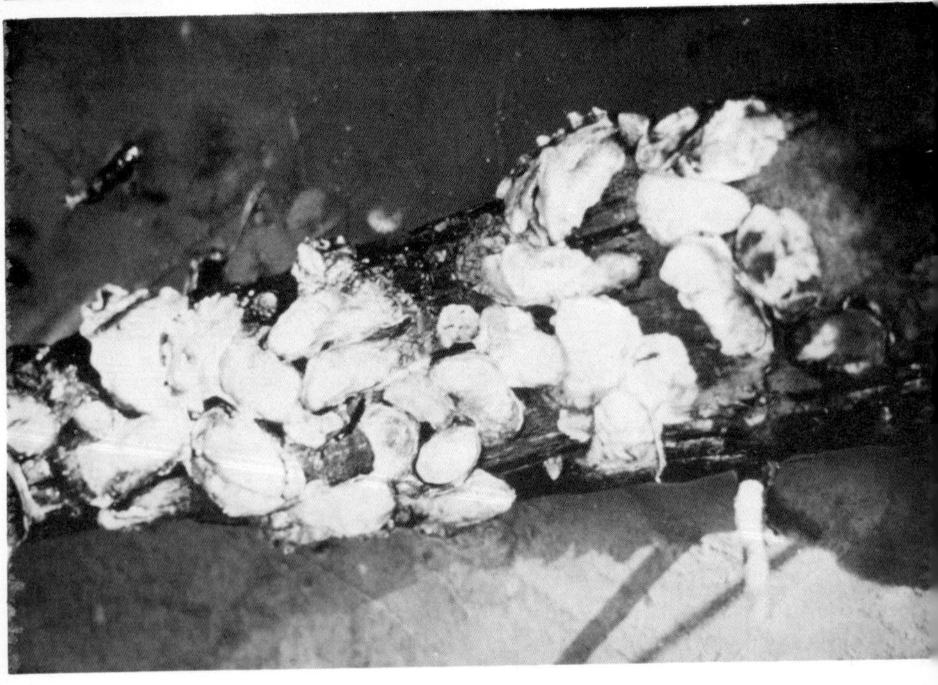

Plate 35 (a) and (b) Encrusting organisms on *Rhizophora* on seaward margin, Arnhem Land. (Photograph by P. Hutchings)

Plate 36 Grazing molluscs, showing trails across the mudflats in front of the mangroves in Arnhem Land. (Photograph by P. Hutchings)

Plate 37 Wood-boring fauna, in northern Australia. Buried log has been uncovered to show *Teredo* burrows. (Photograph by P. Hutchings)

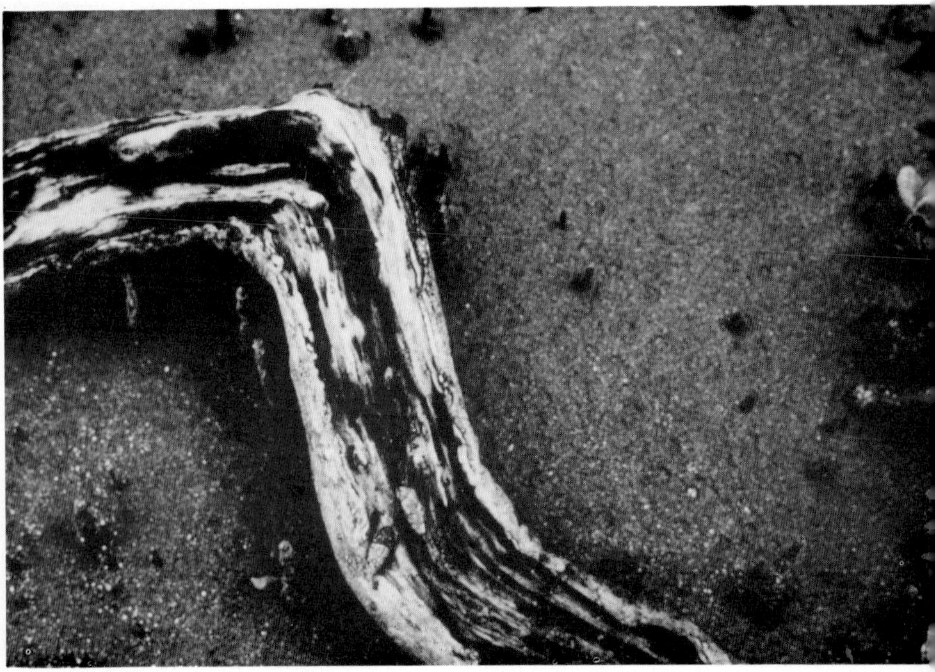

Plate 38 Close up of *Teredo* burrows. (Photograph by P. Hutchings)

Plate 39. The spider *Nephila* sp. in web spun between two leaves of *Avicennia marina*, Careel Bay. (Photograph by P. Whalan)

Plate 40 The green tree ant which has made a nest in the leaves of *Rhizophora*, at Lizard Island. (Photograph by K. Atkinson)

Plate 41 Insect damage caused to leaves of *Rhizophora* at Lizard Island. (Photograph by K. Atkinson)

Plate 42 Unknown insect which has laid its eggs on the undersurface of *Rhizophora* leaves at Lizard Island. (Photograph by K. Atkinson)

Plate 43 The gastropods *Littoraria* complex, two colour phases or perhaps two species, grazing on the surface of *Rhizophora* leaves on Lizard Island. (Photograph by K. Atkinson)

or semi-aquatic species and are discussed later. The majority of terrestrial reptiles occurring in mangroves use them as peripheral habitats. In some areas, mangroves may be important feeding grounds, particularly if nearby terrestrial habitats are depauperate in prey. In other areas, mangroves may provide the only suitable forest habitat and therefore may act as a corridor.

Invertebrates

The only terrestrial invertebrate groups occurring in mangroves are insects and spiders. Like the terrestrial vertebrate fauna, many of the invertebrates found in mangroves also occur in adjacent habitats but, for some, mangroves represent the primary habitat.

Insects

Little work has been done on the insect fauna of Australian mangroves. However, one would imagine that some mangrove characteristics, such as large, almost monospecific stands, abundant detritus and dead wood and some highly specialized habitats such as the red-tufted algae on the pneumatophores in the frontal zone, would provide ideal conditions for the development of endemic species. Certainly, some endemics have been described, and systematists suggest that many more endemic species remain to be discovered.

Traditional insect-collecting techniques are unsuitable for dense tropical mangrove forests, which may explain why few entomologists have ventured into them. Detailed studies have been carried out only on termites, mosquitoes and biting midges which have economic or medical importance.

Termites are an important component of tropical mangroves. Three species of Kalotermitidae occur in the Darwin region, *Neotermes insularis*, *Cryptotermes secundus* and *C. domesticus*. *Cryptotermes domesticus* appears to be restricted to *Ceriops tagal*, and is not known to attack buildings in Darwin; in northern Queensland, by contrast, it is a species virtually confined to domestic situations and is unknown in mangroves. The two populations are interfertile (Gay and Watson 1982). *Cryptotermes domesticus* is distributed widely in southeastern Asia and the western Pacific, in both natural and man-made habitats, but whether these populations are indigenous or were introduced perhaps in the wooden boats of Macassan trepangers probably will never be known. *Cryptotermes secundus* also has been recorded from *Rhizophora* at Kalumburu in Western Australia, and from *Avicennia* at Groote Eylandt; at Cooloola in southern Queensland *Cryptotermes primus* occurs in species of *Avicennia* and

Rhizophora. At Weipa, the mangroves are heavily infested with *Incisitermes barretti*. Another species common on the landward fringe is *Nasutitermes graveolus*, a termite which builds arboreal nests externally on trunks or branches, instead of in the more typical location within trunks. Some species may range below high-tide level, and *Mastotermes darwiniensis* has been recorded in such situations from *Ceriops* in the Darwin region (Miller, pers. comm.). Exactly where this termite goes during high tide is unknown. The same species has been recorded from Corio Bay, Queensland (Ellway 1974), which represents the southern limit of the distribution of *Mastotermes* on the Queensland coast.

A considerable amount of information exists on the mosquitoes and biting midges of mangroves because of their potential for carrying diseases and their nuisance value (Debenham 1978; Lee et al. 1982). Mosquitoes are typically found in the quiet pools at the back of the mangroves, rather than where there is water movement. These pools may be very ephemeral. In the tropics, the life cycle of mosquitoes is about one week from egg to adult; a pool filled at spring tide is adequate even if it dries out soon after the following neap tide. In disturbed mangroves, stagnant pools may be created within the forest as a result of dredging and filling, and these pools are ideal mosquito habitats. The three most common mosquitoes occurring in mangroves are *Aedes vigilax*, *A. alternans* and *Culex sitiens*. The distribution of *A. vigilax* is essentially coastal, with a specialized relationship to the low-lying land of mangrove zones. It is not completely restricted to these environments, however, but occurs inland up the Murray River, at Barrington Tops, New South Wales, and Eidsvold, Queensland (Griffiths, pers. comm.).

Aedes vigilax typically breeds in waters of a very temporary character, and hence it is intermittent in its occurrence. The water must be stagnant and exposed to direct sunshine. Sunshine is essential for the growth of algal plankton on which the larvae feed. Maximum breeding occurs during the summer months in the wet season, and larval development can be completed within ten days. Low temperatures inhibit breeding and greatly prolong duration of the larval stages (Iyengar 1965; Sinclair 1976).

Aedes alternans (Hexham Grey or Scotch Grey mosquito) occurs in all mainland states, and is associated particularly with estuarine areas. It breeds in temporary stagnant pools exposed to the sun, typically in salt marshes, but it also has been recorded from freshwater pools (Hamlyn-Harris 1933). The eggs are laid singly (presumably on mud at the edge of drying pools) and can withstand drying, hatching out when the depression again fills with water (Marks 1966, 1967). The eggs take seven days to hatch under favourable conditions, but below 21°C hatching is delayed. The larvae of *A. alternans* often are found with those of *A. vigilax* in coastal situations. From the second instar stage onwards, the larvae of *A. alternans* are predatory (and at times cannibalistic) on other mosquito larvae. The adults are strong fliers, and are known to migrate with *A. vigilax* in the summer (Hamlyn-Harris 1933). They are vicious biters; biting is mostly diurnal and at sundown.

Culex sitiens occurs from New South Wales northwards and westwards to Western Australia. In the Northern Territory, Hill (1917) found *C. sitiens* near the coast and inland as far as 55 kilometres south of Darwin, breeding in pools, hollow stumps, tins, crab holes, mangrove swamps, weedy lagoons, inland waterholes and shallow wet-season accumulations of water on grassland. Marks (1953) also found it in a variety of habitats, in sunlit, muddy, brackish pools and in deep shade in mangrove swamps, both among roots and in more open water, and in footprints on mud flats.

Three species of *Anopheles* occur at times in mangroves. *Anopheles amictus hilli* prefers breeding in brackish water, and occurs in northern Australia from Western Australia to the Queensland/New South Wales border. *Anopheles annulipes* occasionally breeds in brackish water and occurs throughout Australia. *Anopheles fairauti* is found in the north of Australia from Western Australia to the eastern coast of Queensland. It sometimes breeds in brackish water; in the Solomon Islands, Perry (1946) reported it breeding in extensive brackish-water pools, high in organic debris and subject to tidal fluctuations. Similar habitats in Australian mangroves are likely to be suitable for this species.

Biting midges belong to the family Ceratopogonidae and are often erroneously called sandflies. Associated with mangroves there are probably well over twenty species belonging to the following genera: *Culicoides* (nine described species plus a similar number of undescribed species); *Dasyhea* (an unworked genus in Australia, at least three species); *Forcipomyia* (at least four species); and some *Stilobezzia* spp. (Reye, pers. comm.). Many are still undescribed, and for most there is information only about the adult form. Information is based largely on emergence-trapping, which may or may not mean that breeding occurs nearby. Most genera of biting midges do not require free water at all; the larvae are part of the substrate infauna exclusively above mid-tide level with specific zonation limited vertically by tidal planes and a combination of substrate composition, water movement and ancillary flora and/or fauna (Reye, pers. comm.).

A considerable amount of work has been carried out on the biology of coastal biting midges, with a view to developing satisfactory control techniques. All these data have been summarized by Debenham (1978) and Lee et al. (1984), and it is from these reviews that the following information on the coastal species of biting midges has been extracted. Of the approximately twenty species known, seven are common in mangrove areas. *Culicoides henryi* occurs in coastal southern Queensland and coastal and subcoastal New South Wales; it appears to prefer muddy sand as a larval habitat, possibly with tree cover. The pupae float on the water's surface with their long axis parallel to the surface, and are unable to submerge. The major source of the blood which serves as their food is unknown. Humans are attacked on occasions, and Reye (1972) ranks the species as the seventh most important pest midge of coastal Queensland, noting that it reaches pest proportions only in restricted localities. It seems that the breeding habitat of this species is quite restricted and its nuisance to humans depends upon (1) suitable breeding sites and (2) availability of humans. These two factors rarely coincide, so *C. henryi* probably is less of a pest than implied by Reye's statement. *Culicoides histrio* occurs in Australia from coastal northeastern Australia south to the Sydney region. Immature stages have been collected from a *Juncus* pool at Careel Bay in the Sydney region. At Townsville in northern Queensland, the species seems to favour the mangrove flat well inside the mouth of the Ross River. Populations are maximal during the summer and negligible in winter (Kay and Fanning 1974), and it only has been recorded feeding on birds. *Culicoides magnesianus* occurs on the coast and offshore islands of the Northern Territory and Queensland. It occasionally has been collected biting humans (Lee and Reye 1955), but it is probably normally an avian feeder. Activity is noctural, peak activity in the Townsville area occurring between 2100 and 0300 hours (Reye and Lee 1961). *Culicoides marmoratus* occurs in Australia from northern Queensland to southern New South Wales. Larvae occur in low-lying estuarine zones, often in association with *Sarcocornia*, *Sporobolus virginicus* or *Suaeda australis*. The eggs cannot withstand desiccation, so breeding occurs after a spring tide. This species is known to occur outside its breeding area. It has been reported to bite humans and can be a pest in coastal regions, but Lee, Reye and Dice (1963) suggested that it is an opportunistic feeder, with wallabies being its primary native host. Biting activity is primarily crepuscular, but also diurnal. *Culicoides molestus* occurs throughout coastal eastern Australia. Reye (1972) characterized the larval habitat as clean sand in the open, or among trees disturbed by slight to moderate wave or current action. A wide range of salinities is tolerated. The larvae occur within the top 7.5 cm of sand, and there is evidence that the pupae drift in on a rising tide, strand and emerge there. The species has invaded the sandy banks of canal estates in southern Queensland where banks are at the appropriate level of the intertidal zone (Reye 1971). Kettle, Reye and Edwards (1979), working on canal estates in southeastern Queensland, found that the larvae were concentrated in a narrow zone above mean tide level, and that the density was independent of the distance from the main body of tidal water. It often has been recorded attacking humans but, except in special situation such as canal estates, it is not considered a pest species. *Culicoides subimmaculatus* occurs throughout coastal eastern Australia. Immature stages are found on sandy estuarine foreshores and in the *Sarcocornia* zone. They also have been found in the vicinity of crab holes (often associated with *Avicennia*), and Reye (1969a, 1969b) suggested that the presence of soldier crabs (*Mictyris livingstonei*) is essential for the breeding of this species. Kettle (1977) recorded the larvae feeding on polychaete worms and desmids. Mass emergences appear to be correlated with neap tides. The adults are opportunistic feeders; activity is largely crepuscular but can be diurnal if humidity is sufficiently high and there is little wind.

The last common species is *Culicoides ornatus* which occurs in the Torres Strait Islands and coastal northern Australia, and does not occur south of Tin Can Bay in Queensland (this species often has been misidentified with corresponding erroneous reports of its distribution). The larval habitat is the mean neap tide zone of estuarine areas with muddy substrate, completely sheltered from wave action and with a dense tree cover. It often occurs with *Rhizophora stylosa*, but Reye (1972) suggested that there is also an association with *Aegiceras corniculatum*. Adults are abundant in mangrove swamps, and Reye (1972) regarded this species as the most important Queensland pest species. However, it will feed on flying foxes and birds, as well as on humans.

Much of the rest of the data on the insect fauna is anecdotal, or merely notes the occurrence of particular species. However, it does appear that many orders of insects are represented in mangroves, including Blattodea, Orthoptera, Hemiptera, Coleoptera, Diptera, Hymenoptera, Lepidoptera, Neuroptera, Homoptera and Isoptera; these groups may be extremely abundant either in larval or adult stages (Australian Littoral Society 1977). Below is a brief synopsis of the available information, presented mainly to illustrate the diversity of insects and to illustrate the paucity of detailed data.

The grasshopper, *Valanga irregularis*, has been seen feeding in the mangrove canopy at Corio Bay (Ellway 1974). The pygmy grasshopper, *Coptotettis masticatus*, has been recorded on the mud in mangroves. In New World mangrove areas, many species of crickets and a number of tettigoniids occur (Rentz, pers. comm.).

Diptera occur in mangroves as both free living and parasitic forms, and are a diverse component of the mangrove fauna. Fourteen species were obtained by Hutchings and Recher (1974) at Careel Bay, near Sydney, with minimum collecting. Some Diptera, such as *Melanagromyza avicenniae* which is restricted to mangroves, appear to feed on new shoots. *Copidita nigronotata* is associated with rotting wood; both the adults and larvae are secondary timber borers. Other dipterans specialize on the fruits and propagules of mangroves (Hutchings and Recher 1974; Spencer 1977).

Several species of Lepidoptera are commonly associated with mangroves and they all belong to the Lycaenidae (blue butterflies) (Common and Waterhouse 1972). *Hypochrysops epicurus* occurs from Port Macquarie to Brisbane, and appears to be restricted to stands of *Avicennia marina*. *Hypochrysops apelles* and *H. narcissus* are characteristic elements of the fauna, although their larvae also feed on non-mangrove species. *Hypochrysops apelles*, which occurs along the eastern Australian coast from north of Yeppoon, feeds on *Rhizophora stylosa*, *Bruguiera gymnorhiza*, *Ceriops tagal* and *Avicennia marina*.

Ogyris amaryllis hewitsoni occurs from Maryborough to Cairns and is common around mangroves; its larvae feed on the mistletoe *Amyema mackayense*.

A less common species, *Nacaduba kurava*, occurs from Richmond River to Cape York. It feeds on *Aegiceras corniculatum* but also uses other plants. The larvae and pupae of *Acrodipsas illidgei* have been found in the nests of the black ant (*Crematogaster laeviceps*) near Brisbane. The ant nests in hollow branches of mangroves. Three to four larvae or pupae were present in each colony, and it seems likely that the larvae of *A. illidgei* feed on the immature stages of the ant. Similarly, *Hypochrysops apollo* is associated with ants in mangroves as it breeds in ant-house plants (chapter 4).

The larvae of the tortricid moth, *Procalyptis parooptera*, feed on the leaves of *Ceriops tagal*, firmly joining adjacent leaves with silk to form a shelter in which they live. Pupation occurs in this shelter. After emergence of adults the empty shelter then is used by the larvae of the butterfly *Hypochrysops apelles* as a daytime retreat; they are attended by ants. The caterpillars emerge at night to feed on the surrounding foliage (Common and Waterhouse 1972). Another moth (*Macrocyttara expressa*; family Cossidae) is restricted to mangroves; the larvae tunnel gregariously in the trunks of *Excoecaria agallocha*. The larvae of *Cenoloba obliteralis* (family Oxychirotidae) develop in the cotyledons of fallen seeds of *Avicennia marina* (Common 1970). All the above Lepidoptera occur in eastern Australia.

The ant fauna of Australian mangroves has not been studied well but several species are known. These include *Colobopsis* sp. and *Crematogaster laeviceps*. Dead wood provides suitable habitat for ants (Ellway 1974). No species have been found so far which are restricted totally to mangroves.

Macnae (1968) recorded the weaver ant, *Oecophylla smaragdina*, in *Bruguiera*, *Ceriops* and *Sonneratia* forests in northern Queensland. Its ecology and life history in coconut palm forests have been described by Vanderplank (1960).

Macnae suggested that species of ant which occur below high-tide level in sediment may be able to trap air in their burrows by plugs of mud. Whether ants living in inland areas where freshwater flooding regularly occurs adopt a similar strategy is unknown. If they do not, then the ants found in mangrove muds have developed a unique adaptation enabling them to invade the intertidal environment. One species (not yet identified) occurs at low tide in the Hinchinbrook region and presumably adopts this strategy. Another unidentified species has been seen living in tunnels made by teredos in the prop roots of *Rhizophora stylosa*.

Finally, three species of ant-house plants occur in northeastern Queensland (chapter 4). Their gouty tuberous stems are often hollow and are used by several species of ants. The ant *Pheidole myrmecodiae* is restricted to this habitat, where it obtains protection from desiccation and predation. The relationship between ants and this plant is discussed in chapter 3.

Some of the insect visitors to mangroves may occur seasonally, during the flowering season of particular mangrove species. Some of these insects are important pollinating agents (chapter 4). The honey bee *Apis mellifera* feeds heavily on *Aegiceras corniculatum* and *Avicennia marina* when they are in flower. Native bees also probably visit mangroves, but the species involved have not been documented.

Detailed studies on the insect fauna of Australian mangroves are urgently needed, and the role of insects in the ecosystem needs investigation.

Overseas, Simberloff and Wilson (1969, 1970) and Wilson and Simberloff (1969) provided extensive fauna lists of insects which recolonized small mangrove islands after the islands had been fumigated. By 250 days after all the animals had been experimentally killed by fumigation, the faunas of all the islands except the most distant one from the mainland had regained species numbers and composition similar to those of untreated islands even though population densities were low. As population sizes increased it was expected that competition and predation would become more important. Observed turnover rates showed wide variance, with most values occurring between 0.05 and 0.50 species per day. True turnover rates are probably much higher with 0.67 species per day the extreme lower limit on any island.

Spiders

Until about a decade ago virtually no information was available on the spiders occurring in mangroves. Since then, a series of studies in restricted localities has been carried out (Ellway 1974; Hutchings and Recher 1974; Graham et al. 1975; Australian Littoral Society 1978; Hegerl et al. 1979). Grimshaw (1982) collated all this information with previously unpublished data and produced an extensive list. The majority were identified to species level. The largest number of species occur in the Northern Territory (sixty-nine) followed by northern Queensland (fifty) and southeastern Queensland (fifty-one). No species have been recorded from Victoria and only two from South Australia. Obviously, these numbers are a reflection of the intensity of sampling and may not reflect actual diversity. For instance, only thirty-two species have been recorded from central Queensland, and these are based on a single study. In addition, Grimshaw recorded a number of species which occur in adjacent coastal terrestrial habitats and probably these also occur in mangroves. None of the spiders recorded occur in all of the geographical regions, but a number of species occur from New South Wales to northern Queensland, and some also extend into Northern Territory.

Twenty families of spiders are currently known from mangroves and of these four are well represented: Araneidae (orb web spiders), Saliticidae (jumping spiders), Therdiidae (comb-footed spiders) and Thomisidae (flower spiders).

The orb web builders include both the slant orb web weavers of the genus *Tetragnatha* (the large-jawed spiders) which elsewhere are associated with stream or lake-side habitats, and the vertical orb web weavers of the genus *Eriophora*; both genera are adapted for catching flying prey. Foliage-dwelling spiders, however, seem to be rather uncommon, despite the abundance of habitat for them. This may be correlated with limited prey (such as ants) passing along mangrove trunks and branches, or it may be simply that these spiders are so well camouflaged that they are difficult to see and collect. The wolf spiders of the family Lycosidae (*Geolycosa* sp.) and allied hunting spiders of the family Pisauridae (*Dolomedes* sp.) occur on the ground or among the lowest layer of vegetation. Presumably they are opportunistic hunters, foraging on the exposed mud at low tide. *Dolomedes*, however, often is associated with aquatic habitats and has been observed to move underwater to avoid predators or to take prey.

Freshwater Fauna

Vertebrates

Freshwater vertebrates are restricted to a few species which may occur occasionally at the salt/freshwater interface; they include among the reptiles the freshwater crocodile, *Crocodylus johnstoni*, and occasionally tortoises (Cann 1978). The pitted-shelled turtle, *Carettochelys insculpta*, has been recorded from the Daly, Victoria and Alligator River systems in the Northern Territory (Cogger 1979).

Marine Fauna

Vertebrates

Marine vertebrates include reptiles and fish. Recently, a partial review of selected components of the aquatic fauna of Red Sea mangroves and Brazilian mangroves has appeared (Dov Por and Dor 1984). These articles provide some interesting contrasts to the Australian situation.

Reptiles

The saltwater crocodile, *Crocodylus porosus*, potentially occurs in all river systems from Broome in Western Australia, across the Northern Territory to Maryborough in southern Queensland. Large numbers are present in many Northern Territory rivers with far fewer individuals in the rivers of Queensland (Webb, Messell and Magnusson 1977; Messel and Butler 1977; Messel et al. 1979-82). Crocodiles come into the mangroves to feed on the rising tide. Juveniles feed selectively on sesarmid crabs, prawns, mudskippers and other small fish. As they become larger, their diet shifts towards the large mud crab, *Scylla serrata*, and birds and mammals (Taylor 1979). However, crocodiles do not nest in mangroves (Webb, Messel and Magnusson 1977) but on concave banks where the river comes close to or abuts the adjacent floodplain. Of fifty-two nest sites examined by Magnusson, Grigg and Taylor (1978), fourteen were in adjacent freshwater swamps and thirty-eight were in river-fringing vegetation behind the mangroves. Additional data on nesting sites on the western coast of Cape York are given by Magnusson, Grigg and Taylor (1980).

Sea snakes commonly occurring in estuarine and mangrove areas include: *Ephalophis mertoni*, *E. greyi*, *Hydrelaps darwiniensis* and *Enhydris punctata* in northern Australia and *Hydrophis elegans* and *Aipysurus eydouxii* in central coastal Queensland.

The northern water dragon (*Lophognathus temporalis*) feeds on insects in the mangroves of northwestern Australia and the Northern Territory (Gow 1976; Swanson 1976). Reptiles utilizing mangroves as primary habitat include the little file snake

(*Acrochordus granulatus*) which lives on the mud flats in front of the mangroves in northeastern and northern Australia. It feeds on small crabs and fish. Other mangrove specialists include the bockadam snake (*Cerberus rhynchops*), white-bellied mangrove snake (*Fordonia leucobalia*) and mangrove snake (*Myron richardsonii*), which all occur in northern Australia (Cogger 1979). These mangrove specialists all feed on small fish and/or crabs.

The mangrove monitor (*Varanus indicus*) occurs in mangrove habitats in the Northern Territory, Torres Strait and Cape York, and feeds on insects, fish, reptiles and birds in the tidal mangrove areas. It also occurs in and near forest streams in terrestrial habitats. The rusty monitor (*V. semiremex*) lives in holes in mangrove trees in tidal estuarine situations, or on small islands; it feeds on fish, crabs, insects and lizards (Cogger 1979).

Fish

The use of mangroves as nursery grounds for larval and juvenile fish is well known by professional and amateur fishermen (Milward 1982), although scientific documentation of this is only now being accumulated for Australia. Overseas, the data are more extensive (Austin 1971; Lasserre and Toffart 1977; Yãnez-Arancibia, Linares and Day 1980) and a detailed study of the diet of mangrove fish was carried out by Odum and Heald (1972) in Florida.

Early studies in Australia mainly documented the presence of fish in mangrove creeks. Saenger et al. (1977) listed sixty-eight species. Even so, this must be considered far from complete, and significant gaps are evident (Milward 1982); no records of sharks and rays are included even though they regularly enter mangrove systems. Milward noted that some teleost families such as Apogonidae, Bothidae, Chandidae, Chanidae, Dorosomidae, Elopidae, Engraulidae, Gerridae, Latidae, Leiognathidae, Megalopidae, Platycephalidae, Pseudomugilidae, Soleidae, Syngnathidae and Toxotidae contain species which regularly occur in mangroves, at least in the tropics. He indicated that his (unpublished) data and that of Penridge (1971) add at least another fifty species of fish to the list of Saenger et al. (1977), just for northern Queensland, while Collette (1983) recorded over two hundred species from fifty-eight families from the mangroves of northern Australia (two sampling stations) and New Guinea (twelve sampling stations). Bell et al. (1984) reported forty-six species belonging to twenty-four families (many not previously recorded) from temperate mangroves. A high proportion of these species are economically important, and juveniles of the families Gerridae, Sparidae and Mugilidae dominate the fish community. Obviously, a considerable amount of work needs to be done to fully document the fish fauna of mangroves throughout Australia. Recently a series of studies has been carried out in a temperate mangrove system in Botany Bay on the fish communities occurring in mangroves and associated seagrass beds (Bell 1980; Middleton et al. 1983). These studies are beginning to document the importance of these habitats to fish in the temperate region.

The fish fauna occurring within mangrove environments can be divided into (1) permanent residents, (2) species that visit the mangroves intermittently as adults and (3) those that occur seasonally as eggs, larvae or juveniles. These components vary geographically (Saenger et al. 1977) and also in relation to local topography, tidal range and salinity (Milward 1982).

Much of the mangrove fish fauna also occurs in nearby seagrass beds and estuarine areas, but a small component is restricted to mangroves. The last group are those fish, known as mudskippers, that are adapted to living in tidally exposed parts of mangroves. They belong to the gobiid subfamily Oxcidercinae, which includes about five species in two genera (*Periophthalmus* and *Periophthalmodon*). Mudskippers are common throughout tropical mangroves but are absent from temperate ones (Milward 1974). In Australia they extend only as far south as Hervey Bay in Queensland (lat. 25°40'S). Other members of the subfamily Oxcidercinae are *Boleophthalmus* and *Scartelaos*; these burrow in the mud and occur in the Northern Territory and Queensland.

Other groups of gobies are common in mangroves, but these also occur in other estuarine habitats. They include the genera *Mugilogobius*, *Taenioides* and *Arenigobius* which occur throughout Australia. The oyster blenny, *Omobranchus*, occurs in frontal mangroves where residual water is trapped. Fish of these four genera can withstand some exposure to air but most live in burrows where some water is present even during low tide.

Recent studies have assessed the importance of mangroves and associated seagrass beds and salt marshes to fish. Beumer (1978) investigated the feeding ecology of four species from mangrove creeks in northern Queensland. Two species, the black bream (*Acanthopagrus berda*) and the hair-finned goby (*Ctenogobius criniger*) are predominantly carnivorous; another, the milk spotted toadfish (*Chelonodon patoca*), is largely omnivorous; and the bony bream (*Anodontostoma chacunda*) is detritivorous. These four species, especially *A. berda* and *C. patoca*, use mangroves as nursery grounds and there is a large influx of fry and juveniles in August and September. Beumer (1978) suggested that availability of food is the main reason mangroves provide a suitable habitat for developing fish, rather than lower salinities. Similar conclusions were reached by Blaber (1980) working in northern Queensland; he found that prawns within the mangroves are very important food sources for many of the fish. The fish community was dominated by juveniles and he suggested that mangroves, as well as providing abundant food for juveniles, also provide protection. In general, predation was low as piscivores were under-represented in the Trinity Bay system.

Bell et al. (1984) analysed the fish communities in temperate seagrass beds and associated mangroves in Botany Bay, New South Wales, in terms of numbers of individuals, number of species and stage of maturity at regular intervals over a three-year period. Numerous economically important species were present. Many of the species were found only as juveniles, the adults living elsewhere. Numbers of species and individuals varied seasonally, and it was concluded that mangrove habitats in temperate Australia are important nursery grounds for fishes inhabiting adjacent estuarine and inshore habitats as adults. Other papers have investigated community structure and trophic relationships in seagrass beds both in Botany Bay and nearby Port Hacking (Burchmore, Pollard and Bell 1984; Middleton et al. 1984; Pollard 1984). In saltcouch marshes in Moreton Bay, Moreton (1984) recorded sixteen species of fish and noted that mosquito larvae, copepods and insects were important food items.

Aquatic Invertebrates

Although microfauna, such as nematodes and protozoa are abundant in mangroves (Newell 1974), their presence has been very poorly documented in Australia. Only a single study has been carried out on mangrove nematodes. Dacraemes and Coomans (1978) collected twenty-five species of them from a single sample in the mangroves at Lizard Island, Great Barrier Reef, suggesting that a very rich fauna is awaiting study. Hodda and Nicholas (1985) documented the meiofauna in the mangroves occurring in Fullerton Cove, Hunter River, a temperate mangrove system. Dye (1983) conducted an extensive survey of the composition and seasonal fluctuations of meiofauna in a southern African mangrove estuary.

The aquatic invertebrates of the macrofauna are primarily marine in origin with some freshwater animals occurring at the freshwater/marine interface in the upper reaches of estuaries. In addition, there is a group of marine pulmonate molluscs which are probably closely related to terrestrial and freshwater pulmonate gastropods rather than to other marine molluscs. Zilch (1959) suggested that they have had a long history of independent evolution.

Rather than discussing each aquatic invertebrate group separately, they will be treated by ecological categories. The marine invertebrate fauna of the intertidal zone occupy five different types of habitats. Some of the terrestrial invertebrates discussed above also occur in these habitats.

In addition to specialized intertidal habitats, creeks and lagoons in the mangroves have a characteristic estuarine subtidal fauna dominated by polychaetes, crustaceans and molluscs. Echinoderms also may be represented (Saenger, Stephenson and Moverley 1980). Colonial animals such as sponges and ascidians tend to be restricted to the seaward margins of the mangroves except where tidal creeks flow through them.

In examining the existing data on the marine invertebrate fauna of Australian mangroves, certain factors have to be taken into account; various collecting techniques were employed, the intensity and periodicity of sampling differs and there are many taxonomic problems. Much of the fauna is undescribed or represented by new records. In many cases, a species list is given with no indication of who identified the material or if voucher material has been lodged in a museum for verification of identifications. Some of the faunal records are suspect but cannot be reassessed. These factors, together with the general lack of collecting of the fauna in Australian mangroves, means that there is a serious information deficit. Also, the majority of faunal studies have been carried out near cities or large towns with universities or fisheries institutions. Comparisons between temperate, subtropical and tropical regions are therefore difficult. Virtually no information is available from far northern Queensland or from the Northern Territory except from the Alligator River (Hegerl et al. 1979). Many mangrove studies provide species lists without indicating the specific habitats from which the animals were collected. An example of this is the extensive revisions of spionid polychaetes (Blake and Kudenov 1978) in which several of the species probably were collected from mangroves although detailed habitats are not given. For all these reasons, no species lists are provided here. Some are provided by Saenger et al. (1977), Hutchings and Recher (1982) and Davie (1982) for crabs, but these probably do not reflect adequately the real diversity of the marine invertebrate fauna.

The general lack of data on Australian estuaries makes it difficult to estimate the proportion of fauna restricted to mangroves relative to that which also occurs in other estuarine habitats. However, in general, it seems that much of the marine fauna occurring in temperate mangroves also occurs on adjacent muddy or rocky shores (Branch and Branch 1980; Butler et al. 1977b; Macnae 1966), whereas in tropical mangroves, mangrove specialists occur in addition to general estuarine species. This suggests that tropical mangroves provide habitats in addition to those occurring on other tropical muddy shores. Some of this increased habitat diversity may arise from the retention of water during low tide and by provision of shade. In contrast, in temperate regions mangroves and muddy shores offer similar habitats to other local estuarine situations and this, combined with the factor discussed on p. 163, has resulted in few mangrove specialists having evolved.

The fauna, like the flora, may be zoned within the mangroves (Berry 1963) and along the estuary (figure 47). The zones probably are determined by sediment structure, salinity regimes and tidal patterns (periods of inundation) (Morton 1975). Actively moving animals may modify their zoning during a twenty-four-hour period and during the tidal cycle. For example, fish and prawns move into mangroves at high water and feed on the plankton carried in on the rising tide. Vertical zonation within the sediment and up tree trunks also occurs. Latitudinal effects are superimposed upon these zonation patterns (Wells 1984) (table 33). It should be stressed that for most of the invertebrate fauna little if any information on zonation has been published, although it is quite apparent upon casual observation within both temperate and tropical mangroves.

Encrusting Fauna

Encrusting organisms are those which firmly attach themselves to solid substrates such as trunks of standing and fallen trees, pneumatophores and prop roots. They are restricted to the seaward margins of the mangroves where regular prolonged inundation occurs (Hutchings 1983; Sasekumar 1974). The zone is dominated throughout most of Australia by the oyster *Saccostrea commercialis* a filter feeder. Potter and Hill (1982) studied the effects of high air temperatures on *Saccostrea commercialis* and suggested that oysters growing on mangrove trunks and pneumatophores are partially shaded from the sun, thereby increasing their chance of survival during very hot days when midday coincides with low tide. Once the tide rises, the oysters are cooled by the substantially lower water temperatures. Among the oysters are small gastropods, errant polychaetes, small crabs, amphipods and isopods which form a characteristic community.

The barnacle *Elminius modestus* and the serpulid *Galeolaria caespitosa* occur on the pneumatophores and lower trunks of *Avicennia marina* in the upper part of Spencer and St Vincent's Gulf and at Ceduna in South Australia (Womersley and Edmonds 1958). In Victoria, *Bembicium melanostomum*, *Mytilus edulis planulatus* and the barnacle *Chamaesipho columna* are commonly found on trunks and pneumatophores (Smith, Coleman and Watson 1975). *Balanus amphitrite* is common on *Rhizophora* stilt roots in central Queensland. This species extends up into the estuarine reaches in Queensland and overseas (Fernando and Ramamoorthi 1975). All of these except *Bembicium* are filter feeders and are restricted to feeding during immersion. The breeding periodicity and problems associated with encrusting

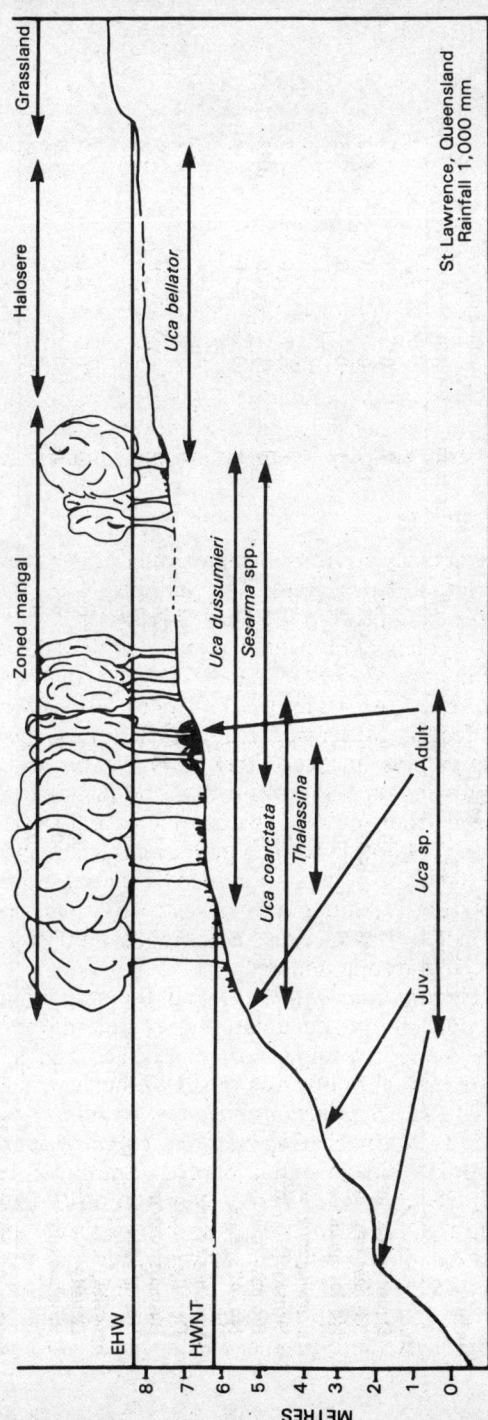

Figure 47 Diagram to show the pattern of distribution of *Uca* crab species in relationship to the zoning of the mangroves in north Queensland (after Macnae 1966).

Table 33 Density, biomass and number of molluscs and crustaceans of either epifaunal, infaunal or arboreal habitats in the Bay of Rest, North West Cape, Western Australia

Habitat	Molluscs				Crustaceans			
	Epifaunal	Infaunal	Arboreal	Total	Epifaunal	Infaunal	Arboreal	Total
Mudflat								
Density (No./m^2)	7.0	10.3	0	17.3	7.6	0	0	7.6
Biomass (mg/m^2)	1038	1460	0	2498	182	0	0	182
No. of species	28	30	0	58	13	0	0	13
Avicennia zone								
Density (No./m^2)	13.4	0	0.5	13.9	1.2	6.7	0	7.9
Biomass (mg/m^2)	2829	7	176	3012	54	978	0	1032
No. of species	16	1	4	21	4	10	0	14
Rhizophora zone								
Density (No./m^2)	0.9	0.1	0.8	1.8	0.5	2.4	0	2.9
Biomass (mg/m^2)	40	1	283	324	5	263	0	268
No. of species	3	2	2	7	8	1	0	9
Backflat								
Density (No./m^2)	0	0	0	0	0	0.6	0	0.6
Biomass (mg/m^2)	0	0	0	0	0	193	0	193
No. of species	0	0	0	0	0	5	0	5

Source: After Wells (1984).

mangrove species of barnacles are discussed by Fernando and Ramamoorthi (1975).

Weate (1975), working in the lower Myall River on the central Nes South Wales coast, found the molluscs *Chizacmaea* sp., *Xenostrobis securis*, *Tatea rufilabris* and *Lasaea australis* in among oysters (*S. commercialis*). Robinson et al. (1983b) also found the bivalve *Trichomya hirsuta* as an encrusting species at Towra Point, Botany Bay.

Saenger, Stephenson and Moverley (1979) studied the fouling communities in the Calliope River, Queensland, and several of these species such as the barnacle *Balanus amphitrite*, the serpulid polychaete *Ficopomatus uschakovi* and the oyster *Saccostrea* also colonize the pneumatophores and tree trunks in the frontal zone of mangroves. *Balanus*, *Ficopomatus* and *Electra* (a bryozoan), constitute a pioneer community which is replaced after three to eleven months by a climax one of *Saccostrea*, *Modiolus auriculatus*, *Balanus* and *Ficopomatus*. A mosaic composed of pioneer and climax species is typically found on pneumatophores and tree trunks.

Two species of oysters have been recorded from the mangrove areas adjacent to the coral reef at Low Isles, and the gastropod *Morula* was associated with them (Stephenson, Endean and Bennett 1958). A recent survey of the fauna of the mangroves along the Alligator River, Northern Territory, by Hegerl et al. (1979) did not indicate specific habitats for the molluscs collected, but *Saccostrea echinata* and *Saccostrea commercialis* were encrusting species. Barnacles were common on the trunks of *Sonneratia*, *Camptostemon* and *Rhizophora*.

The number of encrusting and associated species and the density of individuals decreases with falling salinity (Hutchings et al. 1977). However, factors such as the frequency of tidal inundation and shade are the most critical ones in determining the density of encrusting fauna (Potter and Hill 1982). In areas of dense canopy where low light conditions prevail, epiphytic algae form a dense mat on pneumatophores and around the bases of tree trunks. This mat provides a microhabitat for several species of gastropods and amphipods, and sometimes enough moisture for spionid and nereidid worms.

It appears that the encrusting fauna becomes more diverse with decreasing latitude, at least on the Queensland coast (Shine, Ellway and Hegerl 1973).

Substrate Epifauna

The term "epifauna" refers to the animals, mainly molluscs and grapsid crabs, which live on the surface of mud. The majority of epifaunal molluscs belong to the families Neritidae, Littorinidae,

Potamididae, Cerithiidae and Ellobiidae (Macnae 1968). Scattered information exists on the preferred habitats of some of these species but often species lists are given with no information on habitats. Epifaunal molluscs can be divided into species which prefer hard substrates and those which are found on soft substrates. In New South Wales the following species have been recorded from mangrove trunks, roots, rocks, loose leaves or weed, or on or under rotten logs: *Thalotia comtessei*, *Austrocochlea constricta*, *Bembicium auratum*, *Nassarius burchardi*, *Patelloida mimula*, a complex of species previously referred to as *Littorina scabra*, *Pseudoliotia micans*, *Ranella australasia*, *Assiminea tasmanica*, *Melosidula zonata*, *Ophicardelus ornatus*, *O. sulcatus*, *O. quoyi* and *Onchidium* spp. The last genus may also be found on soft substrates (Loch, pers. comm.; Robinson et al. 1983b). Other species recorded from soft substrates are *Bittium lacertinum*, *Pyrazus ebeninus*, *Salinator solida* and *S. fragilis* (Robinson et al. 1983b). *Terebralia palustris* and *T. sulcata* prefer muddy substrates, whereas *Cassidula* spp. crawls among decaying vegetation. At Low Isles, Stephenson, Endean and Bennett (1958) often found *Onchidium*, *Quoyia* and *Bembicium* concentrated among shingle under *Bruguiera* bushes where some water was retained. Wells and Slack-Smith (1981), working in the Kimberley region of Western Australia, characterized six zones within the mangroves, each with a distinctive molluscan fauna. They determined densities and found maximum values on the mud flat in front of the mangroves and in the *Rhizophora* zone (figure 48, 49). All the species were living on the surface of the mud or on hard substrates; no infaunal species were found.

The grapsid crabs, which are a very abundant and diverse group, were listed by Davie (1982); many species occur over a wide geographical region. Davie himself suggested that his species list was still a preliminary one. Juvenile grapsids cannot be identified to species and extensive field collecting will be necessary to completely document the distribution of these crabs. Molluscs and crabs are zoned in relation to tidal inundation (McCormick 1978; Yates 1978). McCormick found that the frontal zone is the most diverse, although diversity varied among the six sites studied along the New South Wales coast. The sites varied greatly in their salinity regimes and mangrove structure, and there was no trend of increasing diversity from southern to northern New South Wales. He found maximum densities during the winter and spring (figure 50).

Substrate Infauna

The infauna consists of those animals which live within the sediment for all or part of their lives. The term is restricted to the upper layers of the mud and includes both sedentary and mobile animals.

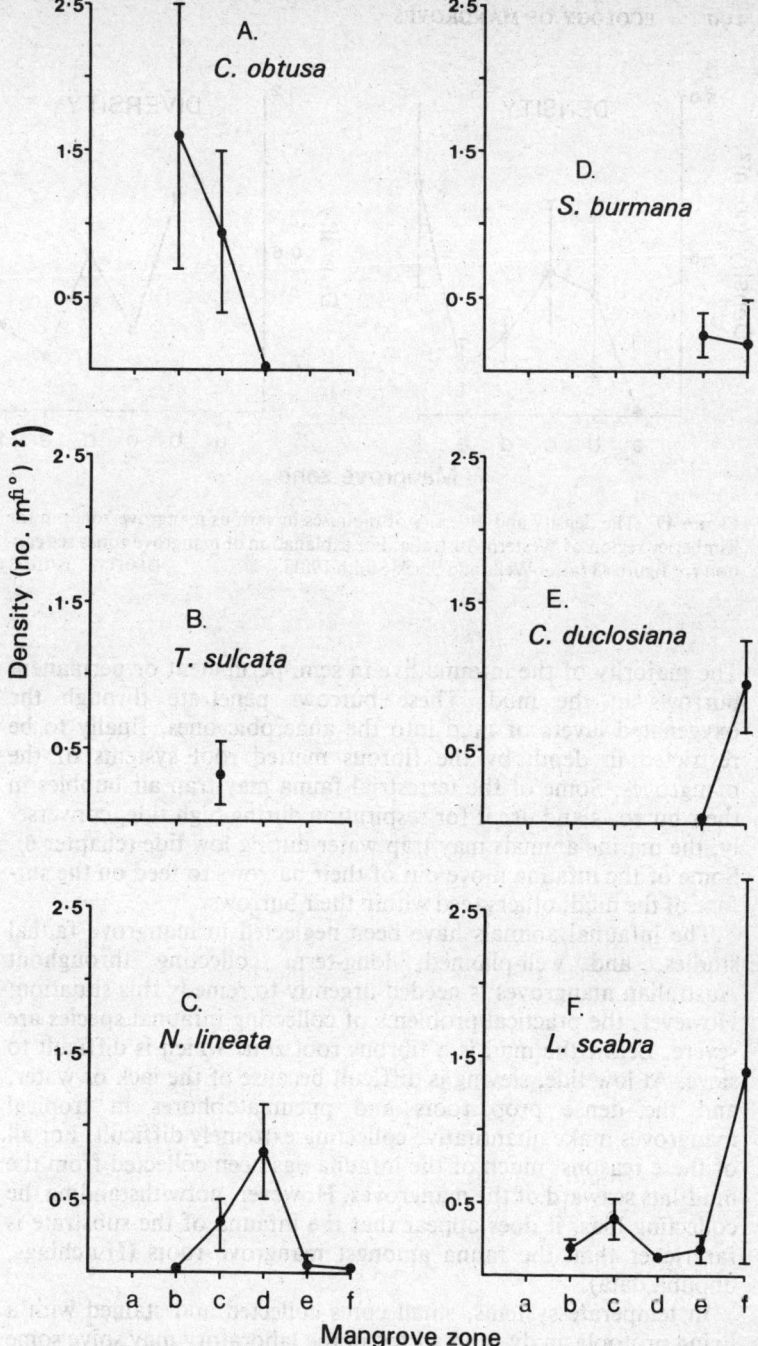

Figure 48 Densities of mangrove mollusc species in various mangrove zones in the Kimberley region of Western Australia. The mean and one standard deviation from the mean are indicated. A = *Cerithidea obtusa*; B = *Terebralia sulcata*; C = *Nerita lineata*; D = *Salinator burmana*; E *Columbella duclosiana*; F *Littorina scabra* (after Wells and Slack-Smith 1981). The mangrove zones are as follows: a. mudflat; b. *Ceriops* zone. c. *Aegialitis annulata* zone; d. *Rhizophora* zone; e. *Sonneratia* zone; f. *Avicennia* zone.

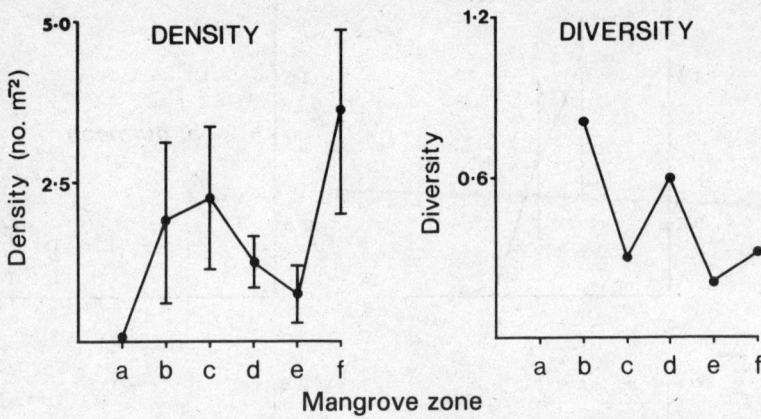

Figure 49 The density and diversity of molluscs in various mangrove zones in the Kimberley region of Western Australia. For explanation of mangrove zones see caption for figure 48 (after Wells and Slack-Smith 1981).

The majority of the infauna live in semi-permanent or permanent burrows in the mud. These burrows penetrate through the oxygenated layers of mud into the anaerobic ones, finally to be restricted in depth by the fibrous matted root systems of the mangroves. Some of the terrestrial fauna may trap air bubbles in their burrows and use it for respiration during high tide; conversely, the marine animals may trap water during low tide (chapter 6). Some of the infauna move out of their burrows to feed on the surface of the mud; others feed within their burrows.

The infaunal animals have been neglected in mangrove faunal studies, and well-planned, long-term collecting throughout Australian mangroves is needed urgently to remedy this situation. However, the practical problems of collecting infaunal species are severe. Below the mud is a fibrous root zone which is difficult to sieve. At low tide, sieving is difficult because of the lack of water, and the dense prop roots and pneumatophores in tropical mangroves make quantitative collecting extremely difficult. For all of these reasons, much of the infauna has been collected from the mudflats seaward of the mangroves. However, notwithstanding the collecting bias, it does appear that the infauna of the substrate is far richer than the fauna amongst mangrove roots (Hutchings, unpubl. data).

In temperate systems, small cores collected and stained with a living protoplasm dye and sorted in the laboratory may solve some of the sampling problems; however, this technique is extremely time-consuming.

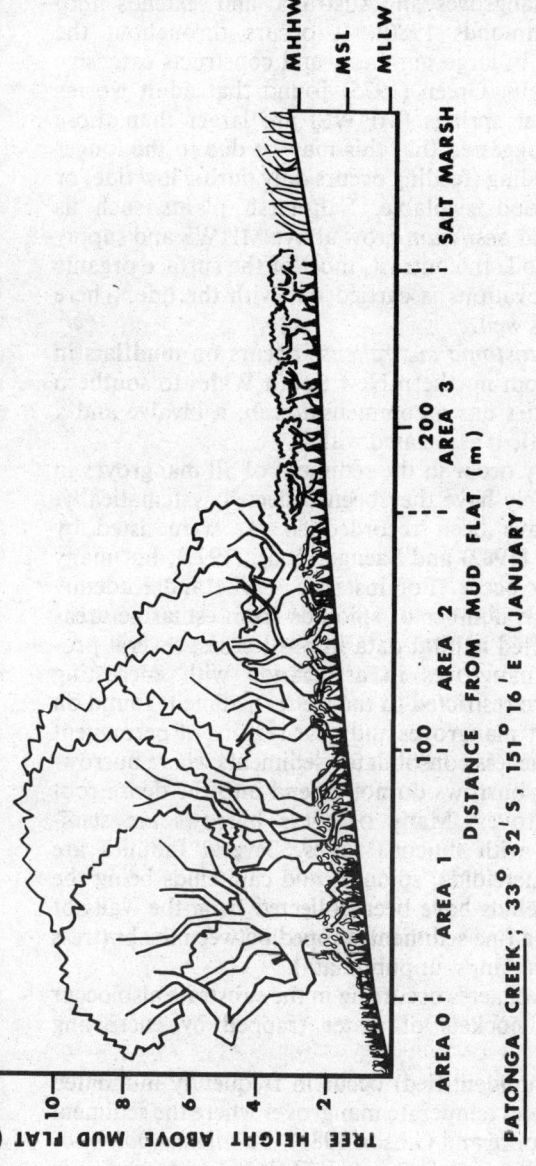

Figure 50 McCormick (1978) studied the horizontal distribution of animals in several areas of mangroves in New South Wales. These data on species numbers for the mangrove stands at Patonga on the Hawkesbury River are presented in this figure. Both epifauna and infauna are richest in species at the lower tide levels where there is regular inundation by the tides. The invertebrate tree fauna is most abundant in the middle zones where refuges during high tide occur, and absent from frontal areas (area 0) which are covered by high tides (after Hutchings and Recher 1982).

The sedentary animals of the infauna are represented by sipunculans, echiuroids, polychaetes and nemerteans. Sipunculans are represented by a single species, *Phascolosoma arcuatum*, which occurs in tropical mangroves in Australia and extends into southeastern Asia (Edmonds 1980). It occurs throughout the mangroves, sometimes in large numbers, and constructs extensive galleries in the substrate. Green (1975) found that adult worms above mean high water springs (MHWS) are larger than those below that level. He suggested that this may be due to the longer period available for feeding (feeding occurs only during low tide) or the quality of the food available. Saltmarsh plants such as *Sporobolus*, *Suaeda* and *Sesuvium* grow above MHWS and supply organic matter to the soil. In contrast, most of the surface organic matter at the lower elevations is carried out with the tide. There may be other reasons as well.

The echiuroid *Ochetostoma australiense* occurs on mudflats in front of mangroves from northern New South Wales to southern Queensland. This species has a commensal crab, a bivalve and a gastropod (all unidentified) associated with it.

Polychaetes probably occur in the sediment of all mangroves in Australia although rarely have they been collected systematically. The species which have been recorded so far were listed by Hutchings and Recher (1982) and Saenger et al. (1977), but many more species probably occur. For instance, Blake and Kudenov (1978) described a large number of spionids from estuarine areas but failed to give detailed habitat data in most cases; several probably occur in the mangroves in association with encrusting oysters. Polychaetes are restricted to the wetter sediment found on the seaward margins of mangroves and near regions of permanent water. They occur in the less consolidated sediments where burrowing is easier, and their burrows do not extend into the dense root systems of the mangroves. Many of these burrows are semi-permanent and lined with mucous tubes. Several families are represented, with the nereidids, spionids and capitellids being the most abundant. Capitellids have been collected from the walls of crab burrows and in the fine sediment trapped between the buttress roots of *Bruguiera* (Hutchings, unpubl. data).

Many species of polychaetes occurring in the substrate also occur in rotten logs or in pockets of water trapped by encrusting organisms.

Nemertean worms (unidentified) occur in frequently inundated areas of both tropical and temperate mangroves where the sediment is not consolidated (Moore and Gibson 1981). *Tubulanus polymorphus* occurs in the sediment of seagrass (*Zostera*) beds and may extend into the frontal zone of the mangroves in Careel Bay near Sydney. Turbellarians (flatworms) also have been recorded in the sediment but they have not been identified.

Within the sediment, large numbers of bivalves (Saenger et al. 1977; Hutchings and Recher 1982; Robinson et al. 1983a, 1983b) and gastropods such as *Conuber sordida*, *Anadara trapezia*, *Tellina deltoidalis*, *Ambuscintilla praemium*, *Glauconome plankta*, *Arthritica helmsi*, *Laternula* sp., *Venerupis crenata*, *Notospisula trigonella* and *Theora fragilis* occur. Overseas, detailed information on the molluscs occurring in mangroves is available for South Africa (Brown 1971), Malaysia and Singapore (Vermeij 1973; Berry 1975).

Macnae (1968) suggested that some of the bivalves occur in particular microhabitats within the substrate, such as in cocoons among the mangrove roots on the seaward fringe. Bivalves are concentrated in the frontal zone of mangroves.

Infaunal crustaceans are represented by amphipods and shrimps, including the snapping shrimp *Alpheus* and the burrowing prawn *Callianassa*. However, the dominant and most conspicuous crustaceans are the crabs, which are represented by large numbers of species in several families (Davie 1982; Hutchings and Recher 1982; George and Jones 1982). Several species have an almost universal distribution throughout Australian mangroves, whereas others are restricted to temperate (e.g. *Macrophthalmus latreillei*, *Mictyris platycheles*) or tropical (e.g. *Macrophthalmus convexus*, *M. latifrons*, *Sesarma darwinensis*) regions. In tropical mangroves the burrows of mud lobsters (*Thalassina anomola*) are very conspicuous as are those of the mud crab (*Scylla serrata*). Mud crabs with a carapace width of 20–90 mm are resident in the mangrove zone; the sub-adults (100–149 mm) move to the intertidal zone to feed at high tide and retreat subtidally at low tide. This movement is accentuated in areas where the density of sub-littoral benthic animals is low, perhaps as a result of prolonged inundation by fresh water during the wet season. Larger specimens (> 150 mm) occur subtidally with an occasional individual being found intertidally at high tide.

Distinct zonation of crabs occurs within mangroves (Warner 1969). In Western Australia, George and Jones (pers. comm.) correlated the distribution of the fiddler crab (*Uca*) with sediment grain size (figures 51a and b). Sediment distribution also has been shown to be important in the distribution of fiddler crabs in Malaysian mangroves (Sasekumar 1974). Tidal exposure and availability of food, together with sediment grain size, probably interact to determine the distribution of crabs. George and Jones (1984) characterized the specific habitats in which thirteen species of crabs from Mangrove Bay, North West Cape, occur. They also tabulated the species according to food, salinity tolerance and type of respiration (tables 34, 35). Wells (1984) (table 36) categorized the feeding types of molluscs collected from four intertidal habitats in Bay of

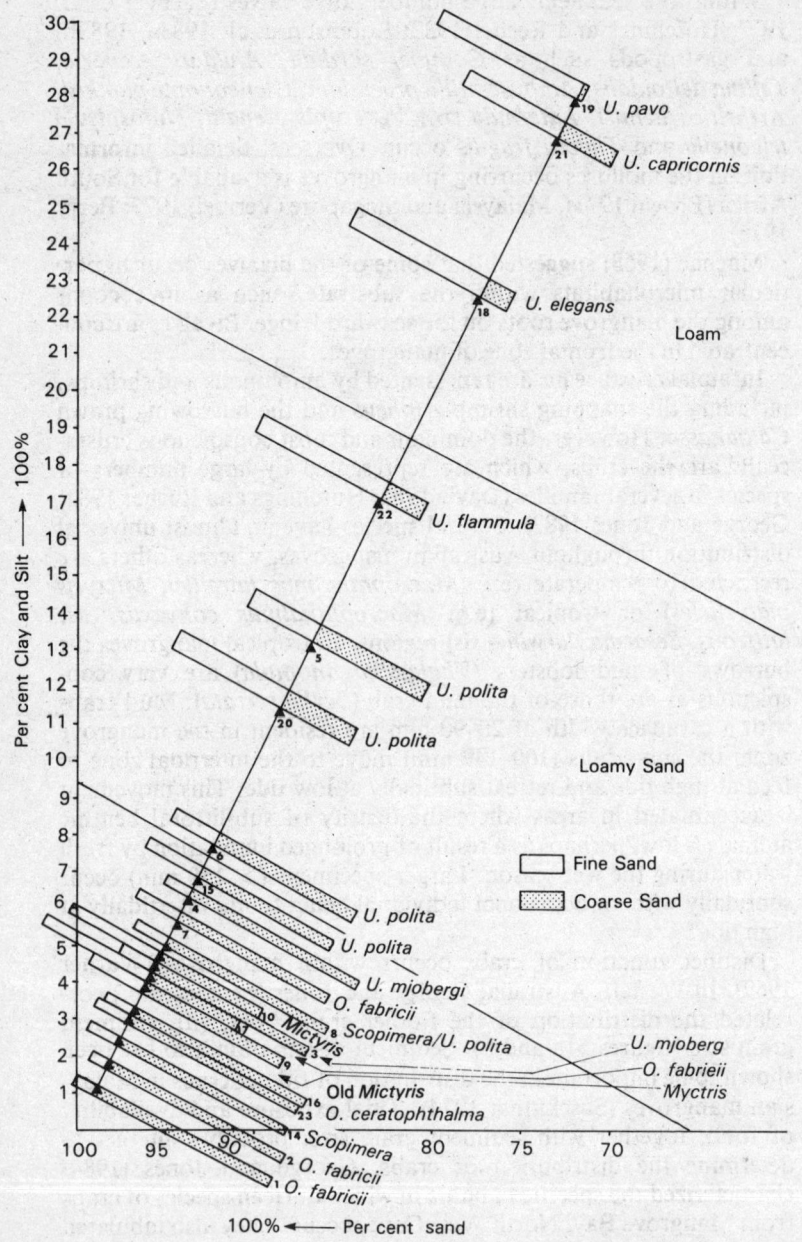

Figure 51(a) The distribution of crabs in relation to the percentage of sand and clay at Bay of Rest, Exmouth Gulf (after George and Jones, unpubl. data).

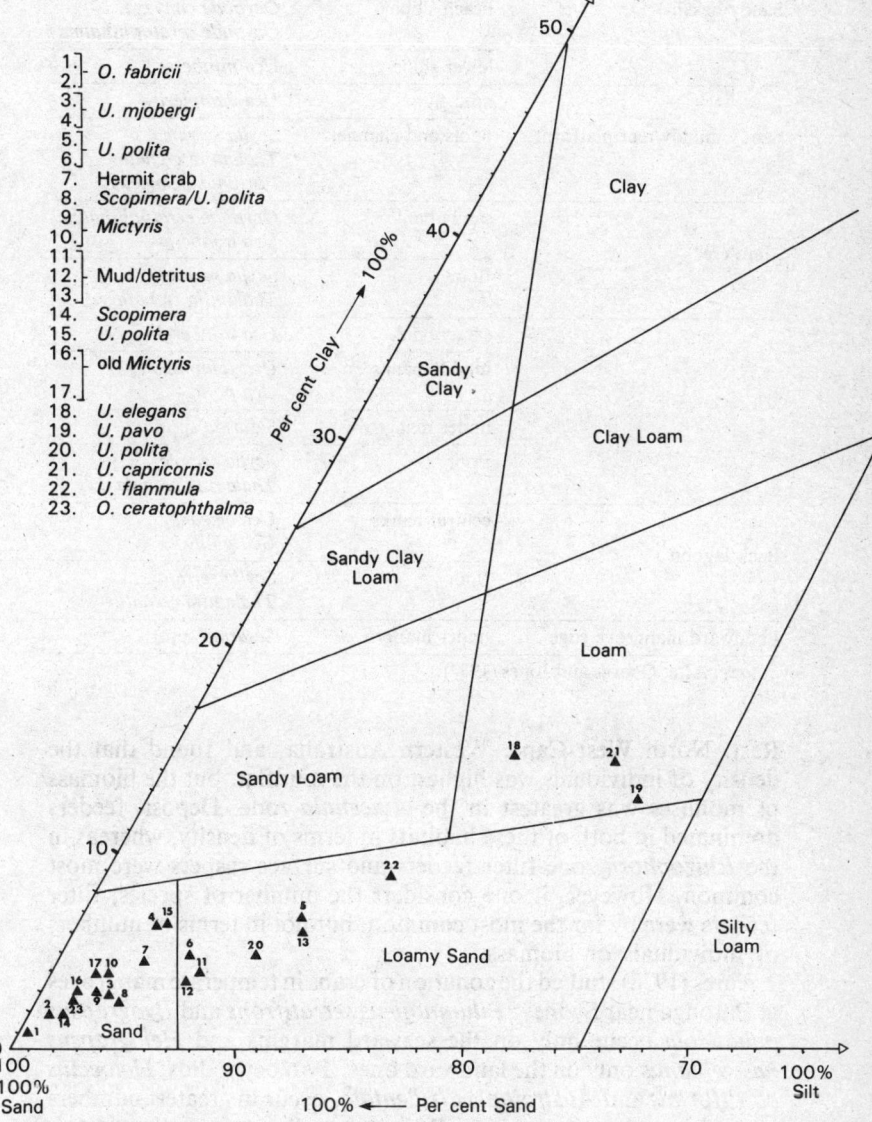

Figure 51(b) The distribution of crabs in relationship to texture of sediment at Bay of Rest, Exmouth Gulf (after George and Jones, unpubl. data).

Table 34 Major crab habitats in Mangrove Bay, North West Cape

Major habitat	Minor habitat	Crab species
Sandy beach	above slope	*Coenobita spinosa*
	beach slope	*Ocypode convexa* *Ocypode ceratophthalma*
Sandy muddy reef platform	lower slope	*Uca mjobergi*
	mud flats	*Uca dampieri*
	pools and channels	*Scylla serrata* *Thalamita crenata* *Portunus pelagicus*
Main creek	sandy bank	*Ocypode ceratophthalma* *Uca mjobergi*
	floor	*Scylla serrata* *Thalamita crenata*
Minor creek	sandy bank	*Uca mjobergi*
	muddy banks	*Uca flammula* *Uca polita*
	under mangroves	*Sesarma* spp.
	floor	*Scylla serrata* *Thalamita crenata*
Back lagoon	central banks	*Uca elegans* *Uca polita*
	floor	*Scylla serrata* *Thalamita crenata*
Landward mangrove edge	under mangroves	*Sesarma* spp.

Source: After George and Jones (1984).

Rest, North West Cape, Western Australia, and found that the density of individuals was highest on the mudflat, but the biomass of molluscs was greatest in the *Avicennia* zone. Deposit feeders dominated in both of these habitats in terms of density, whereas in the *Rhizophora* zone filter feeders and surface raspers were most common. However, if one considers the number of species, filter feeders were by far the most common, but not in terms of numbers of individuals or biomass.

Yates (1978) studied the zonation of crabs in temperate mangroves at Patonga near Sydney. *Pilumnopeus serratifrons* and *Ilyograpsus paludicola* occur only on the seaward margins and *Helograpsus haswellianus* only on the landward ones. Two ocypodids, *Heloecius cordiformis* and *Australoplax tridentata*, occur in greatest numbers in the lowest mangrove zone. By contrast, *Sesarma erythrodactyla* occurs in greatest concentrations in the highest zones. *Paragrapsus laevis* is restricted to the middle regions of the mangroves, and *Helice leachii*, which occurs only in small numbers, is restricted to the

Table 35 Environment, food and type of respiration of crabs in Mangrove Bay, North West Cape

Crab species	Tidal position	Substrate	Food	Respiration	Salinity
Coenobita spinosa	Supratidal	Sand (beach)	Plant, rotting animal	Air	Wide tolerance, fresh to hypersaline
Ocypode convexa	Supratidal	Sand (beach)	Plant, rotting animal	Air	Wide tolerance, fresh to hypersaline
Ocypode ceratophthalma	Supratidal	Sand (beach)	Plant, rotting animal	Air	Wide tolerance, fresh to hypersaline
Sesarma 2 spp.		Mud (salt marsh)	Plant	Amphibious	Wide tolerance, fresh to hypersaline
Uca elegans		Mud (salt marsh)	Plant	Amphibious	Wide tolerance, fresh to hypersaline
Uca mjobergi		Muddy sand (beach)	Blue-green algae, diatoms, bacteria	Amphibious	Wide tolerance, fresh to hypersaline
Uca flammula	Tidal	Sandy mud (banks)	Blue-green algae, diatoms, bacteria	Amphibious	Wide tolerance, fresh to hypersaline
Uca polita	Tidal	Sandy mud (banks)	Blue-green algae, diatoms, bacteria	Amphibious	Wide tolerance, fresh to hypersaline
Uca dampieri	Tidal	Sandy mud (banks)	Blue-green algae, diatoms, bacteria	Amphibious	Wide tolerance, fresh to hypersaline
Scylla serrata			Carnivore		Marine slightly estuarine
Thalamita crenata	Subtidal	Muddy sand (lagoon, sea, creeks)	Carnivore	Water	Marine slightly estuarine
Portunus pelagicus	Subtidal	Muddy sand (lagoon, sea, creeks)	Carnivore	Water	Marine slightly estuarine

Source: After George and Jones (1984).

Table 36 Feeding types of molluscs collected in four intertidal habitats in the Bay of Rest, North West Cape, Western Australia

Habitat	Deposit feeder	Filter feeder	Surface rasper	Carnivore	Carrion feeder	Undetermined	Total
Mudflat							
Density (No./m^2)	8.0	5.6	0.7	0.7	1.4	0.9	17.3
Biomass (mg/m^2)	919	1471	10	10	86	2	2498
No. of species	8	29	2	9	3	7	58
Avicennia zone							
Density (No./m^2)	7.9	3.3	1.7	0	0.5	0.5	13.9
Biomass (mg/m^2)	2466	340	185	0	11	9	3012
No. of species	6	4	7	0	1	3	21
Rhizophora zone							
Density (No./m^2)	0.3	0.8	0.7	0	0	0	1.8
Biomass (mg/m^2)	35	9	279	0	0	0	324
No. of species	2	3	2	0	0	0	7

Source: After Wells (1984).

higher levels of the shore. At Patonga, only two species of mangroves occur, so that the distribution of the crabs cannot be correlated with zonation of mangrove species. In tropical mangroves, some species are restricted to a particular zone of mangroves. *Uca lactea f. annulipes* is almost always associated with *Avicennia*, especially when it is growing in sand, but if the sand is very fine *U. bellator* occurs instead. *Uca bellator* also occurs at the margins of the *Ceriops* zone. The larger seasarmids, being mainly vegetarians, occur among seedlings of *Avicennia, Bruguiera* and *Ceriops*, although *Macrophthalmus depressus*, an omnivore, is restricted to sandy substrates in drainage channels (Macnae 1968). Snelling (1959) investigated the zonation of crabs along the Brisbane River (see chapter 6) (figures 52a and b).

Some mangrove crab populations exhibit seasonal variations, with most species reaching their maximum densities in the warmer summer months, except for *Paragrapsus laevis* which is most abundant in winter (Yates 1978). Not all crabs are seasonal, however. At Gladstone in central Queensland, an eight-year study of mangrove crabs showed that there were virtually no seasonal changes; rather, recruitment and depletion were continuous (Saenger, unpubl. data). Periods of heavy or unseasonal rain resulted in short-term increases in community turnover.

Crabs may show daily activity patterns. Hutchings and Recher (1974) recorded maximum concentrations of crabs at night on a rising tide. Many of the crabs were climbing pneumatophores and trees in search of food. During the day, at low tide, many crabs remain in their burrows, thereby avoiding predation from wading birds. Some species, such as *Sesarma erythrodactyla* and *Paragrapsus laevis*, undergo migrations associated with spawning (Yates 1978).

Mangrove Epifauna

The epifauna contains the non-encrusting animals found on leaves, trunks or pneumatophores of mangroves. The most conspicuous animals, common in all areas from New South Wales northwards, is the gastropod "*Littorina scabra*" which is often extremely abundant, and occurs in all mangrove zones. It was originally thought to be a highly variable species but a recent study by Reid (1985, 1986 in press) revealed that at least ten species have been confused within Australia alone and they all now belong to the genus *Littoraria*.

Other gastropods such as *Potamides obtusa* and *Nerita lineata* occur on the trunks together with *Oncis* sp., a pulmonate slug. Other components of this epifauna are small crustaceans, including crabs, hermit crabs, amphipods and isopods. These are restricted to the frontal zone of the mangroves where they occur on the

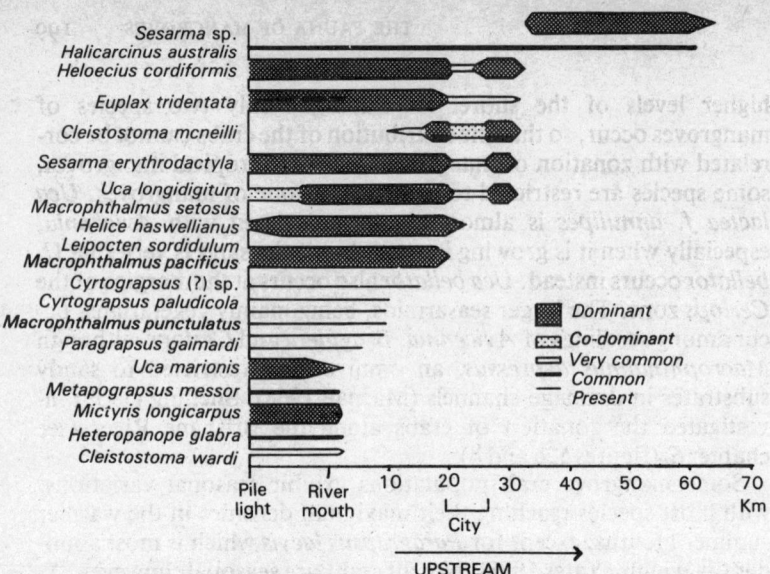

Figure 52(a) Diagrammatic representation of abundance and penetration of crabs along the Brisbane River during normal weather conditions. The number of dominants at each locality is due to the diversity of habitats (after Snelling 1959).

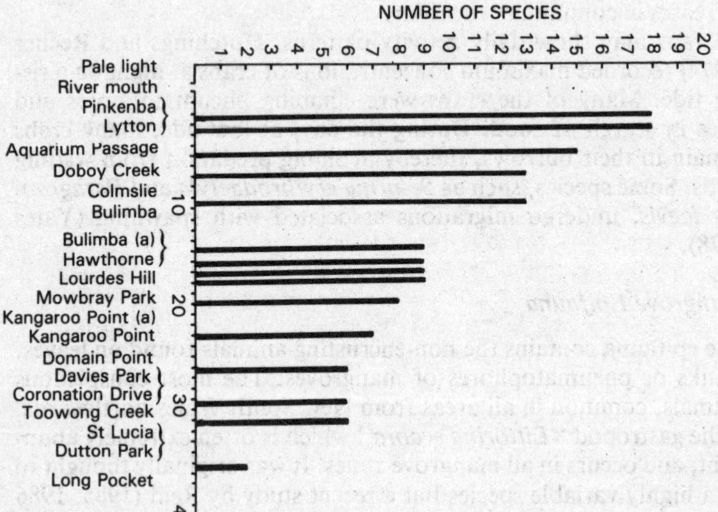

Figure 52(b) Histogram showing numbers of species of crabs at selected localities along the Brisbane River (after Snelling 1959).

pneumatophores and bases of the trunks. Many of these animals are active only at low tide during the night or on dull days when desiccation and predation are reduced. At times, these epifaunal animals may become associated with the encrusting, substrate epifaunal or infaunal communities.

Wood-boring Fauna

Wood-boring animals are not restricted to mangroves but, because fallen logs tend to accumulate there, mangroves characteristically have many wood-boring species. Animals such as ship worms (teredinids) are very important in the breakdown of wood.

In Papua New Guinea, Cragg and Swift (unpubl. data) examined the patterns of colonization of dead wood by decomposer organisms. Their preliminary data suggested a close relationship between terrestrial decomposers (particularly fungi, termites and beetles) and marine timber borers (teredinids and isopods). Elsewhere, non-marine borers have been implicated also (Woodruff 1970). Cragg and Swift (unpubl. data) suggested that infection by fungi may be a prerequisite for successful boring by teredinids. It seems likely that a similar relationship may exist in Australian mangroves, although it has not been investigated yet.

The marine timber borers are dominated by the teredinid bivalves (figure 53). Thirty-one species in eight genera occur (Turner et al. 1972; Turner and McKoy 1979). Other borers include isopods; recently Holdich and Harrison (1980) described several new species of *Gnathia* from dead wood substrates in Queensland mangroves. Other species certainly occur but they rarely have been identified. In Australia, no species of isopod which attacks live mangrove roots has been found. However, in Florida an isopod *Sphaeroma terebrans* commonly attacks the live aerial roots of *Rhizophora mangle* at the air/sediment interface. This was originally considered to be highly detrimental to the mangrove, but now is thought to be beneficial (Simberloff, Brown and Lowrie 1978), because the isopod damage, together with that done by insects, stimulates root branching, so that for every root produced aerially by the tree an average of 1.4 roots reach the substrate (figure 54).

Associated with wood-boring organisms are a range of non-boring animals such as barnacles, limpets, crabs, amphipods, nemerteans and polychaetes. This fauna is most diverse in and under logs trapped in the frontal margins of the mangroves which are frequently inundated. A single species of nemertean (*Pantinonemertes winsori*) has been recorded so far from tropical mangroves (Moore and Gibson 1981). It occurs beneath bark or in cavities in rotten fallen timber of *Avicennia marina* and *Ceriops*

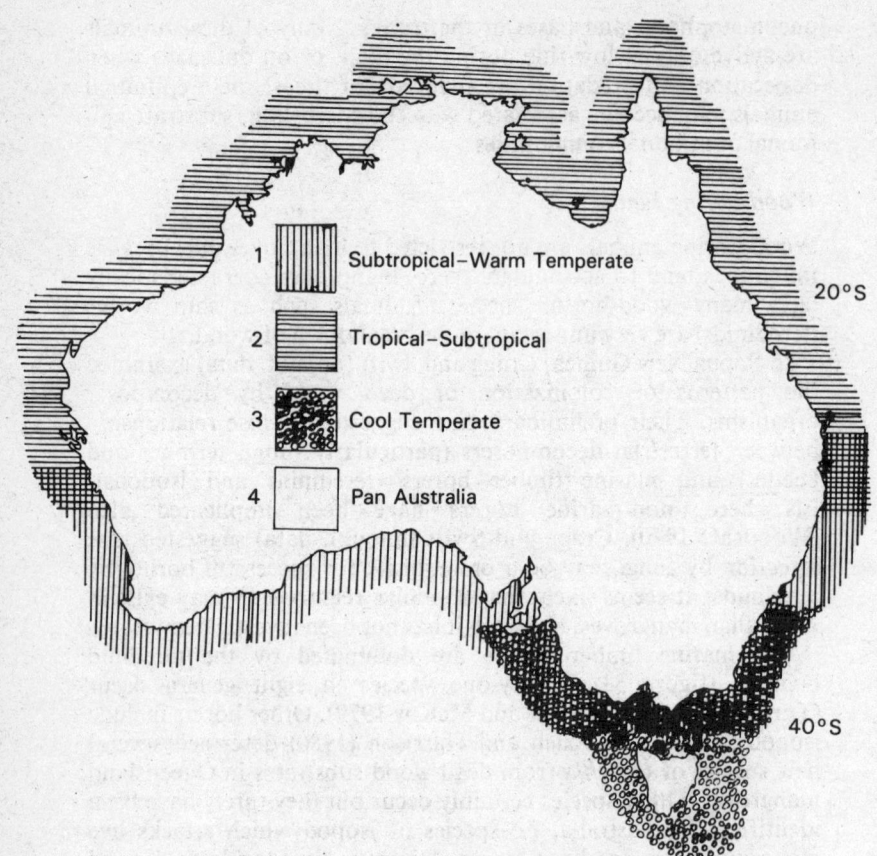

Figure 53 General distribution patterns of some Australian molluscan wood borers. All are Teredinidae, except *Martesia striata* (Linnaeus) (marked with an asterisk) which is in the family Pholadidae. *Nausitora sauli* is marked with a question mark because of taxonomic problems. The full list of names is as follows:
Group 1: *Bankia australis, Teredo navalis, Teredo fragilis, Lyrodus medilobata*
Group 2: *Dicyathifer manni, Bactronophorus thoracites, Teredo matacotana, Bankia rochi, Bankia nordi, Teredo mindanensis, Teredo furcifera, Lyrodus bipartita, Lyrodus massa, Martesia striata**
Group 3: *Nausitora sauli?, Teredo navalis, Bankia australis*
Group 4: *Lyrodus pedicellatus, Teredo clappi, Lyrodus tristi, Nototeredo edax*

tagal at the upper tidal levels. Graham et al. (1975) found thirty-four species of molluscs associated with logs in Trinity Bay, Cairns. These molluscs use the tunnels and cavities created by wood-boring teredinids. In addition, some species such as *Ellobium aurisjudea* and several microgastropods like *Assiminea* spp. and *Iravadia* spp. seek refuge in the logs during the day, and feed on the muddy substrates in the *Rhizophora* and *Bruguiera* zones at night or on cloudy, rainy days.

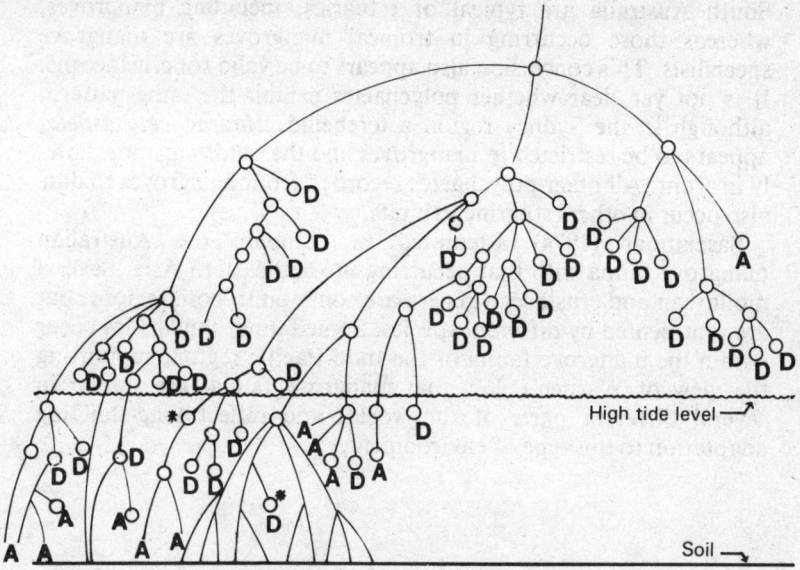

Figure 54 Branching pattern for a single *Rhizophora mangle* root from Clam Key, Florida. A = alive; D = dead; O = bored by *Ecdlytolophora* sp. (caterpillar); O (above water) = bored by unknown insect; O (below water) = bored by *Sphaeroma terebrans*;* = bored by *Teredo* sp. (after Simberloff 1978).

The crab fauna associated with rotting logs has been documented well for the Alligator River, Northern Territory (Hegerl et al. 1979), and around Gladstone in central Queensland (Saenger unpubl. data); however, it is not clear whether these species prefer to live in this specialized habitat or whether they also occur in nearby muddy substrates.

Distribution of the Marine Fauna

As already mentioned, there is a tendency for the number of species of all mangrove animals to increase with decreasing latitude, and although in most cases this correlates with increasing diversity of the mangroves themselves, there are some discrepancies with birds (Ford 1982; Schodde, Mason and Gill 1982) and crabs (George and Jones 1982).

Increasing diversity may be exhibited even within a genus. For example, the molluscan genus *Nerita* is represented by a single species, *N. atramentosa*, in temperate mangroves, whereas six species occur in tropical and subtropical mangroves.

Macnae (1968) suggested that molluscan species occurring in temperate mangroves such as Westernport Bay and the gulfs of

South Australia are typical of estuaries, including mangroves, whereas those occurring in tropical mangroves are mangrove specialists. This conclusion also appears to be valid for crustaceans. It is not yet clear whether polychaetes exhibit the same pattern, although in the Sydney region a terebellid, *Hadrachaeta aspeta*, appears to be restricted to mangroves and the mudflats immediately in front. All other polychaetes recorded from mangroves to date also occur in other estuarine habitats.

Sasekumar (1974) attempted to compare the Australian mangrove fauna with that occurring in southeastern Asia. Several molluscan and crustacean genera are common to both regions, but are represented by different species. Some faunal similarities occur within the mangrove fauna of the Indo-Pacific region, supporting the view of Warner (1969) that mangrove faunas are similar in several different parts of the world, and reflect long-standing adaptation to this type of environment.

6 Adaptations of the Mangrove Fauna

Mangroves are forests with a muddy intertidal substrate, and many of the terrestrial and marine animals are merely adapted to either a forest or to a muddy intertidal habitat and otherwise show no specific adaptations to mangroves. However, mangroves do possess some unique features to which some species show special adaptations.

Mangroves are normally inundated daily by the tide for varying periods of time depending on the tidal cycle (minimum period of inundation during neaps and maximum during springs). Mangroves above high-water neaps are inundated less regularly. They may also be inundated by fresh water during floods. Several problems for the terrestrial fauna arise from such inundation. Their movement across the forest floor may be restricted and their foraging or seeking of prey on the ground or on the lower levels of tree trunks may be inhibited. Ground movement may also be impeded in dense tropical mangroves even at low tide by above-ground roots. Terrestrial animals may be preyed upon by marine predators which come into the mangroves on a rising tide.

Perhaps the major problem for the terrestrial fauna is the absence of permanent fresh water. Several mangroves have a high tannin content in their leaves, which may make them unpalatable to a number of herbivores. In addition, mangrove species generally have high concentration of salt in their tissues which also may render them unattractive to herbivores.

For the marine fauna, mangroves provide additional habitats and less harsh conditions than occur on other muddy shores. Epifaunal and encrusting species are more numerous as more habitats are available in mangroves (such as tree roots). The dense canopy of the forest provides some protection against predators and desiccation. Parts of the mangrove environment are less rigorous in terms of salinity and temperature changes compared with open muddy shores (Macnae 1967). As a result of these conditions, some marine animals that can not survive in open muddy habitats persist in mangrove habitats and are restricted to them. It also seems that some marine animals extend their vertical range up the beach in a mangrove situation in comparison to a muddy shore. Because mangroves offer these benefits, competition and predation pressures are possibly greater within mangroves than on nearby open mud, although recently Branch and Branch (1980) suggested that predation is not important in controlling the distribution and density of the herbivorous gastropod *Bembicium auratum*, but that food supply and availability of hard substrates such as oysters in the frontal zone of the mangroves are the controlling factors. The substrate has a high organic content and low oxygen levels, which may modify burrowing activity.

Some species of the mangrove fauna are adapted to an intertidal existence or to living in a multi-layered dense forest rather than exhibiting specific adaptations to mangroves (facultative mangrove-dwellers). Others are adapted specifically to mangroves (obligatory mangrove-dwellers). A full assessment of the adaptations of the mangrove fauna would necessarily involve comparisons with species occuping other muddy habitats. That has not been achieved yet.

Terrestrially Derived Fauna

Mammals

For the majority of terrestrial vertebrates, mangroves simply provide an extension of their range or possibly act as a substitute forest if no terrestrial forest exists in the vicinity, or they may be a bridging habitat between two terrestrial habitats; consequently, few adaptations are necessary, especially for the arboreal mammals which rarely traverse the muddy floor of the forest. An exception is the small grey false water-rat, *Xeromys myoides*, for which mangroves are a very important habitat. This rat feeds on crabs and has developed a means of incapacitating them even when they are larger than itself. The rat bites at the crab's centrally positioned eye stalks. The crab then moves its claw anteriorly to defend itself and, avoiding the claw, the rat lunges forward and bites the basal attachment of the claw, thus severing it from the body. The process is then repeated with the other claw. By then, the crab cannot defend itself and the rat turns the crab on its back and removes and eats its legs and body (Magnusson, Webb and Taylor 1976). *Xeromys* has a water-repellent fur but has not been seen swimming, so presumably, during high tide, it rests in the trees, which it can climb with agility. The Black Rat (*Rattus rattus*), a species introduced into Australia, elsewhere in its range occupies many habitats including mangrove islets lacking even brackish water or dry land. This species can not survive on full-strength sea water, and consequently must depend on water in its food, or perhaps temporary fresh water during rains (Dunson and Lazell 1982).

Birds

The group of terrestrial vertebrates showing the most adaptations to the mangrove habitat are the endemic bird species. Their adaptations are mainly concerned with feeding. The mangrove robin (*Eopsaltria pulverulenta*), white-breasted whistler (*Pachycephala lanioides*), mangrove fantail (*Rhipidura phasiana*), dusky gerygone (*Gerygone tenebrosa*), mangrove gerygone (*G. levigaster*) and the red-headed honeyeater (*Myzomela erythrocephala*) all have longer bills than closely related terrestrial species. In some species these long bills may prevent the clogging of bristles around the mouth and muddying of the face while foraging on surface mud, while in other species the function of longer bills is unknown. Subspecies occurring in the mangroves, such as the little shrike thrush (*Colluricincla megarhyncha aelptes*) and the black butcher-bird (*Cracticus quoyi spaldingi*), have much longer bills than those subspecies inhabiting rainforests in northeastern Australia and New Guinea. The white-breasted whistler has a strongly hooked bill which may help in the cracking of crustacean carapaces. Crabs are their staple food. Schodde, Mason and Gill (1982) suggested that the more rounded wing and tail of the mangrove robin compared with those of other species of the genus *Eopsaltria* confer greater manoeuvrability as it flies through the mangrove canopy in search of food.

Amphibians

The Indo-Malaysian frog *Rana cancrivora* which lives in mangroves is euryhaline (tolerant of a wide variety of salinities) throughout its life (Gordon, Schmidt-Nielsen and Kelly 1961; Gordon and Tucker 1965). Adults retain urea within their body fluids and thereby raise the osmotic pressure nearer that of sea water. In contrast, tadpoles are good osmoregulators and may excrete excess salts by some extrarenal pathway. They can tolerate salinities up to 40 per cent sea water indefinitely and even have 50 per cent survival in water of 80 per cent sea water (Dunson 1970, 1974, 1978).

Reptiles

A few reptiles inhabit mangroves either exclusively or in addition to other brackish or marine habitats. They are included among the "terrestrial" animals here as they are terrestrially derived. However, some of the species discussed have semi-aquatic or nearly completely aquatic habits. They include the acrochordid snake *Acrochordus granulatus*, the colubrid snakes *Cerberus rhynchops*,

Fordonia leucobalia and *Myron richardsoni*, the goannas *Varanus indicus* and *V. semiremex*, the skink *Emoia atrocostata* and the saltwater crocodile *Crocodylus porosus* and the freshwater crocodile *C. acutus*. In addition, various sea snakes (families Hydrophiidae and Laticaudidae) sometimes inhabit mangroves, especially the banded sea krait, *Laticauda colubrina*. Many terrestrial and arboreal species make forays into mangroves.

Valvular nostrils and/or a laterally compressed tail are common aquatic specializations that are displayed by some of these species (Loveridge 1946, Heatwole 1978).

The saline conditions experienced by such reptiles pose an osmoregulatory problem as the reptilian kidney is incapable of excreting urine with a higher salt concentration than that of the blood. This problem has been met through the development of special salt-secreting glands. Salt glands have developed independently at least five times in reptiles. In lizards it is a nasal gland which secretes brine into the nasal cavity from which it is sneezed; in sea snakes and file snakes it is a sublingual gland which empties its secretions into the tongue-sheath from which they are expelled by protruding the tongue; in homalopsine colubrid snakes it is probably a premaxillary gland; in some crocodiles there are a number of small salt glands located on the tongue; and finally, in sea turtles the salt gland is a modified tear gland (lachrymal gland) associated with the eye (Dunson 1973; Heatwole 1976, 1978; Dunson and Dunson 1979; Taplin and Grigg 1981; Taplin et al. 1982; Taplin, Grigg and Beard 1985).

The rusty goanna (*V. semiremex*) from mangrove forests can secrete brine from the nasal gland up to concentrations of 748 mM for sodium, 800 mM for chloride and 76 mM for potassium. It can adjust the Na/K ratio over a wide range and has a maximum secretory rate of 34 μmol $(100\ g)^{-1}\ hr^{-1}$ for sodium and 3 μmol $(100\ g)^{-1}\ hr^{-1}$ for potassium. In its ability to excrete electrolytes extrarenally, it is intermediate between terrestrial lizards and the marine iguana, and can excrete sodium at a slightly higher rate than some sea snakes (Dunson 1974).

The little file snake (*A. granulatus*) inhabits a variety of aquatic habitats ranging from freshwater to estuaries and mangrove swamps to the sea; it is capable of extrarenal salt excretion via a sublingual salt gland. It can secrete brine up to 600 mM of sodium and 20 mM of potassium and can eliminate the anion chloride at a rate of up to 60 μmol $(100\ g)^{-1}\ hr^{-1}$, higher than in some strictly marine snakes (Dunson and Dunson 1973; Heatwole 1976).

The dog-faced water snake, *C. rhynchops*, inhabits mangroves. It drinks fresh water when it is available, but when it is not it can excrete sodium at about 15 μmol $(100\ g)^{-1}\ hr^{-1}$, a rather feeble performance in comparison to other marine reptiles, but clearly of advantage in a salty environment when fresh water is unavailable (Dunson and Dunson 1979).

It was originally thought that crocodiles could inhabit salt water only temporarily and then by virtue only of their large size and their concomitant low surface-area-to-volume ratio, or by having periodic access to fresh drinking water (Dunson 1970); young *Crocodylus acutus* are known to selectively drink water of lower salinities (Mazotti and Dunson 1984). However, more recently it has been shown that even hatchlings of the saltwater crocodile, *Crocodylus porosus*, can survive indefinitely and grow at a variety of salinities ranging from fresh water to sea water and beyond, without drinking fresh water, even on a diet of marine crustaceans, which are high in electrolytes (Grigg et al. 1980). They can sustain salt and water balance and maintain a high degree of plasma homeostasis (Grigg 1981; Taplin 1984). Although K, Ca and Mg are excreted in the urine, the kidney is not a vehicle for the excretion of Na or Cl in sufficient quantities to account for osmoregulation (Grigg 1981), and it was found that extrarenal excretion of these ions was via lingual salt glands with secretory rates of Na of up to 49 μmol $(100 \text{ g})^{-1}$ hr^{-1}, the secretions having a concentration up to more than 1,000 mOsm (sea water is about 1,000 mOsm) (Taplin and Grigg 1981).

The crocodilians vary in their ability to cope with salt water. Not only do *C. porosus* (Australia) and *C. acutus* (America), the two species inhabiting salt water, have functional salt glands on the tongue, but so does *C. johnstoni*, a primarily freshwater species that only occasionally occurs in salty situations. *C. johnstoni* has secretory rates (1-2 μmol $(100 \text{ g})^{-1}$ hr^{-1}) and concentrations (320-420 mM Na) somewhat lower than the mean values for the saltwater species when in fresh water (6-20 μmol $(100 \text{ g})^{-1}$ hr^{-1}; 450-600 mM Na) but still sufficient to permit occupancy of salt water (Taplin et al. 1982). A survey of crocodile tongues revealed that all the members of the subfamily Crocodylinae tested had functional lingual salt glands, whether from fresh water or salt water, but that members of other subfamilies had only lingual salivary glands of no osmoregulatory significance. It is likely that crocodylids as a group had their origin in fresh water but that the subfamily Crocodylinae represented a marine radiation (Taplin, Grigg and Beard 1985). Those species inhabiting mangroves and other salty habitats have probably retained the ancestral habits of the subfamily, whereas the freshwater species represent a return to fresh water. In this context, the mangrove species are not newly specialized to the mangrove habitat; rather, the freshwater crocodylids have abandoned it.

The permeability of the skin to water and salt is important to aquatic reptiles; those species of snakes inhabiting marine and estuarine habitats, including mangroves, have skins that are relatively impermeable compared with those of freshwater reptiles (Dunson 1978; Dunson and Dunson 1979). Indeed, an American colubrid snake that lacks a salt gland is able to inhabit mangroves by virtue of its very low skin permeability (Dunson 1980a).

The extent of respiration through the skin varies among aquatic snakes. *Crocodylus rhynchops* has skin which is relatively impermeable to the passage of respiratory gases in comparison to sea snakes (Heatwole and Seymour 1978). In mangrove swamps where this snake lives, low cutaneous permeability is probably an advantage, as the water there is often low in dissolved oxygen and rich in carbon dioxide. *Acrochordus granulatus*, also a mangrove species, has low cutaneous uptake of oxygen but in this species, which also lives in oxygen-rich environments, its low skin respiration probably reflects a low metabolic rate rather than low skin permeability as the proportion of its total oxygen requirements met by cutaneous respiration is higher than in some sea snakes (Heatwole and Seymour 1975, 1978).

Breathing characteristics vary among snakes: terrestrial species breathe frequently (land rhythm), whereas sea snakes hold their breath for very long periods and when they do breathe they take only one to three breaths before again holding their breath (aquatic rhythm) even when at the surface or in air. *Cerberus rhynchops* from mangroves and the sea krait *L. colubrina* are intermediate in that they have both types and can switch from one to the other (Heatwole 1977, 1981).

Insects

Many of the mangrove insects lay their eggs in the fruit of particular mangroves, and these species must time their reproductive activity to coincide with fruiting of their food plant. This adaptation clearly is not restricted to insects living in mangroves, but occurs in all insect species which lay their eggs in the fruit of particular trees (Hutchings and Recher 1974). Macnae (1968) reported that some mangrove insects lay their eggs in pockets of water trapped in holes or forks in mangrove trees. Some mosquitoes lay their eggs in crab holes where permanent water is present (Bright 1977).

Clements (1963) found smaller anal papillae in larval mosquitoes living in saline areas, including mangroves, than those found in freshwater species. This suggests that the size of anal papillae is a function of the volume of fluid excreted.

Dunson (1980b) studied osmoregulation in larvae of a dragonfly (*Erythrodiplax berenice*) from mangroves. This species was resistant to salinity changes and is the only dragonfly known to be highly tolerant of hypersaline conditions.

Finally, some species of ants which live on the floor of the mangrove forest have adapted to the daily tidal inundation by plugging their burrows with mud and trapping bubbles of air inside the burrow. Other arborescent species of ants survive by sheltering in the hollow twigs of *Rhizophora* and *Ceriops* (Taylor, pers. comm.).

Marine-Derived Fauna

In contrast to the terrestrial fauna, many of the marine species show adaptations to the mangrove habitat, or at least to an intertidal one. These adaptations can be divided into three main types: morphological, behavioural and physiological.

Morphological Adaptations

Mudskippers are fishes related to gobies and, like gobies, are characterized by fused pelvic fins (Milward 1974). They are characteristic of tropical mangroves and are well adapted to alternating periods of exposure to air and submersion (Gordon et al. 1969). Their eyes are highly mobile and compensate for the lack of a neck; they are set in turrets and are protected from desiccation by secondary spectacles (figure 55). These spectacles are produced by a

Figure 55 Schematic drawing of *Periophthalmus* to show adaptations of the eyes and pelvic fins (after Berjak et al. 1977).

fusion of the overlying skin and the cornea. Walls (1942) noted that the lower half of the retina is far richer in cones than the upper half which has more rods. The upper half is used for locating prey on the surface of the mud and the lower half for watching a rival in display. The eyes are perched on top of the head which may increase the field of view. The tissue of the sides and back of the buccal cavity serve as a respiratory epithelium (Schottle 1932). Schottle considered the gill surfaces to be reduced in comparison to fully aquatic gobies. Accessory respiratory surfaces are also present on the fins and in the nasal sac diverticula (Macnae 1968). However, these additional surfaces have not been shown experimentally to be respiratory and perhaps, instead, they are associated with salt regulation. Mudskippers respire under water like other fish. However, when they come out of the water, they gulp air by rapidly distending the branchial cavity. The small valvular, opercular openings are then closed, trapping water and oxygen which can be used for respiration while on land. Opercular movements cease during this period (Teal and Carey 1967b).

When submerged, all mudskippers swim relatively easily in a normal fishlike way (Stebbins and Kalk 1961). On land they proceed by a series of skips and jumps. Some species also can climb trees using their pelvic fins as suckers and their pectoral fins as "arms" which embrace the twig or root. The ability to jump and skip is the result of modifications to the skeleton and musculature. Eggert (1929) studied the movement of several mudskippers and compared them with gobies in an attempt to trace the adaptations associated with migration onto land. Subsequently Petit (1920), Van Dijk (1960) and Harris (1960) continued these investigations. Three types of locomotion have been recognized:

1. Ambipedal progression or "crutching". In this mode the pectoral fins are used as crutches (figure 56); they are stretched forwards and then the fish swings on them and at the completion of the stroke the weight is transferred to the pelvic fins. No sinuous movement of the body occurs.
2. Skipping on land. This is typically an escape reaction, but is used also during feeding. During skipping, the tail is bent fowards and to one side, with the caudal fin rays digging into the mud. When the body is straightened the fish is projected forwards and upwards into the air. The vertical component of the skip is provided by the thrust of the pelvic fins, with the pectoral fins acting as stabilizers.
3. Skimming on the water. A mudskipper can skim across the water in a series of bounds. Each bound is preceded by a short burst of swimming and during this the propulsive forces for the bound or leap are produced. The tail, second dorsal and anal fins produce the forward thrust; the pectoral fins are held to

ADAPTATIONS OF THE MANGROVE FAUNA 213

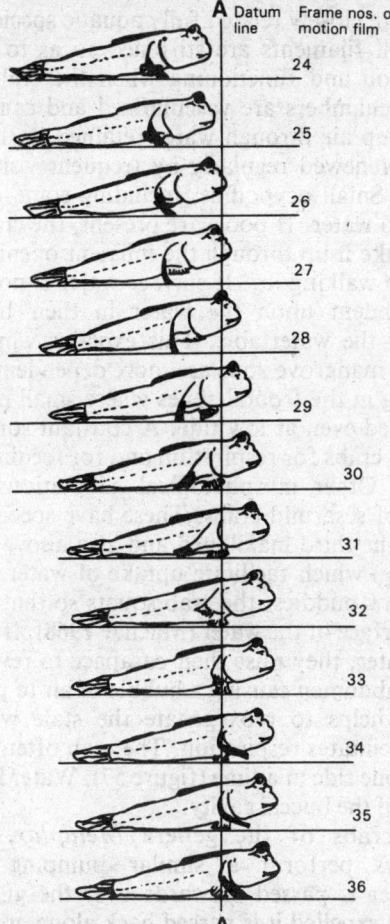

Figure 56 Cycle of fin movements during locomotion on land by crutching of *Periophthalmus*. Traced from a motion film. Each frame represents a time lapse of 0.3 seconds (after Macnae 1968).

the side of the body during flight which lasts about a twentieth of a second. This type of movement is similar to the way a flying fish moves through the air on a glide.

Many of the crabs living in mangroves exhibit morphological adaptations for breathing while on land. Gray (1957) suggested that the size of crab gills is correlated with habitat and metabolic activity. Intertidal species in temperate areas generally have reduced gill areas in comparison to fully aquatic species. This appears to hold true for mangrove species of *Ocypode* and *Uca* which have fewer gill filaments than closely related fully aquatic species (table 37). In addition, the gill filaments are stiffened so as to maintain their shape, orientation and functioning when the crab is out of the water. The gill chambers are vascularized and can act as a lung. These crabs pump air through water retained in the gill chamber which must be renewed regularly by frequent visits to the water (Macnae 1968). Small ocypodids, including some *Uca*, need continuous access to water. If pools are present, the crab will squat in the water and take it up through the inhalent opening between the third and fourth walking leg. If surface water is not available, the crabs are dependent upon the water in their burrows, which penetrate below the watertable. This explains why species living higher up in the mangrove zone are more dependent upon burrows than those living in the frontal zones where small pools of surface water are retained even at low tide. A constant source of water is necessary for all crabs for respiration and for feeding in some, as is discussed later. Other morphological adaptations are found in several species of sesarmid crabs. These have specialized openings (at the base of the third maxilliped and also above the fourth and fifth walking leg) which facilitate uptake of water into the buccal cavity. In shallow puddles, the crab squats so that these openings are under the surface of the water (Macnae 1968). If the crabs are in very shallow water, they raise their carapace to reveal a slit above the edge of the abdomen causing a bubble of air to pass into the gill chamber. This helps to reoxygenate the stale water inside the chamber and facilitates respiration. The crab often performs these movements on one side at a time (figure 57). Water is exhaled along the upper edge of the buccal cavity.

Species of crabs of the genera *Metaplax*, *Ilyoplax* and *Macrophthalmus* perform a similar pumping behaviour to sesarmids. Water is passed forwards over the gill as usual, but instead of being expelled it is passed back along grooves under the edges of the carapace to enter the gill chambers again posteriorly. While flowing back in a thin layer over the carapace the water is reoxygenated, and this reduces the necessity for continual replenishment of water.

Some of the polychaete worms commonly found in mangroves (such as those belonging to the families Spionidae and Eunicidae) and some of the nereidid worms have gills along their bodies which facilitate respiration (figure 58). However, these families are also well represented in other estuarine habitats so that gills are not exclusive to mangrove species. Storch and Welsch (1972) made a study of the epidermis of polychaetes found in mangrove swamps in Sumatra and were able to show from the ultrastructure and histochemistry that some species are able to absorb oxygen directly

Table 37 Crabs, with their average gill areas per gram, arranged by habitat

Aquatic		Low tide		Intertidal		Above tide	
Callinectes	1367	Menippe	887	Uca pugnax	770	Sesarma cinerea	638
Areneus	1301	Panopeus	874	Uca pugilator	624	Ocypode	325
Ovalipes	1288			Sesarma reticulata	579		
Hepatus	1099			Uca minax	513		
Portunus gibesii	1003						
Portunus spinimanus	901						
Libinia dubia	748						
Libinia emarginata	566						

Source: After Gray (1957).

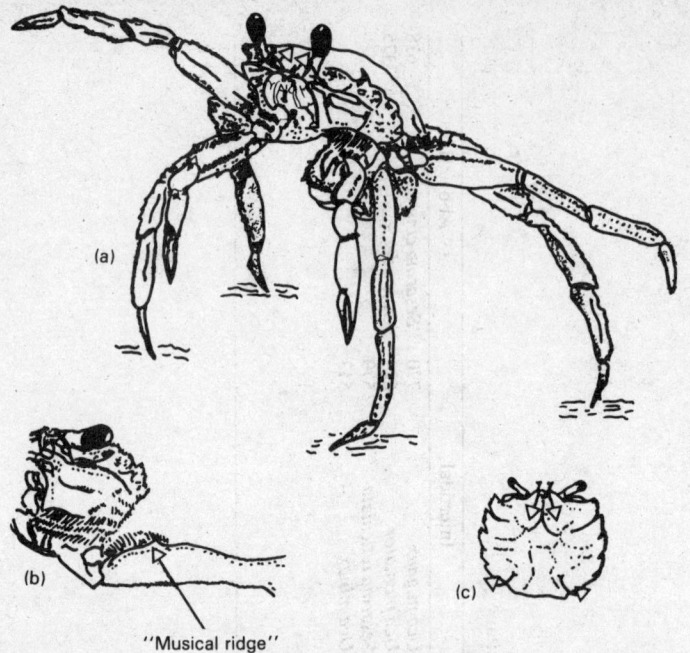

Figure 57 *Metaplax crenulatus*: (a) Female in characteristic pose. The arrows indicate the direction of water flow while pumping and reoxygenating the water. This process is almost constant while the crab is out of water. (b) Base of cheliped and pterygostome of a male showing the "muscle ridge" on the pterygostome below the orbit and the "bow" on the merus of the cheliped. (d) Dorsal view showing direction of water flow over the back while the crab is out of water. (After Macnae 1968).

from air. For details of the structure of the epithelium see the photomicrographs of Storch and Welsch (1972).

Most of the molluscs living at high levels in the mangroves have some respiratory modification. The ellobiids are air breathers and some of the tree-living prosobranchs have their mantle cavities modified as lungs.

The mangrove fauna also faces problems of general body desiccation and many of the molluscs and crabs have thickened shells or carapaces which reduce water loss by evaporation. The thickened skeletons also presumably afford some protection from predation by wading birds.

Warner (1969), working on crabs in a Jamaican mangrove swamp, found that within a species certain sizes were restricted to a particular mangrove zone. The largest individuals occurred higher up the shore than smaller individuals of the same species (figure 59). Smaller crabs were believed to be less resistant to desiccation than larger animals because of their greater surface-area-to-volume

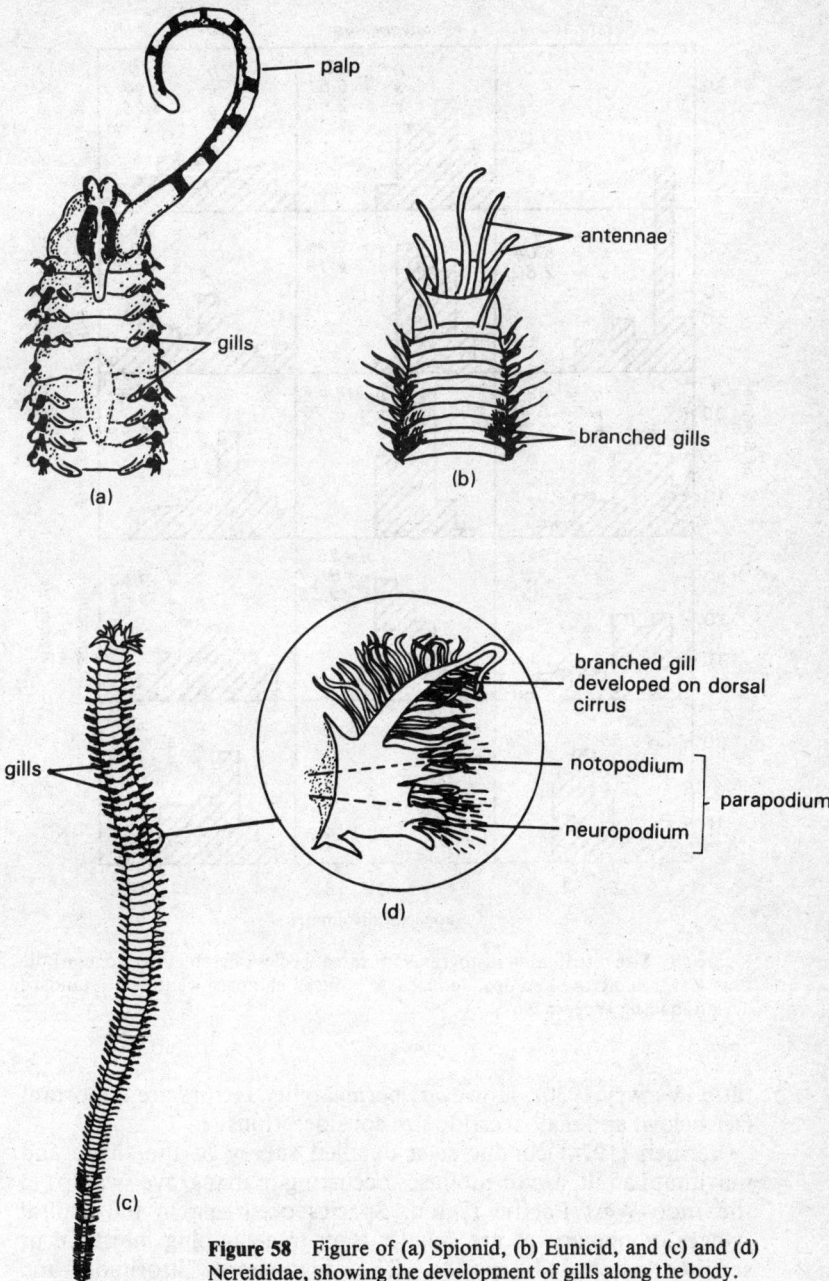

Figure 58 Figure of (a) Spionid, (b) Eunicid, and (c) and (d) Nereididae, showing the development of gills along the body.

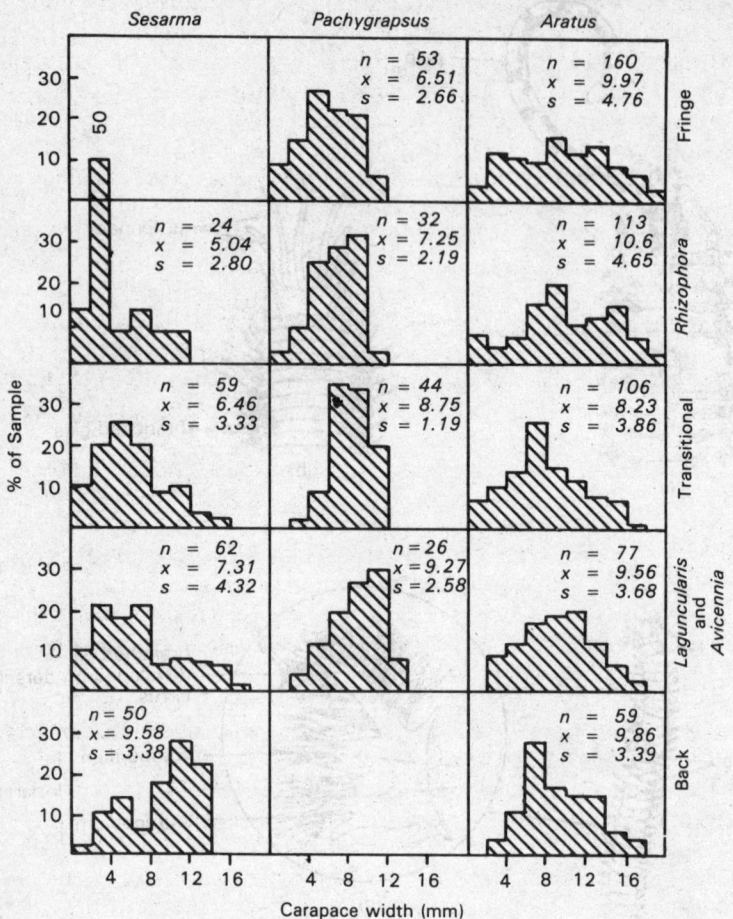

Figure 59 Size distribution histograms for three species of crab in five zones of the Port Royal mangrove swamps, Jamaica. $\bar{x}$ = mean carapace width; s = standard deviation (after Warner 1969).

ratio (Verwey 1930). However, permeability factors are important (see below) and may override size considerations.

Vermeij (1974) conducted a detailed survey of the shape and maximum adult size of molluscs occurring in mangrove swamps in the Indo–West Pacific region. Species occurring in the littoral fringes of mangroves are smaller than tree-climbing intertidal or sediment-associated species. The high-shore littorinids and potamidids are larger and more slender, and neritids are larger and more globose than their lower-shore counterparts. These interspecific trends parallel those among rocky-shore gastropods, and reflect adaptation to extremes in temperature and desiccation (Vermeij 1973b). Shells of species in several families in the littoral fringe exhibited either internal whorl respiration or apical decollation, possibly resulting from acidic conditions during anaerobic respiration while the animals were exposed to air ((Vermeij 1974).

The zonation of many of the smaller ocypodid crabs is related to the particle size of the substrate on which they feed and in which they burrow (Crane 1975). Frith and Frith (1978) suggested that substrate grain size and organic content, the presence or absence of mangrove vegetation, relative salinity and degree of tidal wetting are the main factors affecting the distribution, zonation, density and sympatry of *Uca* species. George and Jones (1984) were able to predict which species of *Uca* would occur in particular places by the distribution of sediments. All appear to feed on diatoms or particles of detritus associated with the sediment. The crabs sort the sediment and ingest some of it and discard the rest in the form of rounded pellets (Macnae 1968). *Uca* males use only the small chelae to pick up the sediment whereas *Uca* females and both sexes of other genera (which have both chelae similar in size) use them alternately. These chelae are slightly hollowed and function as "spoons" to scoop up the sediment which is then transferred to the first and second maxilliped (mouth parts). These maxillipeds have highly specialized setae (especially on the meropodites which are used to sort the sediment) (figure 60). Some of these setae are feathery, others are spoon-shaped. The setae on the outer surface of the first maxilliped form a stiff brush. If the species lives in coarse sand, then the spoon-shaped setae are more numerous and larger than those of species living in fine sediment (figure 60). A current of water pumped out of the branchial chamber assists in the sorting of the sediment. The sediment picked up by the chelae is rolled between the specialized setae of the meropodites of the first and second maxilliped. The spoon-shaped setae trap the sand grains while the "brush" of the first maxilliped scrapes off any diatoms on the surface of the sand grains. Detritus associated with the sediment and the released diatoms are carried by the water current towards the mouth. Sand, together with other large or heavy sediment particles, sinks to the base of the buccal cavity where it is mixed with mucous to form pellets which are discarded at intervals. The water is returned to the branchial cavity through an opening at the base of the outer edge of the third maxilliped. Species like *Uca dussumieri* and *U. urvillei*, which feed on fine mud, have a mass of hair-like setae forming a filter at the base of the third maxilliped, which prevents sediment from entering and clogging the branchial cavity (Macnae 1968).

Obviously, these adaptations to feed are related to the fact that

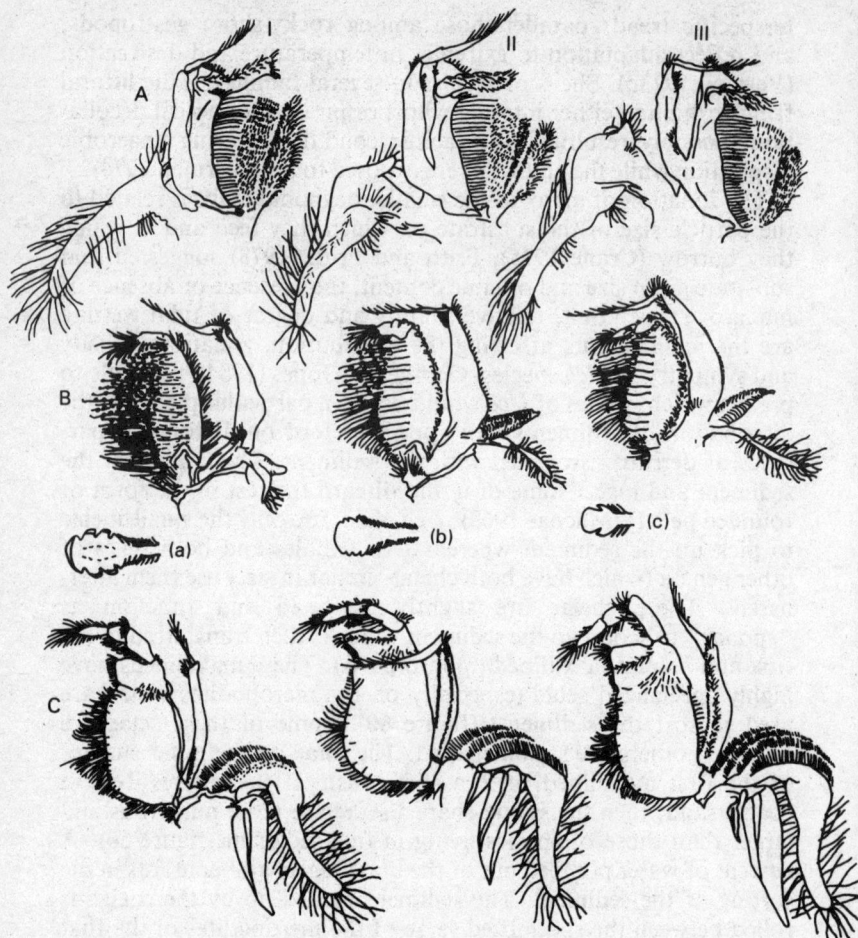

Figure 60 Maxillipeds of three species of fiddler crabs which inhabit different substrata. Line I: *Uca lactea f. annulipes* — sand; line II: *Uca urvillei* — find mud; line III: *Uca dussumieri* — which feeds off very find mud. Row A: outer view of first maxilliped; row B: inner row of second maxilliped; row C: outer view of third maxilliped. (a)–(c): typical spoon-shaped bristle from respective species.

the animals feed on sediment rather than that they inhabit mangroves. However, sediments suitable as food for particular species of crabs tend to accumulate in mangroves (George and Jones 1984).

Macrophthalmus depressus, unlike other species in the genus, does not live in discrete burrows; it lies buried in the sand with only its eye stalks showing. The maxillipeds function as a sieve, rather than as sorting agents, trapping detritus and plankton in the water. This species also scavenges on carrion.

Hermit crabs, such as *Clibanarius taeniatus* and *C. virescens* which live on mudflats associated with mangroves, feed on detritus and use modifed setae on the mouth parts for filtering, and plates of heavy teeth for masticating their food. In contrast, hermit crabs which live on rocky intertidal shores, such as *Paguristes squamosus* and *Dardanus setifer*, feed on larger items and use their chelae to grasp and shred the food (Kunze and Anderson 1979) (figure 61-64). The mud crab (*Scylla serrata*) uses its mouth parts to crush food such as bivalves, gastropods or small crabs. The large powerful chelae are used to crush any prey too large to be broken up by the mouth parts.

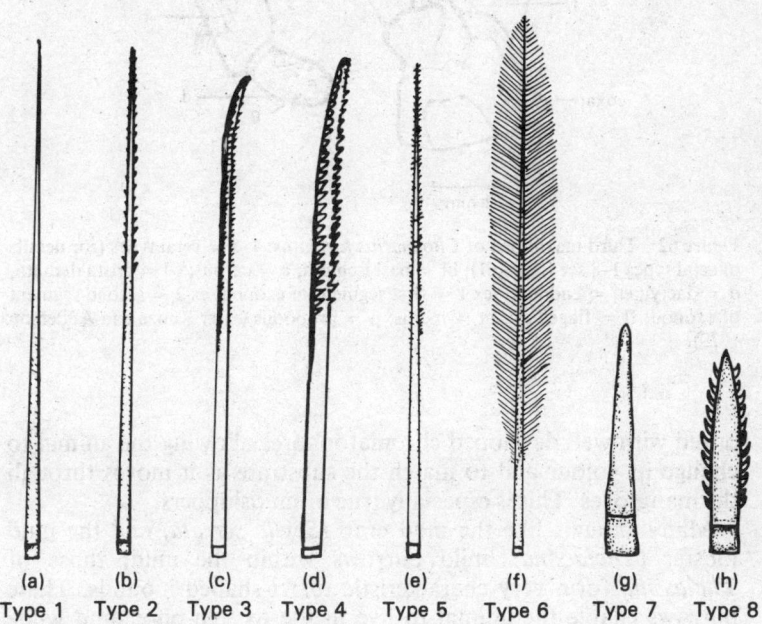

Figure 61 Types of setae found on the mouthparts of hermit crabs: (a) type 1, simple; (b) type 2, spinose; (c) type 3, comb; (d) type 4, serrate; (e) type 5, papillose; (f) type 6, plumose; (g) type 7, stout simple; (h) type 8, stout serrate (after Kunze and Anderson 1979).

Behavioural Adaptations

Many species of the marine fauna are cryptic, living in burrows or under logs, and thereby may avoid desiccation and at least partly escape predation from wading birds. Many are also cryptically col-

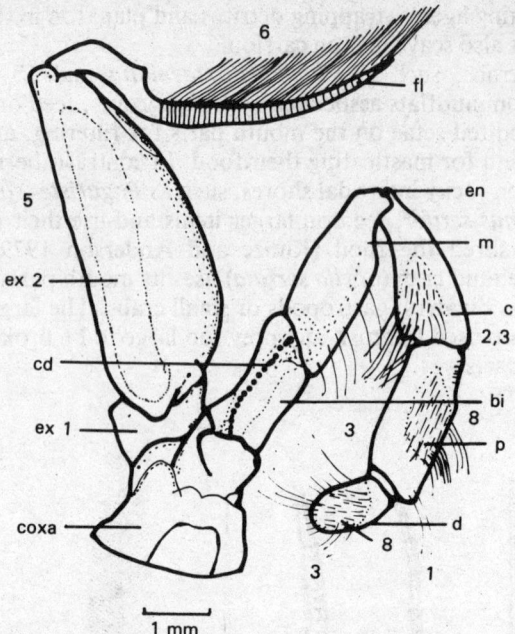

Figure 62 Third maxilliped of *Clibanarius taeniatus*: 1–8 = setal types (for details of setal types 1–8 see figure 61); bi = basi ischium; c = carpus; cd = crista dentata; d = dactyl; en = endopod; ex 1 = first segment of exopod; ex 2 = second segment of exopod; fl = flagellum; m = merus; p = propodus (after Kunze and Anderson 1979).

oured with well-developed chromatophores allowing the animal to change its colour and to match the substrate as it moves through the mangroves. This is especially true of mudskippers.

Many animals like the mud crab (*Scylla serrata*) and the mud lobster (*Thalassina*) build burrows within the mud; those of *Thalassina* form very characteristic turret-shaped mounds. These burrows enable the animal to live in the oxygen-poor mud while still maintaining an opening to the surface through which oxygenated water or air can pass. The burrows extend below the watertable so that a pool of water is always present at the bottom of the burrow which can be used for respiration and feeding. By this means, many species can live further up the shore than otherwise would be possible. This adaptation also occurs in other muddy habitats.

Many of the burrowing animals are more active at night than during the day; this is especially true of crabs (Hutchings and Recher 1974). Yates (1978) has shown that many crabs are more

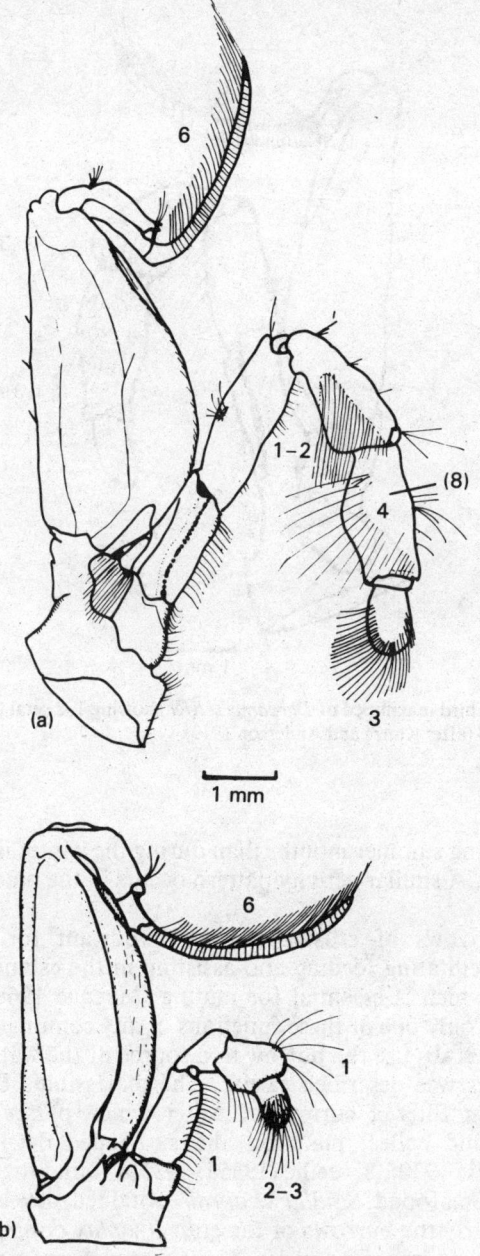

Figure 63 Third maxilliped of (a) *Clibanarius virescens* and (b) *Paguristes squamosa*, showing setal types 1–8 (for details see figure 61) (after Kunze and Anderson 1979).

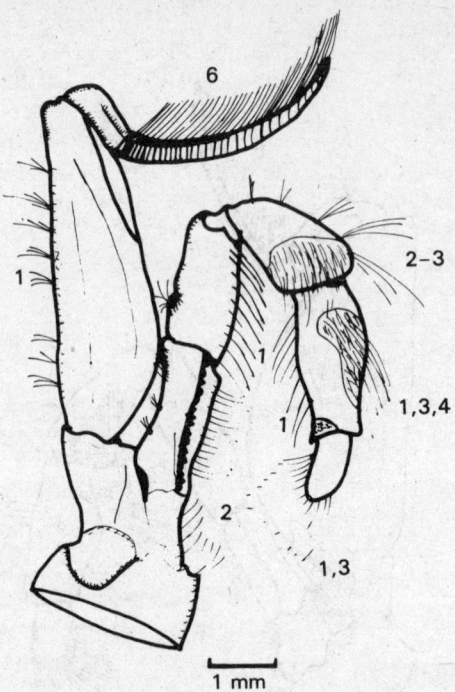

Figure 64 Third maxilliped of *Dardanus setifer* showing 1-8 setal types (for details see figure 61) (after Kunze and Anderson 1979).

active during summer months than during the winter in a temperate mangrove. A similar activity pattern occurs in the mud crab, *Scylla serrata*.

The burrows of crustaceans are important for providing a refuge, facilitating feeding and assisting in the establishment of a territory which is essential for mating (Macnae 1968). A burrow may serve only one of these functions or any combination of them. Ocypodid crabs use the burrow as a source of the water needed for feeding, as was described earlier. The thalassinids *Upogebia* and *Callianassa* filter a current of water which passes through the burrow, and collect plankton and suspended detritus for food (MacGinitie 1930). Tweedie (1935a, 1935b) found that the burrows of the stomatopod *Squilla choprai* contained developing larvae, and similarly the burrows of the crab *Ilyoplax delsmani* contained all stages of the life cycle. These records suggest that some mangrove crustaceans may pass their entire life cycle within their burrows.

McKillup and Butler (1979) studied the burrowing activity of the crab *Helograpsus haswellianus* in South Australia. They suggested that crab holes are a limited resource and that this limitation is self-imposed. Although the density of crabs sometimes exceeds the density of holes, substrate breakdown through overdigging is never observed. They postulated that each different kind of substrate has a characteristic density of crab holes that it can support, which they term the substrate tolerance limit. Above this density the substrate collapses. Exposed crabs are preyed upon by several species of birds and fish, and crabs in holes have a selective advantage over those on the surface. Crabs on the surface are seen commonly only when their density exceeds that of holes or when individuals are in transit from one hole to another. McKillup and Butler (1979) therefore postulated that once the maximum number of holes has been dug in a particular type of substrate, a crab will no longer dig a hole for itself but enter an occupied one. The entrant is then either evicted or it evicts the original resident and the evicted crab quickly enters another hole. Such a control on the number of crab holes may be widespread, and in areas where predation is high it may be a factor controlling density. The mechanism by which individuals recognize when the substrate tolerance limit is being approached and therefore do not dig is unknown.

Uca musica terpsichores dislodge their neighbours and fill in their burrows with sand. Burrow infilling is performed only by sexually mature males which will display later on during low-tide. Zucker (1977) suggested that burrow filling is a primary means by which the males establish and maintain their territories. Similarly, three species of *Ocypode* (ghost crabs) have been observed to fill in neighbouring burrows that lie within the minimum required space around a burrow. Lighter (1974) suggested that such a mechanism may be important in controlling the spacing and distribution of ghost crabs.

The fiddler crabs (*Uca*), which are common in mangroves and associated mudflats, establish territories and defend them rigorously. *Uca* males have developed a complex system of signalling with their enlarged, brightly coloured claws (figure 65). Salmon (1984) discussed the behaviour, aggression and mating system of a "primitive" fiddler crab, *U. vocans*, which occurs in tropical Australia. He concluded that the previously held views on "primitive" and "advanced" fiddlers needs to be re-examined (Crane 1975) and that the behavioural patterns in these two groups may have evolved independently. He proposed that ideas concerning the evolution of behaviour in both groups should be re-examined, stressing in particular ecological variables that affect male and female reproductive success. In these crabs the mating system is promiscuous (each sex mates with multiple partners) and

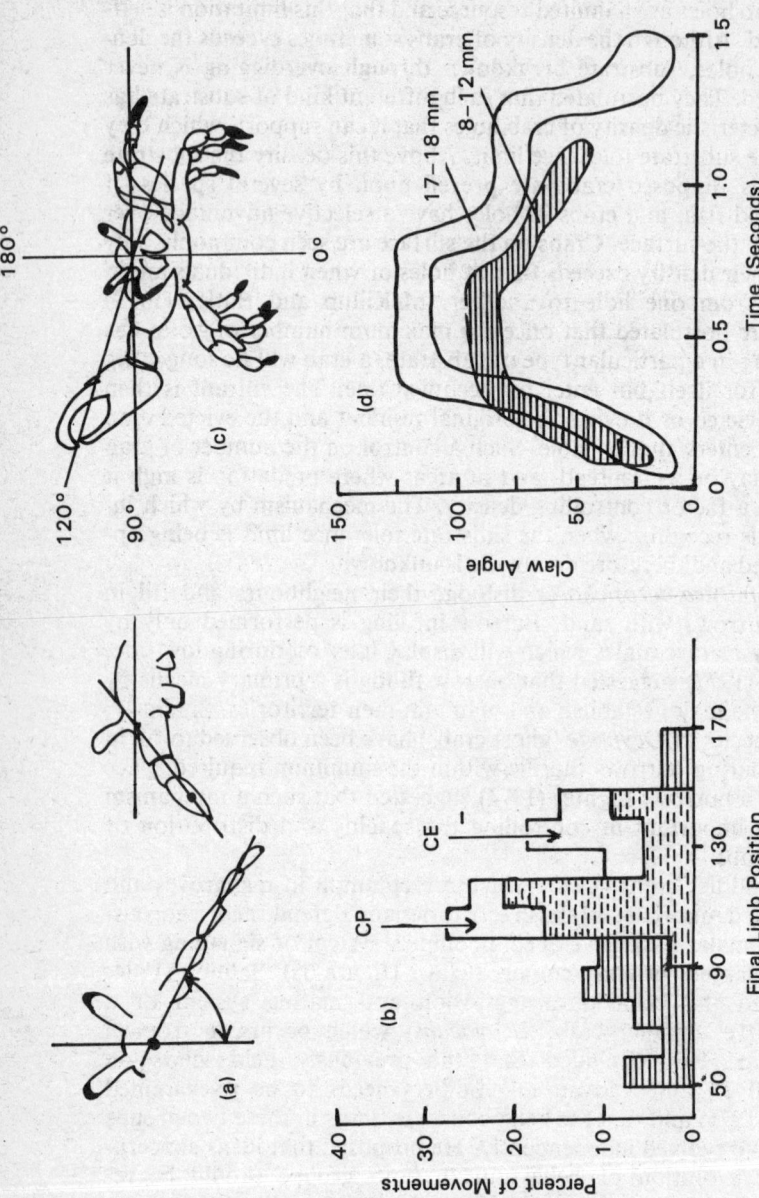

Figure 65 Crab displays analyses by cine film: (a) ambulatory raise (left) and cheliped movement (right) during hermit crab agonistic displays; (b) distribution of angles during non-displays (lined) and display (clear) movement; overlap areas are dashed lines; CP = cheliped presentation; CE = cheliped extension; (c) major chela angle during waving display of a male fiddler crab; (d) distribution of angles for several waves shown by smaller (8, 12 mm carapace width) and larger (17–18 mm) males of *Uca pugilator* (after Salmon and Hyatt 1983).

resource-free (neither sex provides the other with any resource other than sex cells). In contrast, in other species the mating system is polygamous (one male mates with several females but females mate only once) and resource-based (each male provides his mate with a safe burrow for mating, ovulation and incubation). The development of the signalling patterns in *Uca*, which are species-specific, may have allowed many closely related species to coexist in a very similar environment. For example, the males of the sibling species *Uca pugilator* and *U. panacea* show differences in visual and acoustic displays. These differences are exaggerated in the overlap zone. Occasional forced matings can be produced in the laboratory, but with high mortality of larvae. Thus, both premating (behavioural) and postmating (higher larval mortality) barriers occur which prevent interbreeding (Salmon et al. 1978).

Zucker and Denny (1979) suggested that at least two different male courtship strategies have evolved among tropical *Uca* species. Most species are referred to as "non-herders". In these species, males wave near their burrow entrance and enter their burrow as soon as a female wanders into the vicinity. The female is free either to enter the burrow where mating takes place or to ignore the male. "Herders", on the other hand, attract a female to the vicinity of their burrows by their waving display, but then attempt to envelop her with the large claw and drag her into the burrow where mating occurs.

Zucker (1978) studied the time of courtship in *Uca* and found it often coincided with low tides occurring in the late morning. At this time, the substrate is often dried out by high temperature, low humidity and the extreme low tide which thoroughly drains the flat. The dried mudflat may not be appropriate for efficient feeding. These species feed by filtering organic matter left on the mud by the ebbing tide. If, as seems likely, feeding can occur only when the substrate is wet, courtship is restricted to times when the substrate is too dry for feeding. Thus members of the genus *Uca* efficiently partition their time.

Heloecius cordiformis, a common crab in temperate mangroves, also has a courtship display closely resembling that of some fiddler crabs (Griffin 1968).

The burrows of mudskippers serve numerous functions including refuge from predation, refuge from high tide and desiccation and as an observation post (Nursall 1981); their burrows are also essential for reproduction (Brillet 1970, 1975).

Complex behavioural patterns effectively isolating species are found in mudskippers where several species occur in the same general habitat (Whyte 1979). Nursall (1981) described four species of mudskippers living together in a small creek in the mangroves near Townsville. These species showed intra- and interspecific antagonistic behaviour, and had microhabitat preferences which separated them completely (figures 66 and 67). Nursall (1981) tested the interaction between these sympatric species both in the field and in the laboratory.

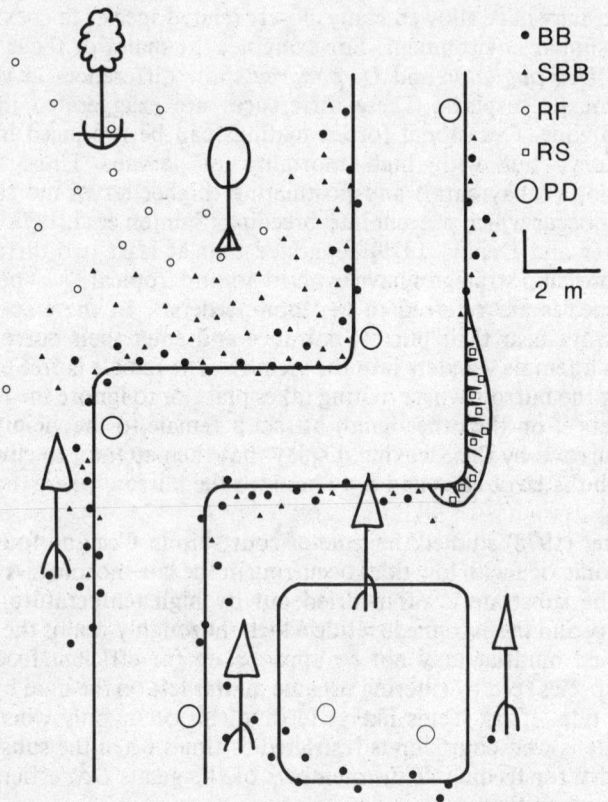

Figure 66 Schematic distribution of mudskippers along Three Mile Creek, north Queensland. BB = *P. vulgaris*; SBB = *P. gracilis*; RF = *Periophthalmus* sp.; RS = *P. expediionium*; PD = *Pn. schlosseri*. *Rhizophora* is shown with prop roots, *Ceriops* with buttresses and *Avicennia* with pneumatophore. The creek meanders across the diagram; standing water is shown by a square surrounded by *Rhizophora*. RS are located on a steep bank of the creek (after Nursall 1981).

Although mudskippers are territorial during the breeding season (Kobayashi, Dotsu and Takita 1971), there is little evidence that territories are defended during the non-breeding season. Instead, there is aggressively forced spacing with size and species hierarchies. The animals move about taking their territories with them.

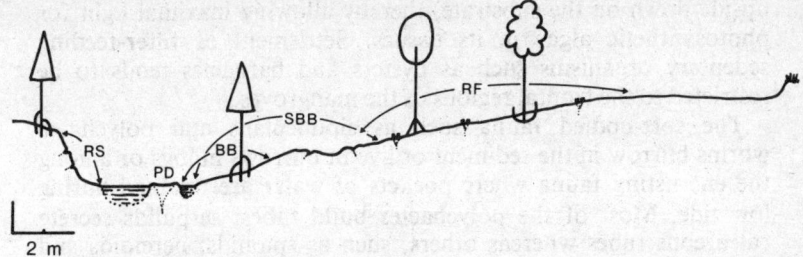

Figure 67 Diagrammatic elevation of the mangroves in Pallarenda swamp, near Townsville, north Queensland, to show distribution of mudskippers. BB = *P. vulgaris*; SBB = *P. gracilis*; RF = *Periophthalmus* sp.; RS = *P. expeditionium*; PD = *Pn. schlosseri*. For key to mangrove species, see caption to figure 66 (after Nursall 1981).

Once Nursall had recognized the behavioural patterns and the dominance hierarchy within the various populations, he was able to find morphological differences between the coexisting species and identify them as *Periophthalmus vulgaris*, *P. gracilis*, *P. expeditionium* and *Periophthalmodon schlosseri* and an undescribed species of *Periophthalmus*, which morphologically could not be distinguished easily from *P. vulgaris*. However, this new species could be separated on behavioural patterns and habitat requirements (Nursall 1981) (figures 66 and 67).

Soldier crabs (*Mictyris longicarpus*) aggregate in large numbers and move in a single large army across the mudflat at low tide; they rapidly burrow simultaneously to escape from predators (Cameron 1966). They also have a well-developed carapace which may give them some protection from desiccation and predation. Juvenile *Mictyris* are found lower down the shore among seagrass beds (Hutchings and Recher 1974) and presumably migrate up the shore as they mature.

Several other mangrove crabs show migratory behaviour including juvenile mud crabs, *Scylla serrata*, which settle in spring and early summer among the mangrove pneumatophores or in seagrass beds. They make small burrows and forage for food in the vicinity. By the end of summer they have grown much larger. During the winter they tend to shift from the intertidal to the subtidal area, but they may move back into the intertidal region briefly during high tide for feeding (Hill, Williams and Dutton 1982; Hyland, Hill and Lee 1984).

Some of the gastropod molluscs living in old burrows made by teredos (bivalve molluscs) are active only at night or during cloudy days when desiccation is reduced (Graham et al. 1975; Little and Stirling 1984).

The jelly fish *Cassiopea* has adopted a sedentary lifestyle, resting upside down on the substrate, thereby allowing maximal light for photosynthetic algae in its tissues. Settlement of filter-feeding sedentary organisms such as oysters and barnacles tends to be restricted to the frontal regions of the mangroves.

The soft-bodied fauna such as sipunculans and polychaete worms burrow in the sediment or live in burrows in logs or among the encrusting fauna where pockets of water are retained during low tide. Most of the polychaetes build tubes; serpulids secrete calcareous tubes whereas others, such as spionids, nereidids and terebellids, build semi-permanent tubes of mucous and sediment. These tubes retain some moisture during low tide which prevents desiccation and is essential for respiration. The sipunculan *Phascolosoma arcuatum* burrows at least 25 cm below the surface of the mud, and excavates extensive galleries. Green (1975) measured the fluctuation in temperature during the day at this depth and found it varied little. He also suggested that the corresponding salinity variation is likely to be small. *Phascolosoma arcuatum* feeds at night during low tides on the moist surface mud by completely everting its introvert which exposes its ciliated tentacles. The cilia create water currents which draw fine particles from the surface of the sediment on to the tentacles, which in turn carry the sediment to the mouth.

This species can survive extended periods of starvation. In tropical estuaries, which carry a large freshwater run-off during the wet season, the worms may be subjected to extended periods of immersion during which time they cannot feed. Animals kept in the laboratory did feed after twenty-four days of continual immersion (Green 1975), but in the field worms were never observed feeding except at night when exposed at low tide.

Reproductive Adaptations

As already mentioned in the behaviour section, fiddler crabs (*Uca*) perform extremely complicated behavioural patterns associated with mating, which permit a number of species to share the same habitats. Mudskippers have an elaborate courtship routine and nuptial parade (Brillet 1970) which lead the female to the male's burrow where mating occurs (figure 68). However, whether sympatric species perform slightly different courtship routines has not been investigated.

Some species exhibit behaviour after mating that ensures that the larvae are released into the sea and thereby dispersed. For example, the crabs *Sesarma erythrodactyla* and *Paragrapsus laevis* migrate to the lower zones of the mangroves to spawn (Yates 1978). The mud crab, *Scylla serrata*, undertakes extensive migration during its life cycle. After mating in the summer, the females move offshore from

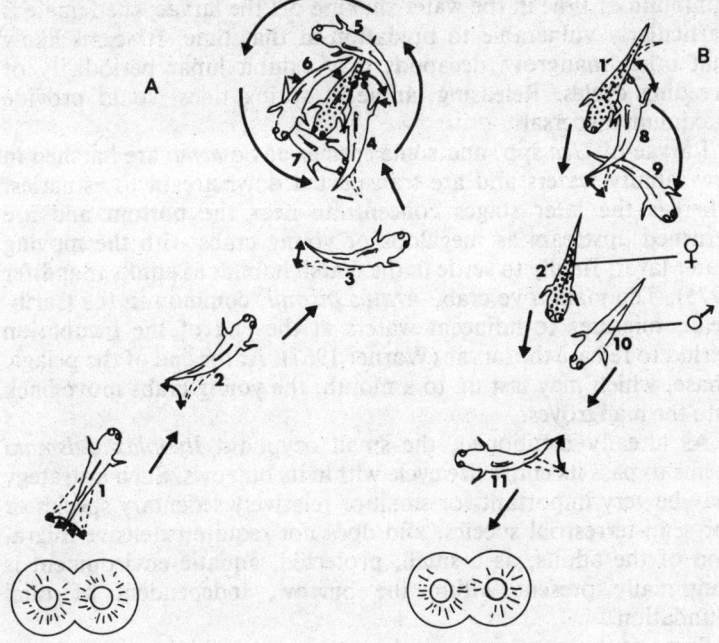

Figure 68 Nuptial dance of *Periophthalmes* (after Brillet 1970).

the shallow coastal water to areas up to 30 kilometres from land and down to 300 metres in depth. As in all decapods, a considerable number of eggs are spawned and then carried on the underside of the abdomen until they hatch and are released into the plankton. In *S. serrata* there are five larval stages, consisting of four zoeae and one megalops stage; these last for about four to six weeks. The megalops stage is the one which moves back into the estuaries and mangrove swamps and metamorphoses into juvenile *S. serrata*. The migration by the female may assist in the dispersal of the larvae, and may be related to differential predation pressures between offshore deep waters and the estuarine ecosystem (Hill 1982; Hyland, Hill and Lee 1984). The majority of marine decapods, including those species found in swamps, have planktonic larval stages.

The mangrove tree-climbing sesarmid of North and South America, *Aratus pisonii* synchronizes breeding and migration to the sea with the lunar cycle (Warner 1967). Eggs develop attached to the pleopods of the female and hatching is synchronized with new or full moon and hence spring tides. When the eggs are about to hatch, the females migrate to the lower seaward zones of the mangroves and release the newly hatched larvae into the water. All the eggs hatch simultaneously which enables the female to spend a minimum of time in the water shaking off the larvae; the female is particularly vulnerable to predation at that time. It seems likely that other mangrove decapods may exhibit lunar periodicity of breeding cycles. Releasing larvae at spring tides would provide maximum dispersal.

Larvae of *Uca* spp. and some species of *Sesarma* are hatched in low-salinity waters and are transported downstream to estuaries, whereas the later stages concentrate near the bottom and are returned upstream as megalops or young crabs with the moving water layer, finally to settle in the marsh habitat as adults (Sandifer 1975). The mangrove crab, *Aratus pisonii*, common in the Caribbean, migrates to adjacent waters at the end of the incubation period to release the larvae (Warner 1967). At the end of the pelagic phase, which may last up to a month, the young crabs move back into the mangroves.

As already mentioned, the small ocypodid *Ilyoplax delsmani* seems to pass its entire life cycle within its burrows. Such a strategy may be very important for small or relatively sedentary species or for semi-terrestrial species, and does not require extensive migration of the adults, as a small, protected, aquatic environment is continually present within the burrow, independent of tidal inundation.

Some of the mangrove polychaete worms exhibit brooding — for example, the spirorbids and the spionids — but these families occur in a wide variety of habitats apart from mangroves; in these cases brooding cannot be considered a special adaptation to life in mangroves. *Ceratonereis aequisetis*, a common estuarine and mangrove nereidid in New South Wales and Queensland, is often found with juvenile worms in the adult tube, suggesting that the worm gives birth to miniature worms, eliminating the planktonic larval stage (Hutchings and Glasby, 1985). The breeding strategies of most estuarine and mangrove polychaetes have not been investigated.

Like the decapods, some of the molluscs produce a free-swimming larval stage. This is true for the majority of tropical gastropods and bivalves, and presumably also occurs in mangrove species. However, the larval stage (veliger) has been suppressed in ellobiids such as *Ophicardelus*, *Cassidula*, *Ellobium* and *Pythia* (Macnae 1968). The life history of a member of the *Littorina scabra* complex (currently being described by Reid 1986), a very common mollusc in temperate and tropical mangrove areas, has recently been studied by Muggeridge in the Sydney region (1979). This particular population spawns for two to three days around the new or full moon from November to April. This ensures that the release of veligers coincides with the maximum tidal heights for each lunar cycle. The eggs develop to the veliger stage in the mantle cavity of the female which acts as a brood pouch. Each female can breed several times during the breeding season and may spawn monthly or bi-monthly. About twelve veligers are released at each spawning. In contrast, another littorinid, *Bembicium auratum*, spawns throughout the year with peaks from August to January. The spawn is laid in bean-shaped jelly masses in microhabitats protected from desiccation (Muggeridge 1979).

Pulmonate snails are hermaphroditic, several of them alternately producing spermatozoa and eggs with changes in the cycles of the moon or tide. Berry (1958), working on a Malaysian mangrove ellobiid snail, *Cassidula auris-felis*, found that like many other molluscs it exhibits changes of sex during the course of the life cycle. Typically, the majority of snails had developing male gametes, but few oocytes. Abundant oocytes occurred only in a substantial proportion of the population after periods of more than eight days without tidal inundation. In addition, more snails had abundant oocytes in the lunar "weeks" following new and full moon than after half moons. Following the development of oocytes, rates of copulation increased and gametogenesis in young snails began. Berry speculated on why continuous exposure may stimulate the production of oocytes. It may affect oviposition, or possibly the microflora on the organic material of the mangrove mud builds up during exposure but is adversely affected by tidal cover. Since *Cassidula* is a deposit feeder, such a mechanism might control the quality or quantity of the snails' food supply and hence the production of oocytes. Little is known about the breeding of other pulmonate snails, but most have internal fertilization achieved by copulation and the majority do not develop through the typical planktonic veliger; instead they develop directly into miniature snails within gelatinous egg masses or, in a few cases, viviparously within the female snail (Berry 1972). These features of reproduction are of advantage to animals which are isolated from the open sea by considerable distances for many days at a time.

Some work has been done on the dynamics of fouling communities on mangrove roots (Saenger, Stephenson and Moverley 1979). Sutherland (1980), working in Venezuela, found no seasonality in the recruitment of such encrusting organisms; rather, recruitment occurred throughout the year in low numbers. The main feature of the community was that species composition differed among roots. He attributed these differences to chance resulting from the low recruitment rate of most species and the low rate of supply of new roots. Once present on a given root, species seem to live for many years and resist the invasion of most other larvae. The initial differences in species composition are maintained over time. Salinity fluctuations were suggested by Sandison and Hill (1966) as the major factor limiting distribution of encrusting fauna in Nigerian mangroves. The Sydney Rock oyster, *Saccostrea commercialis*, can survive for higher temperatures when it is covered with a fine layer of mud or is in the shade (figures 69 and 70).

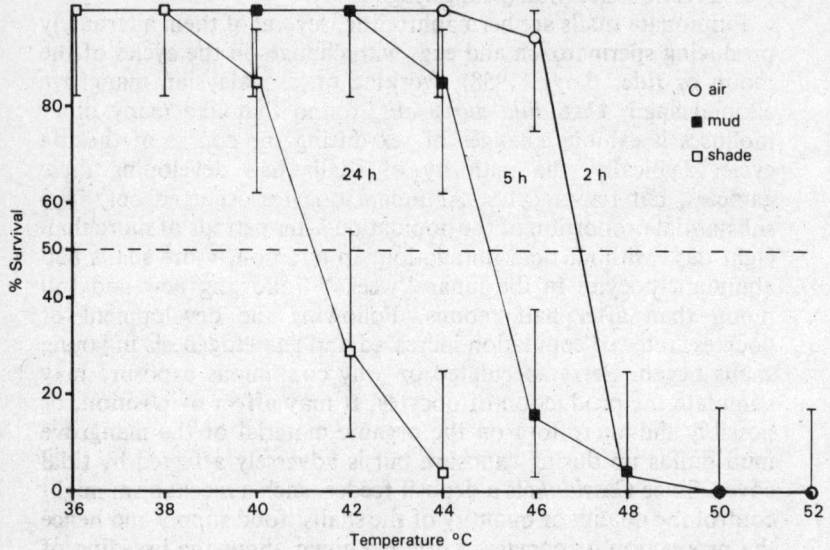

Figure 69 Survival of Sydney rock oyster *Saccostrea commercialis* exposed to high temperatures in heated sea water for periods of 24, 5 and 2 hours (n = 20 at each point; vertical lines indicate 95% confidence units) (after Potter and Hill 1982).

The sipunculan *Phascolosoma arcuatum* spawns in the summer from December to January in the Brisbane region. The gametes develop in the body cavity (coelom) over a period of five to seven months. This is slightly faster than in the temperate species *Golfingia vulgaris* and *Phascolosoma agassizii* (Green 1975). Green suggested that the spawning period of *P. arcuatum* varies within its geographical range, and that spawning often coincides with flood conditions, dispersing the larvae widely.

Physiological Adaptations

Nutrition

Most of the standing crop and primary production in a mangrove is in the form of angiosperm tissue (Golley, Odum and Wilson 1962). This can serve as food for the mangrove fauna in several ways:

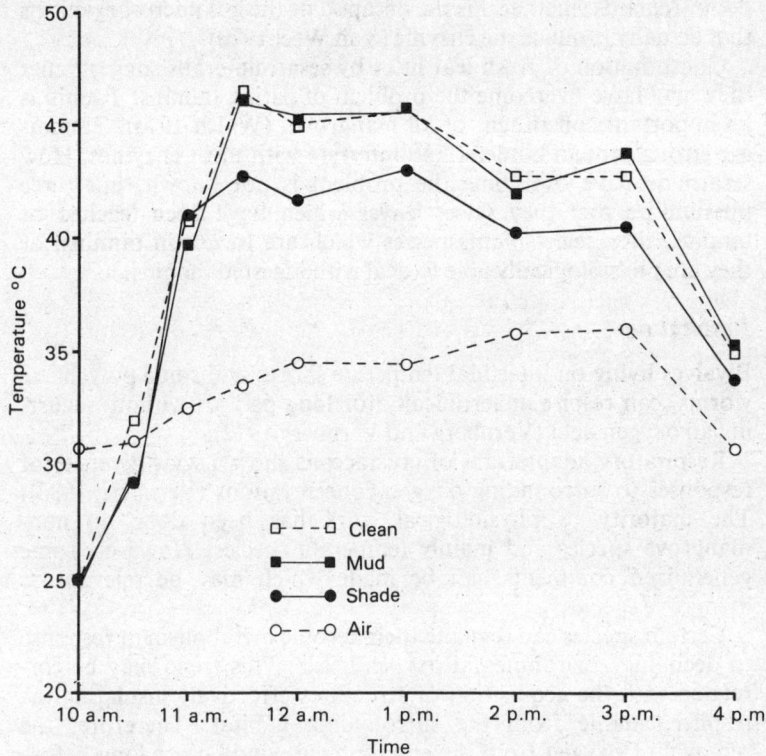

Figure 70 Tissue temperature of Sydney rock oysters (*Saccostrea commercialis*) exposed directly to sunlight in air, starting at 10 a.m. and terminating at 4 p.m. Temperatures of oysters covered with a fine layer of mud, shaded oysters, cleaned oysters and air temperature are shown (n = 5 at each point) (after Potter and Hill 1982).

directly by grazing, by consumption of freshly fallen litter or after its decomposition to detritus (Malley 1977). Two sesarmid crabs, *Chiromanthes onychophorum* and *C. dussumieri*, occurring in Malaysian mangrove swamps feed almost entirely on fallen mangrove leaves. However, it is not known whether they are obtaining nutrition from the leaves or from the microflora associated with the decomposing leaves (Malley 1977). Some Australian sesarmids pick up recently fallen leaves and take them into their burrows, suggesting that they are herbivorous or at least omnivorous. No detailed physiological studies have been undertaken on mangrove herbivores or detritivores to determine how they obtain their nutrients. There have been reports that the enzyme cellulase is secreted by the hepatopancreas of some species of decapods, but doubt remains whether it is the decapod or the gut micro-organisms that actually produce the enzyme (Van Weel 1970).

Consumption of fresh leaf litter by sesarmid crabs suggests that they may have overcome the problem of eating tannins. Tannin is an important constituent of all mangroves (Walsh 1974). Tannins are strong protein binders, and interfere with most enzymes. How sesarmids have overcome the problem is not known, but three possibilities are: they select leaves which have been leached of tannins, they select plant species which are lowest in tannins, or they are physiologically able to deal with ingested tannins.

Respiration

Bivalves living on intertidal temperate shores and some polychaete worms, can respire anaerobically for long periods without incurring an oxygen debt (Vernberg and Vernberg 1972).

Respiratory adaptations of crustaceans show a wide diversity of responses to surrounding oxygen concentrations (Vernberg 1983). The majority of physiological work has been done on non-mangrove species and mainly temperate species. However, some generalized comments can be made which may be relevant to mangrove species.

Certain species can regulate their aerobic metabolism in response to declining environmental oxygen levels. This trend may be correlated with the acquisition of structures effectively insulating the respiring tissue from the surrounding habitat. Therefore, the removal of oxygen from the environment would cease long before the available supply of oxygen was exhausted. Such species appear to have long-lasting internal oxygen reservoirs such as gas bubbles, or pools of respiratory pigments with high oxygen affinity. Species lacking these oxygen stores may switch to anaerobic pathways (Mangum and Van Winkle 1973).

The fiddler crabs *Uca pugilator* and *U. pugnax* live in burrows in muddy intertidal temperate shores, and as they do not pump water through the burrows even when the burrows are inundated, they may experience low oxygen tensions. These crabs appear relatively insensitive to anoxia and their critical oxygen tension is low. They can continue to use oxygen down to very low levels compared with non-burrowing species (Teal and Carey 1967a), and can survive anoxic conditions in the laboratory for 32–138 hours. Following return to oxygenated conditions, oxygen consumption is elevated above pre-anoxic conditions, suggesting that an oxygen debt had developed (Thompson and Pritchard 1969). These species produce high levels of lactate but also appear to produce CO_2, which suggests that metabolic pathways beyond glycolytic fermentation are being used for anaerobosis (Cousens 1974). Typically, the oxygen consumption rate of these animals is highest during that phase of the tidal cycle when greatest locomotory activity occurs (fiddler crabs are relatively inactive in their burrows when covered by sea water).

It seems likely that two distinct mechanisms may exist in mangrove animals for coping with low O_2 levels: one in which the metabolic rate is reduced and no O_2 debt is incurred and the other in which an O_2 debt is incurred. In the latter situation the animals must return frequently to oxygenated conditions to eliminate the debt. The intertidal crab *Carcinus maenas* switches from aerobic to anaerobic metabolism (as indicated by increased lactate production) (Theede 1973; Breteler 1975; Spaargaren 1977). It regulates oxygen uptake down to pressures of about 20 mm Hg. As oxygen pressure increases, the crab can maintain its gill ventilation rate and heart beat until O_2 tension reaches 60–80 mm, and then they decrease. The crab can survive anoxia for about two days, after which irreversible damage occurs. Although there is a reduction in energy production during anoxia, ionic regulation can be maintained through anaerobic pathways. Under these conditions the crab moves very slowly and is therefore very susceptible to predation.

Osmoregulation

The osmoregulation of relatively few mangrove species has been studied in detail. However, it is likely that mechanisms which operate for other intertidal species also apply to those from mangroves. Consequently, this section draws heavily upon studies of intertidal organisms generally.

Animals living in mangrove swamps experience fluctuations in salinity during a tidal cycle. Salinities may exceed that of sea water on a hot day during low tide, and may be fresh during rain or flood conditions. In tropical mangroves, in northern Queensland and the Northern Territory, flood conditions may last for several weeks during the wet season. In contrast, in South Australia and probably in northwestern Western Australia, where there is little or no freshwater run-off into the mangroves, very high ground salinities may be experienced. How do the animals cope with these fluctuations? Deep burrowing by some species, such as bivalves and sipunculans, perhaps avoids saline conditions. Green (1975) suspected that at a depth of 25 cm little fluctuation in salinity occurs, and Saenger (unpubl. data) found little fluctuation in salinity at these depths in Queensland clayey mangrove soils.

Bivalves may simply close their valves tightly when salinities are unfavourable. Obviously, under such conditions no feeding can occur, and this response of necessity must be short term. Rayner (1979) studied the salinity ranges tolerated by a variety of teredinids found in New Guinea. She found that salinity tolerances of adults and possibly of presettlement larvae were not the most important factors limiting their distribution in an estuary.

During heavy flooding, a considerable percentage of the mangrove fauna may be killed or washed out of the swamp into the estuary. McCormick (1978) suggested that canopy faunas were depauperate in certain New South Wales mangrove swamps because of recent flooding. Salinity fluctuations and associated flooding may not completely eliminate a population but rather considerably reduce its size.

A great deal of work has been done on the osmoregulatory mechanism employed by mangrove crustaceans. There is a tendency for the permeability of crab shells to decrease in estuarine and terrestrial species in contrast with fully aquatic species (figure 71).

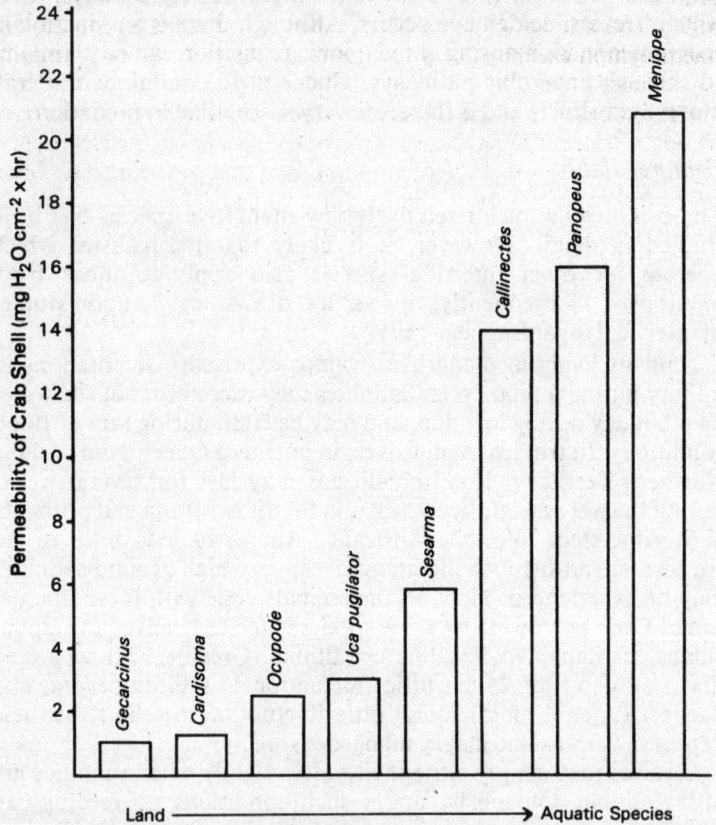

Figure 71 Permeability of crab shells to water in air. Land crabs are on the left, aquatic species on the right. An average of six determinations were made for each species. The average standard deviation among the measurements for each species was ± 30% of mean value (from Vernberg and Vernberg 1972).

Mantel and Farmer (1983) summarized all available data on estuarine species, including some mangrove species; a brief synopsis is given below.

Barnacles, which are sessile organisms, cannot avoid changes in salinity by moving elsewhere as do many other crustaceans. However, when exposed to high or low salinities they can shut their valves and remain inactive for several hours. If prolonged exposure to high or low salinities occurs, the valves remain shut for several days; when they are open, both the haemolymph (blood) and mantle fluid are isosmotic to the new external salinity. The exact way in which this is achieved is still unknown (Foster 1970; Fyhn 1976).

Mangrove species of gammarid amphipods have not been studied, but estuarine species in general are hyperosmotic in dilute salinities and become isosmotic with the surrounding sea water at higher salinities. The urine produced is normally isosmotic to the haemolymph except in very low salinities when a more dilute urine is produced.

Regulators are those species which maintain their internal osmotic concentration relatively constant, either higher (hyperregulators) or lower (hyporegulators) than that of the medium over part or all of their range (figure 72). Intertidal isopods are hyper- and hyporegulators but the mechanisms for regulation have not been studied (Segal and Burbanck 1963). An interesting study on *Sphaeroma* showed that, as the population "ages" over a year, individuals regulate less strongly at both low and high salinities; this is thought to be a sign of senescence (Charmantier 1973; Charmantier and Trilles 1973). Whether these changes coincide with population migrations of *Sphaeroma* is unknown.

The decapods are in general long-lived, mobile animals, able to tolerate some changes in salinity. Many undertake daily, seasonal or yearly migrations related to feeding or reproduction, which may expose them to large changes in salinity. This group of crustaceans have evolved many types of mechanisms to cope with changes in salinity. Also, species less tolerant may avoid stress behaviourally by burrowing.

Snelling (1959) studied the distribution of intertidal crabs in the Brisbane River and found that during drought conditions marked changes in the zonation of crabs occurred. However, a freshwater flood of short duration subsequently did not alter the zonation; thus, some crabs can withstand a sudden but short decrease in salinity. Barnes (1967) continued this study by investigating the osmoregulatory abilities of one component of the fauna, the grapsid crabs. All these crabs are capable of some degree of hypo-osmoregulation. He found that their capabilities for osmoregulation do not always correspond to the environmental conditions in

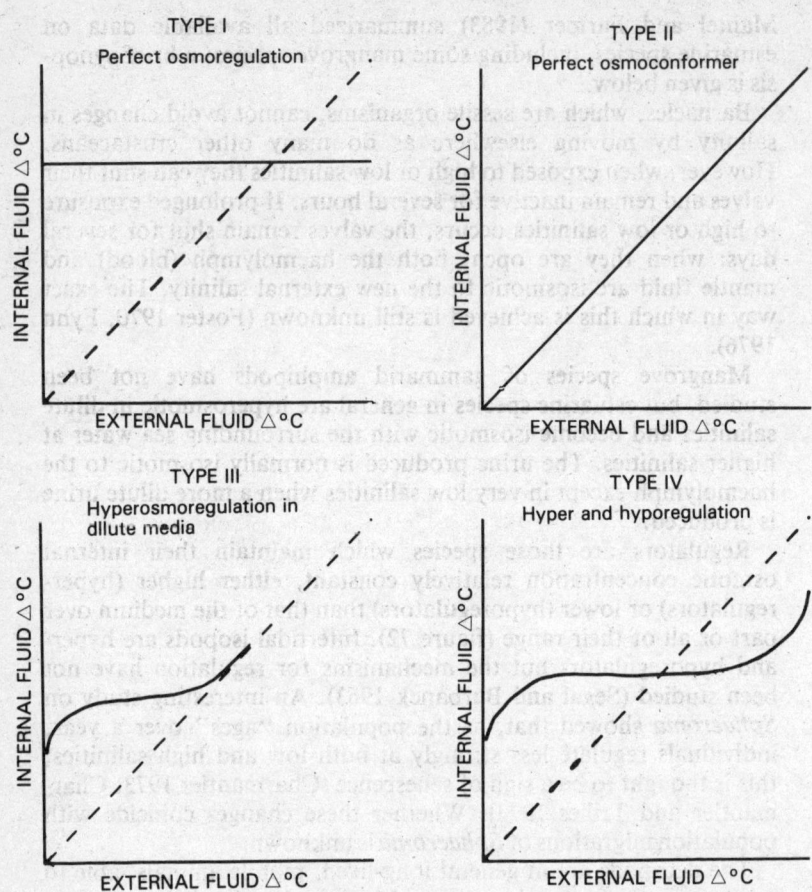

Figure 72 A generalized representation of the comparative osmoregulatory response of marine, estuarine, freshwater and terrestrial animals (after Vernberg and Vernberg 1972).

which they are found in the field, suggesting that these species have tolerance ranges which exceed the usual range of fluctuation experienced in their immediate environment. However, if the grapsid crabs are listed in order of decreasing ability to hyperosmoregulate (*Paracleistostoma mcneilli*, *Australoplax tridentata*, *Mictyris longicarpus*, *Macrophthalmus setosus* and *M. crassipes*) (figures 73 and 74), it corresponds with their ability to penetrate up the Brisbane River; *Paracleistostoma mcneilli* occurs furthest upstream. The one exception is *M. longicarpus* and Barnes (1967) suggested that preference for sandy beaches, rather than salinities, restricts this species to the lower reaches of the river (figure 75).

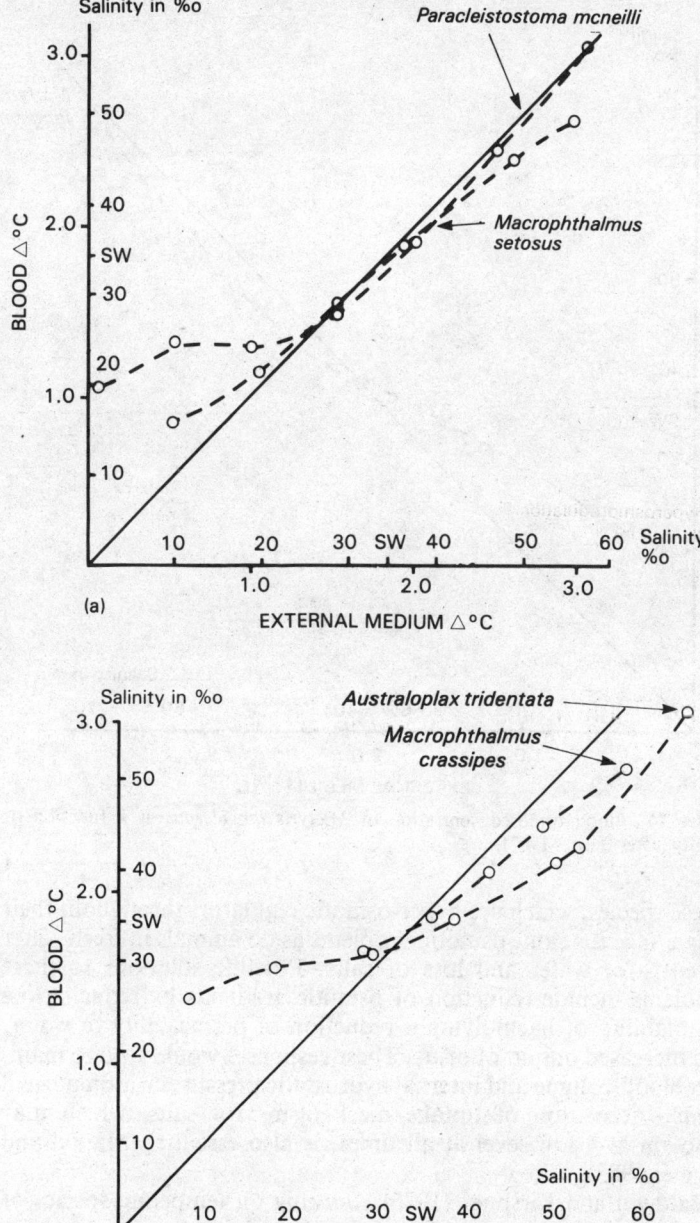

Figure 73 Blood osmo-concentration as a function of salinity of (a) *Macrophthalmus setosus* and *Paracleistostoma mcneilli* and (b) *Australoplax tridentata* and *Macrophthalmus crassipes* (after Barnes 1967).

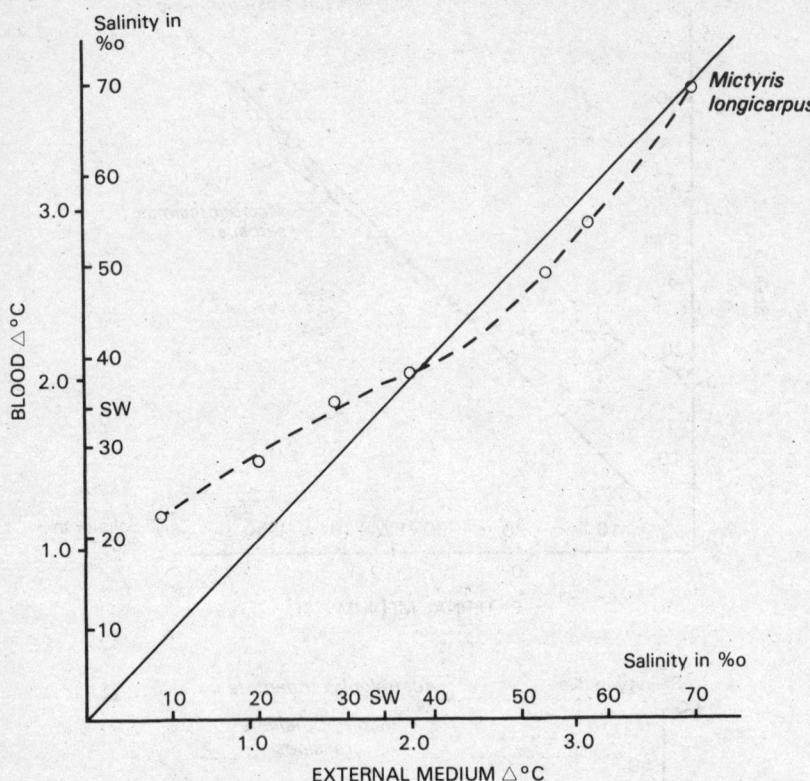

Figure 74 Blood osmo-concentration of *Mictyris longicarpus* as a function of salinity (after Barnes 1967).

These species, which are hyper-osmotic regulators throughout their range, face the same osmotic problems as do animals in fresh water — entry of water and loss of salts. Possible solutions to these problems include reduction of osmotic gradients by reducing the osmolability of haemolymph, reduction of permeability to water, and increased output of urine. These responses would tend to maintain blood volume and internal hydrostatic pressures within normal limits. Activation of uptake mechanisms for salts, which may function at a low level at all times, is also essential (Mantel and Farmer 1983).

Baldwin and Kirshner (1976), working on temperate species of *Uca* on the western coast of America, found that these crabs maintain the ionic concentration of their body fluid below that of sea water. They suggested that this is beneficial because the crabs do not experience such large or frequent changes in blood ionic concentrations when there are changes in dilution and concentrations

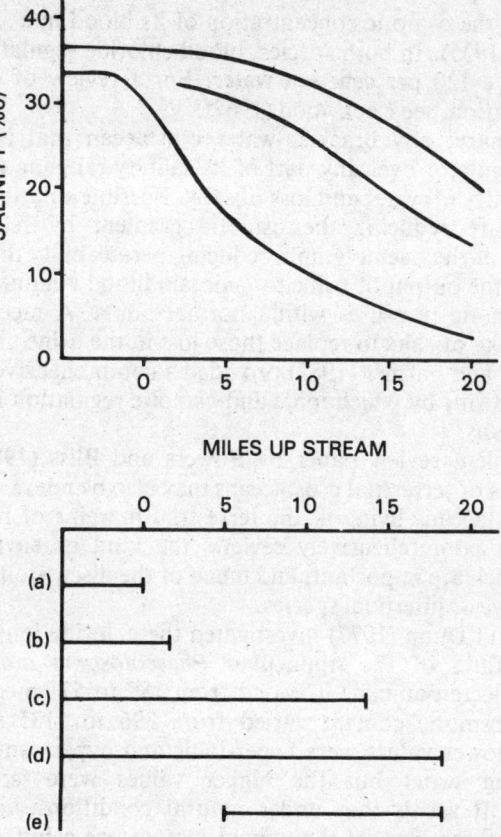

Figure 75 The approximate distribution of crabs along the Brisbane River, in relation to salinity and miles upstream: (a) *Macrophthalmus crassipes*; (b) *Mictyris longicarpus*; (c) *Macrophthalmus setosus*; (d) *Australoplax tridentata*; (e) *Paracleistostoma mcneilli* (after Barnes 1967).

of the water. Maintenance of a more uniform internal ionic environment, however, must have an energetic cost.

The mangrove crab, *Goniopsis cruentata*, which occurs in the mangroves of Venezuela, is a brackish-water species with limited powers of ionic and osmotic regulation. The crab maintains its blood chloride concentration at an almost constant level when exposed to external concentration ranging from 20 per cent to 120 per cent that of sea water. However, this crab cannot survive in fresh water. It maintains its blood concentration by regulating the amount and concentration of urine produced by the antennary glands; water is conserved by an excretion of unwanted ions (Zanders 1978). The Australian crab *Heloecius cordiformis* appears to regulate the osmotic concentration of its blood in a similar way (Edmonds 1935). In both species, blood chloride regulation breaks down above 120 per cent sea water. For a review of crustacean osmoregulation, see Lockwood (1967).

In summary, any brackish-water crustacean that is a hyperosmotic regulator over any part of its salinity range faces the problems of entry of water and loss of salts. Possible solutions to these problems are reducing the osmotic gradient by reducing the osmolality of the haemolymph, reducing permeability to water and increasing the output of urine to maintain blood volume and internal hydrostatic pressures within normal limits. A mechanism of active uptake of salts to replace those lost in the urine also is needed. Mantel and Farmer (1983) provided a comprehensive review of the mechanisms by which ionic and osmotic regulation is achieved in crustaceans.

An excellent review paper by Powers and Bliss (1983) on the adaptations of terrestrial crustaceans may also be relevant for some of the crustaceans living at the terrestrial margins of mangroves. This paper comprehensively reviews the kind of environmental factors which are important, and much of the discussion is relevant to mangrove or intertidal species.

Green and Dunn (1977) investigated the chloride ion content of coelomic fluid of the sipunculan *Phascolosoma arcuatum*. In freshly collected animals it varied from 189 to 571 meg/l Cl$^-$ and the total osmotic content varied from 396 to 1135 mOsm per litre. The lower values were hyper-ionic and hyper-osmotic to the surrounding water but the higher values were isoionic and isosmotic. It seems that under natural conditions, mud in the worm's gut and around the animal may act as a buffer, thereby allowing the animal to maintain an ionic and osmotic state differing from that of the general environment. Such a mechanism is clearly a successful adaptation for a species that is found throughout the mangroves and over a wide geographical area where it is subjected to extremely variable and fluctuating salinities.

It is also possible that a widespread species such as *Phascolosoma arcuatum* may occur as different physiological races, with the races in northern Queensland that are subjected to regular freshwater flooding having different osmoregulatory capacities from those occurring in subtropical areas where flooding is less common. Whether mangrove faunas show any physiological capacity to withstand desiccation or just avoid desiccation by behavioural and morphological adaptations is unknown. The nature of the excretory products produced by most of the mangrove fauna is also unknown. Much work remains to be done on the adaptations exhibited by the fauna in this fluctuating environment.

7. Productivity of Mangrove Ecosystems

Mangroves contribute significantly to estuarine and inshore productivity, although this has been recognized only since the late 1960s with Heald's (1969) pioneering work on Florida mangroves. Odum (1971) and Odum and Heald (1975) and others have shown in broad terms how production is linked to estuarine food chains via an energy flow based on detritus (see chapter 8). Most of the data on primary production have been collected overseas predominantly from southern Florida where only four species of mangroves occur, dominated by *Avicennia germinans* and *Rhizophora mangle*. These overseas studies form the basis of this chapter, so care must be taken in extrapolating to Australia where a greater number of species of mangroves occur over a much wider geographical area that differs markedly in structure, climate, geomorphology and hydrology.

In order to provide a context for the following discussion, primary production of a variety of terrestrial and marine communities, based on a review by Larkum (1981) is presented in table 38.

Definition of Primary Production

Net primary production (NPP) has been defined by Whittaker, Likens and Lieth (1975) as "that part of the total or gross primary productivity of photosynthetic plants that remains after some of this material is used in the respiration of these plants". The respiratory component of net primary production is thus that of the whole plant, including leaves, stems and roots. This can be summarized as the total accumulation of new organic matter in plant tissues in excess of respiration per unit area per unit time (usually measured as dry weight of organic matter). Although the main contributors to primary production in the mangrove ecosystem are the mangroves themselves, epiphytic plants such as algae on the pneumatophores, bases of trees and mud surfaces also contribute to this productivity. Productivity can be thought of as the recombination of mineral elements into organic matter formed as a result of photosynthesis and associated metabolic activities. The degradation of organic matter is accompanied by a release of elements

Table 38 Primary productivity estimates for plant communities in Botany Bay, New South Wales

	Annual production (t ha^{-1} yr^{-1})	% area	Area (ha)	Total primary production (t yr^{-1})	% cont.
Posidonia australis	7	7.22	500	3,500	
Zostera capricorni	1.8	4.11	284	512	25.1
Epiphytic algae	1.2	11.34	784	941	
Mangroves	15	5.78	400	6,000	30.4
Salt marsh	6	2.17	150	1,200	6.1
Benthic algae	3	2.89	200	600	3.0
Phytoplankton	1.5	66.49	4,600	7,000	35.4
			6,918	19,753	100.00

Source: Larkum (1981).

which can be used again. These two processes (recombination and degradation) in the cycling of matter can be evaluated by determining the rates of fixation and release of carbon (the basic building block) in terms of production and respiration.

Methods of Measurement and Results

Measurement of total net primary productivity is extremely difficult. Instead, most workers have attempted to measure some feature related to primary productivity which can then be used as a basis for making comparisons of productivity.

Biomass

Biomass is one such attribute. It is the total weight of organic matter in the community, or of particular species in the community. It is not a direct measurement of productivity, and indeed may be a very poor indicator of productivity. Productivity is the rate at which matter is produced whereas biomass is the amount present at any one time. For a given biomass, productivity can be high (if turnover is rapid) or very low (if turnover is low). For valid comparisons, rates of biomass turnover must be taken into account. Nevertheless, biomass has been used as a basis for making crude comparisons of productivity, and for evaluating worldwide productivity patterns (Clough and Attiwill 1982).

It has been suggested that biomass estimates tend to underestimate the impact of animals on the mangrove system, and certainly none of the figures discussed below include an animal component. Lugo and Snedaker (1974) summarized existing data on biomass (table 39) from Florida, Panama, Puerto Rico and the Philippines, and more recently from southern Thailand. The majority of studies have concentrated on above-ground biomass, and in most cases the values appear to be very similar, ranging from 5.4 to 18.4 kg m^{-2}. The exceptions are mangroves from Panama and Florida scrub and successional mangrove communities from Florida where the following respective values have been recorded: 28.3, 0.9 and 1.0 kg m^{-2}. Recently, in Malaysia, standing biomass (above ground) has been measured for areas of mangrove forest of known age. The trees in the study area are logged on a thirty-year cycle and thus it is possible to determine the age of the different groups and compartments (table 40). Biomass does not increase uniformly with time; initially growth is very rapid and then gradually declines. The only figures available from Australia are

Table 39 Estimates of biomass for non-Australian mangroves. The data from Panama, Puerto Rico, the Philippines and Florida are from Lugo and Snedaker (1974) who cite the original sources; data from Thailand are from Christensen (1978). Estimates are expressed as kg m^{-2}.

Locality/Type	Leaf	Above-ground	Total above-ground biomass	Below ground	Total (where all components measured)
Panama	0.4	27.9	28.3	19.0	47.3
Puerto Rico	0.8	6.3	7.1	5.0	12.1
Philippines	1.3	4.6	5.9	ne	
Florida overwash	0.7	13.0	13.7	ne	
	0.7	12.0	12.7	ne	
Florida riverine	0.4	9.8	10.2	ne	
	1.0	17.4	18.4	ne	
Florida fringe	0.6	8.6	9.2	ne	
	0.6	11.8	12.4	ne	
	0.7	15.3	16.0	ne	
Florida scrub	0.1	0.8	0.9	ne	
Florida island	0.5	4.9	5.4	0.8	6.2
Florida succession	0.2	0.8	1.0	1.4	2.4
Thailand	0.7	15.9	16.6	ne	

ne = not estimated.
Source: After Clough and Attiwill (1982).

Table 40 Figures from a study carried out by the Forest Research Institute, Kepong, and University Pertaman Malaysia, for BIOTROP (the Southeast Asian Ministers of Education Organization's Regional Center for Tropical Biology) unpublished report, June 1982

Age of stand	Biomass
5 years	8.91 kg m^{-2} yr^{-1}
18 years	17.5 kg m^{-2} yr^{-1} — 97% increase in 13 years
28 years	21.18 kg m^{-2} yr^{-1} — 20% increase in 10 years

from temperate monospecific stands of *Avicennia marina* (Briggs 1977a; Clough and Attiwill 1982; Field et al. 1983). The values (table 41) are of the same order of magnitude as those reported from other parts of the world.

Table 41 Estimates of biomass for mangroves in temperate Australia (data are expressed as kg m^{-2})

Compartment	Lane Cove, NSW, Lat. 33°50'S	Koorangang Is., Lat. 31°51'	Hunter River, Lat. 31°51'	Westernport Bay, Victoria, Lat. 38°S
Leaves	1.35			0.6
Branchwood		19.35	16.85	6.4
Trunk	11.5			2.0
Roots	15.35	1.61	0.08	14.6
Total	28.2	20.96*	16.93*	23.6

* The figures for roots include pneumatophores only. These figures are for the healthy, relatively undisturbed sites at Koorangang, and consist only of *Avicennia marina*.
Source: After Clough and Attiwill (1982), and from Field et al. cited in Moss (1983).

Values of biomass for *Avicennia marina* in the Lane Cove River (Briggs 1977a) are in the lower range of values reported for mature forests, but are higher than the usual values for woodlands, shrublands and grasslands (Whittaker, Likens and Lieth 1975).

Briggs (1977a) compared the biomass of *Avicennia* in the Lane Cove River with those of *Rhizophora mangle* communities studied by Golley, Odum and Wilson (1962). The total biomass of the *A. marina* community growing at 31°S latitude was 28.3 kg m^{-2}, more than twice that of *R. mangle* growing at 18°N, 11.28 kg m^{-2}. Part of this difference may be a latitudinal effect. However, it may also reflect differences in maturity of the two communities; the *A. marina* community was more mature than the *R. mangle* stand. Biomass of woodlands increases with maturity and this may be true of mangrove communities as well. This may be an exception; as can be seen from table 39, similar biomasses above ground have been reported from a range of mangrove communities occurring in different climates and localities, and which are of various ages and past histories. This is partly due to the more general observation that, in well-developed mangrove communities, tree size and tree density often tend to vary inversely with consequent variations in canopy (Clough and Attiwill 1982). Although above-ground biomass appears similar in a wide variety of mangrove habitats, the few figures available for biomass of roots and pneumatophores show considerable variation (see table 39). Root biomass often constitutes more than 20 per cent of the total biomass in many temperate forest species (Clough and Attiwill 1982).

Litter Production

Another method of assessing productivity is to measure a component of net primary production such as litter production and compare the results obtained from different systems. Goulter and Allaway (1979) measured the rates of production of leaf and other kinds of litter (seeds, flowers, twigs, and so on) in *Avicennia marina* woodland in Middle Harbour, Sydney. The values they obtained are rather high in comparison to other areas including tropical mangrove stands both in northern Australia and elsewhere (table 42). Briggs (1977a), also working in Sydney, estimated that if the trees in an *Avicennia* woodland replaced their leaves and petioles annually, litter production would be 1.2–1.5 kg m^{-2} yr^{-1}. This range is larger than those reported for *Rhizophora mangle* forests of 0.9 kg m^{-2} yr^{-1} (Heald and Odum 1970) and 0.5 kg m^{-2} yr^{-1} (Golley, Odum and Wilson 1962). Considerable variation occurs in maximum litter production recorded from Hinchinbrook, tropical Queensland, but the figure of 28.1 dry t ha^{-1} yr^{-1} is the highest figure so far recorded from any mangrove community. So both in Australian temperate and tropical mangroves a large amount of leaf litter is potentially available to the estuarine ecosystem.

Gas Exchange of Leaves

Several workers (Golley, Odum and Wilson 1962; Miller 1972; Lugo et al. 1975) measured rates of photosynthesis and respiration of individual leaves or small branches as an indication of primary production (figure 76). From these figures they calculated the net primary production of individual trees and subsequently the net primary production per unit area. Lugo et al. (1975) suggested that as much as 4–10 per cent of gross primary production may be lost through respiration of stems or surface roots. Many studies have not considered this loss. However, Lugo and his co-workers may still have underestimated the loss, as they based their figures on the gas exchange of surface roots (prop roots and pneumatophores) only and these may reflect only part of the loss from the entire root system.

The values of net primary production obtained by extrapolating from individual leaves or small branches are plausible in that they fall within the range of woody plants with the C_3 pathway of carbon fixation. In chapter 2, the evidence that C_4 metabolism is used (Joshi et al. 1975) is considered, but the evidence leans towards the more usual C_3 pathway.

Extrapolating from individual leaves presupposes that within a species similarly sized leaves and rates of productivity occur. This is

Table 42 Mangrove litter production at various localities in the world

Species	Locality and latitude	Litter production (dry t ha^{-1} yr^{-1})	References and comments
Rhizophora apiculata	Phuket Is., Thailand (8°N)	6.7	Christensen (1978); dead leaf production
R. mangle	Puerto Rico (18°N)	4.8	Golley, Odum and Wilson (1962); dry organic matter in litter
Mixed mangrove forest	Hinchinbrook Is., Qld (18°S)	3.7–28.1	Bunt (1978); litter fall
Mostly *R. mangle*	Southern Florida, USA (25°N)	8.8	Odum and Heald (1975); total litter production
R. mangle, Laguncularia racemosa and *Avicennia nitida* (syn. *germinans*)	Southern Florida, USA (26°N)	0.8–12.7	Pool, Lugo and Snedaker (1975); litter fall
Principally *R. mangle* and *L. racemosa*	Southern Florida, USA (26°N)	1.3–10.7	Teas (1976); dry organic matter in litter
A. nitida	Southern Florida, USA (26°N)	4.9	Lugo and Snedaker (1974); total leaf fall
A. nitida	Southern Florida, USA (26°N)	2.9	Teas (1976); dry organic matter in litter
A. marina	Roseville, NSW (34°S)	5.8	Goulter and Allaway (1979); total litter fall
A. marina	Westernport Bay, Vic. (38°S)	1.62	Clough and Attiwill (1982); total litter production
A. marina	Lane Cove, Sydney (33°50'S)	12–15	Briggs (1979); estimated from biomass
A. marina	Salt Pan Creek, Sydney ((33°51'S)	6.8	Love (1981)
A. marina	Botany Bay, Sydney (34°S)	7.0	Love (1981)
A. marina	Tuff Crater, Auckland (36°48'S)	3.65–8.10	Woodroffe (1982); total litter fall

Note: A more detailed table giving additional figures is provided by Woodroffe (1982).

not necessarily true. Lugo and Snedaker (1974) reported that leaf size is smaller in mangroves subjected to thermal stress (table 43).

Kolehmainen (1973) indicated that mangroves in areas of thermal loading (5°C above ambient) developed greater densities of

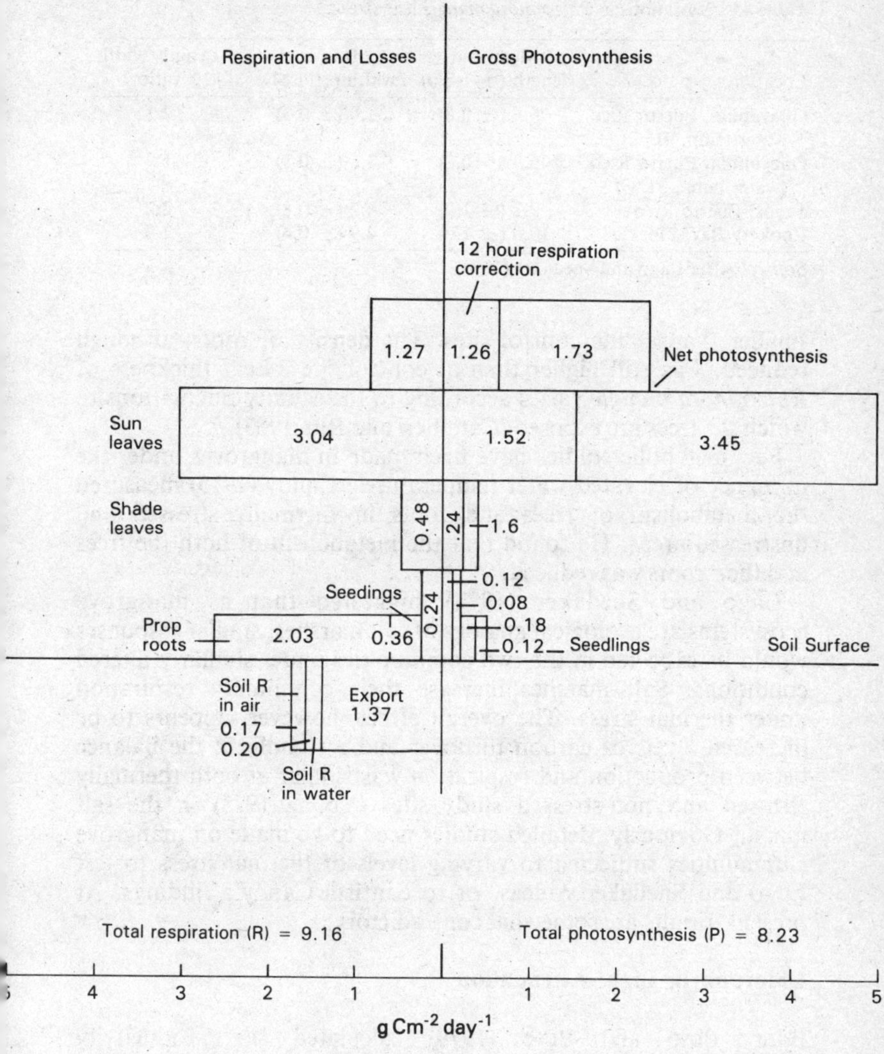

PRODUCTIVITY OF MANGROVE ECOSYSTEMS 251

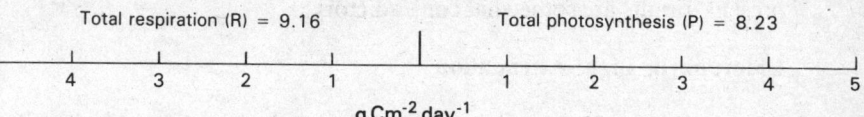

Figure 76 Rates of photosynthesis, respiration and export in the Puerto Rican mangrove forest components in g C m^{-2}) day^{-1} in May (after Golley, Odum and Wilson 1962).

prop roots. Canoy (1975) studied the mangroves along the southern coast of Puerto Rico which are subjected to warm water (30–40°C) from a power station which had been operating for seventeen years prior to the study. He found that, with some thermal stress, the size of the leaves and numbers of adventitious roots per square metre increased, but towards temperatures of 40°C the leaves were

Table 43 Variations in *Rhizophora mangle* leaf sizes

Location	Mean leaf length (± 1 SD)	Mean leaf width (± 1 SD)	Length/width ratio
Guayanilla, Puerto Rico water temp. 41°C)	6.4 (± 0.8)	3.0 (± 0.3)	2.1
Guayanilla, Puerto Rico (water temp. 31°C)	6.9 (± 0.8)	3.6 (± 0.3)	1.9
Jobos, Puerto Rico	9.2 (± 1.9)	4.2 (± 0.8)	2.2
Rookery Bay, Florida	10.0 (± 1.0)	4.9 (± 0.4)	2.0

Source: After Lugo and Snedaker (1974).

smaller than at the control sites. The density of roots, although reduced, was still higher than at control sites. Leaf thickness of *Rhizophora mangle* varies according to the salinity fluctuations to which the trees are exposed (Camilleri and Ribi 1983).

Few metabolic studies have been made in mangroves under the influence of elevated water temperatures. Canoy (1975) measured the metabolism of trees and roots in thermally stressed and unstressed areas. He found that the metabolism of both the trees and their roots was reduced.

Lugo and Snedaker (1974) suggested that as mangrove ecosystems are ecological analogs of salt marshes, similar responses would be expected in the two communities under similarly altered conditions. Salt marshes increase their community respiration under thermal stress. The overall effect, however, appears to be increased levels of carbon turnover and metabolism; the balance between production and respiration was similar at both thermally stressed and non-stressed study sites (Young 1973) in the salt marsh. Obviously, detailed studies need to be made on mangrove communities subjected to varying levels of thermal stress to test Lugo and Snedaker's ideas, or to confirm Canoy's findings. At present, results are somewhat contradictory.

Chlorophyll; Light Attenuation

Bunt, Boto and Boto (1979) calculated the productivity of a mangrove forest in northern Queensland by measuring the light attenuation through the forest canopy and assuming that this was attributable to photosynthetic utilization. This assumption was confirmed by assays of leaf pigments. Productivity was calculated by assuming similar rates of assimilation to those published in the literature. Production estimates ranged between 16 and 26 kg ha^{-1} day^{-1}. Their estimates of photosynthetic production were slightly higher than accounted for by litter-fall production but obviously some of the photosynthetic production is used in root and trunk growth which was not included in the litter data; thus, the two methods generally agree. Bunt, Boto and Boto (1979) believed their method to be of value in obtaining estimates of primary production for a wide range of mangroves found in northern Queensland, and that differences owing to varying environmental conditions, such as rainfall, freshwater inflow, soil nutrients and human impact, could be compared.

Utilization of All Components

Christensen (1978) used several components of primary productivity in his calculations for *Rhizophora apiculata* in southern Thailand (table 44). He collected all the biomass above ground at various 1-metre horizontal levels, and measured the rate of leaf production. Combining these data with the annual increment in the form of trunks, branches, and prop roots, he calculated that the total net production was 27 t (dry matter) ha^{-1} yr^{-1} or 6.9 g ash-free dry matter m^{-2} day^{-1} (figure 77).

In table 44 the breakdown of the individual components is shown and leaves constitute less than 25 per cent of the total annual productivity. However, leaves have a much higher rate of turnover than trunks or roots so that the most continuous source of plant matter for the rest of the mangrove community must be the leaves. Christensen (1978) calculated that the average period of life for leaves of *Rhizophora apiculata* was about eighteen months. Leaves also degrade rapidly in contrast to trunks and prop roots which may take up to several years to break down completely.

Factors Influencing Primary Production

The environmental factors affecting mangroves, including their productivity, have been discussed in detail in chapter 3 and will not be repeated here. It should be reiterated, however, that many variables are involved. These can be divided into external factors and inherent factors and may include physical factors (such as light, temperature), spatial effects (such as latitude) and temporal effects (such as season and inter-year differences). Characters of the canopy such as leaf area index, leaf inclination and the distribution of leaves within the canopy have a marked effect on primary productivity by modifying leaf temperature and boundary layers. These factors are not unique to mangroves but occur in forests generally. In figure 78, the vertical distribution of leaf biomass, leaf area, chlorophyll *a* and light intensity is shown for a Puerto Rican mangrove system indicating the variation in productivity within the canopy (Golley, Odum and Wilson 1962).

Table 44 Net primary production of *Rhizophora apiculata* in Phuket estimated from observations of marked shoots and biomass values

Component	Dry matter (g m^{-2} yr^{-1})	% Contribution	Ash free dry matter (g m^{-2} yr^{-1})	% Contribution
Leaves, bud, scales	670	24.77	600	23.70
Buds, flowers	18	0.66	16	0.63
Propagules	16	0.59	15	0.59
Trunks, branches, prop roots	2,000	73.96	1,900	75.06
Total	2,704		2,531	
Total in t ha^{-1} yr^{-1}	27		25	
Total in g^{-2} day^{-1}			6.9	

Source: After Christensen (1978).

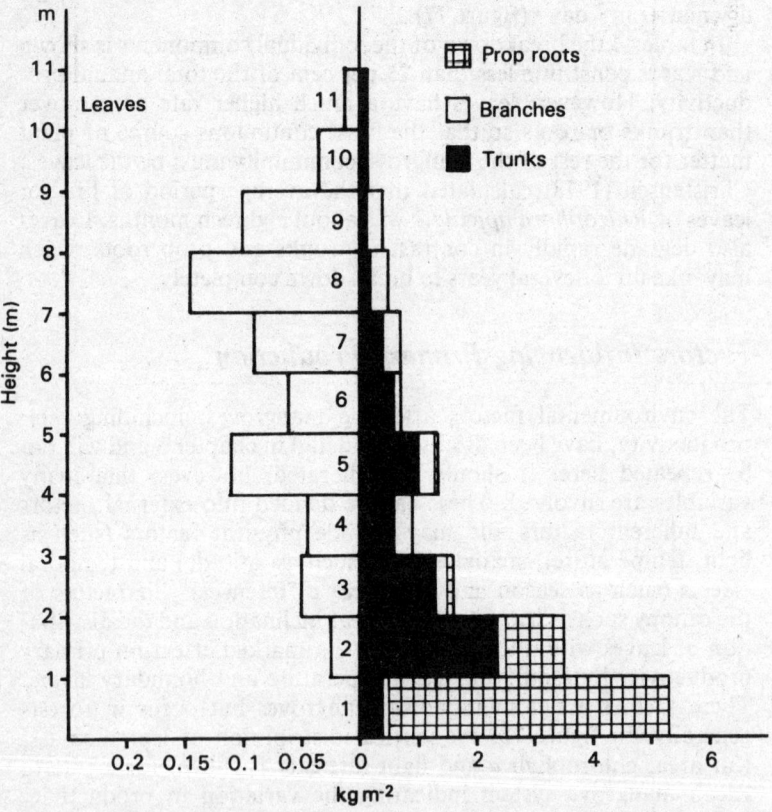

Figure 77 Vertical profile structure of the biomass of *Rhizophora apiculata* in Phuket, February 1976 (after Christensen 1978).

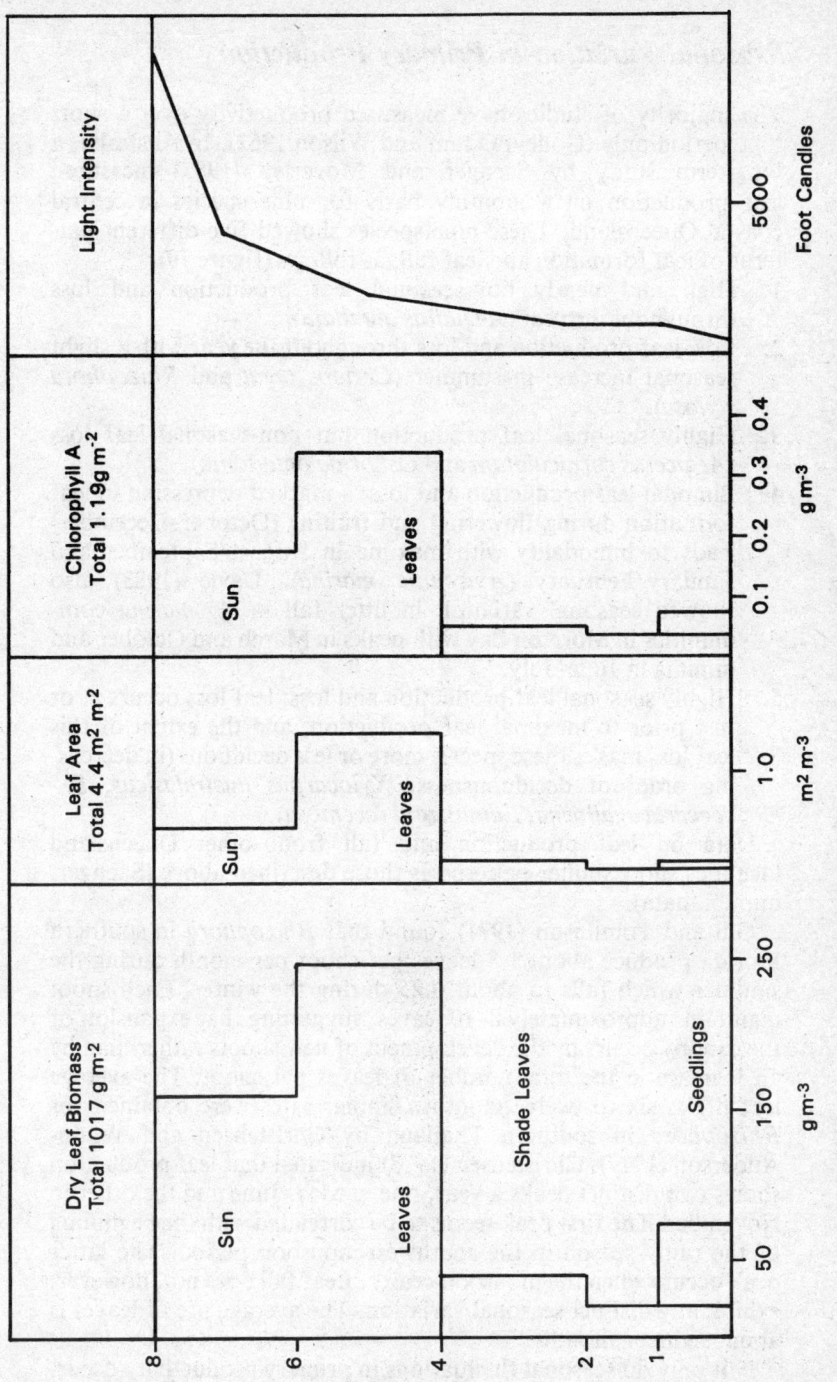

Figure 78 Vertical distribution of leaf biomass, leaf area, chlorophyll and light intensity in the red mangrove forest of Puerto Rico (after Golley, Odum and Wilson 1962).

Seasonal Variation in Primary Production

The majority of studies have measured productivity over a short time period only (Golley, Odum and Wilson 1962). In Australia, a long-term study by Saenger and Moverley (1985) measured leaf production on a monthly basis for nine species in central coastal Queensland. These nine species showed five different patterns of leaf formation and leaf fall, as follows (figure 79):

1. High and nearly non-seasonal leaf production and loss throughout the year (*Aegialitis annulata*).
2. Low leaf production and loss throughout the year with a slight seasonal increase in summer (*Ceriops tagal* and *Rhizophora stylosa*).
3. Highly seasonal leaf production but non-seasonal leaf loss (*Aegiceras corniculatum* and *Osbornia octodonta*).
4. Bimodal leaf production and loss; a marked depression in leaf formation during flowering and fruiting (October-December) leads to bimodality with maxima in August/September and January/February (*Avicennia marina*). Davie (1983) also showed seasonal variation in litter fall in *A. marina* communities in Moreton Bay with peaks in March and October and minima in June-July.
5. Highly seasonal leaf production and loss; leaf loss occurs at or just prior to maximal leaf production, and the extent of this leaf loss makes these species more or less deciduous (in decreasing order of deciduousness: *Xylocarpus australasicus*, *Excoecaria agallocha*, *Lumnitzera racemosa*).

Data on leaf production and fall from other Queensland localities show similar patterns as those described above (Saenger, unpubl. data).

Gill and Tomlinson (1971) found that *Rhizophora* in southern Florida produce about 1.8 leaves per shoot per month during the summer which falls to about 0.25 during the winter. Each shoot maintains approximately 8-10 leaves, suggesting that expansion of the canopy occurs by the development of new shoots rather than by an increase in the total number of leaves per shoot. The average leaf life is six to twelve months. Similar rates were obtained for *Rhizophora* in southern Thailand by Christensen and Wium-Anderson (1977). Christensen (1978) indicated that leaf production shows two distinct peaks a year, one in May-June and the other in November. The first peak seems to be correlated with the beginning of the rainy season in the southwest-monsoon period. The latter peak occurs when the monsoon ceases. Leaf fall does not, however, exhibit any distinct seasonal variation. The average life of leaves is about eighteen months.

Not only do seasonal fluctuations in primary productivity occur,

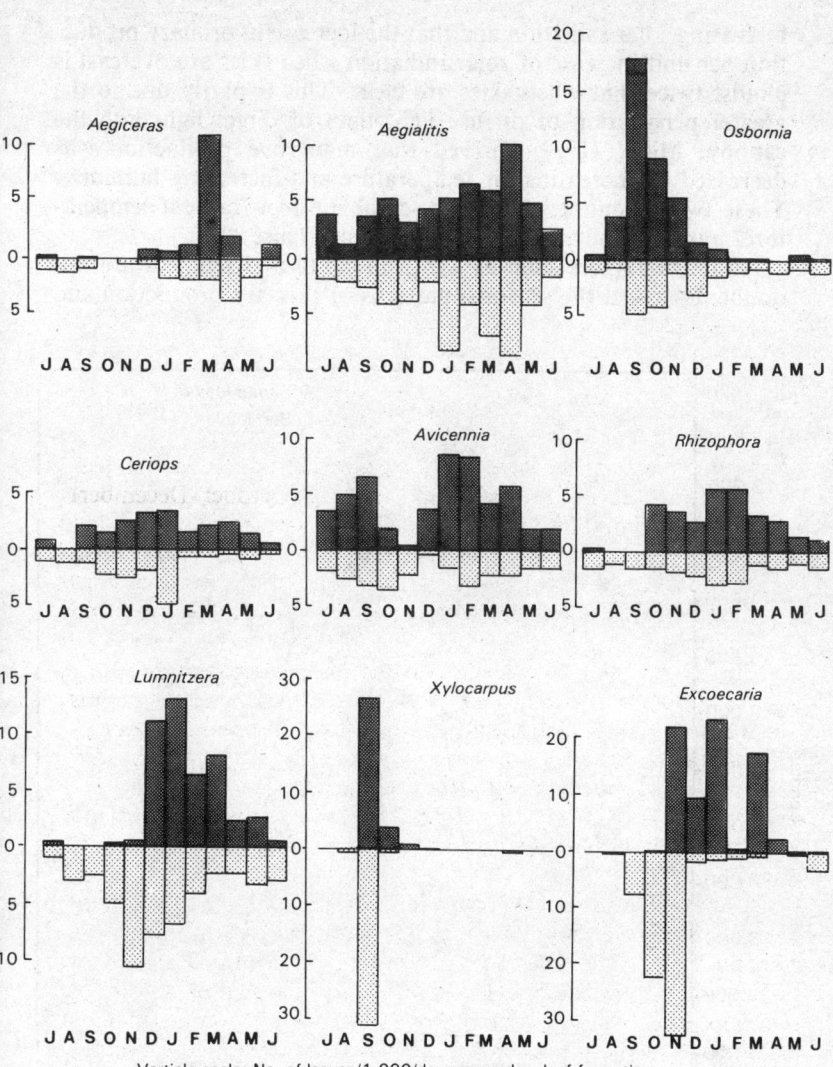

Figure 79 Leaf production and leaf drop in nine species of mangroves in central coast Queensland, measured over a twelve-month period (Saenger and Moverley 1985)

but there are diurnal fluctuations in rates as photosynthesis is limited to daylight and will vary according to local weather patterns.

De Witt (1965) showed that primary production increases with increasing solar radiation and that the increase in primary production per unit increase of solar radiation when skies are overcast is almost twice that when skies are clear. This is partly due to the greater penetration of diffuse light than of direct light into the canopy. Miller (1972) showed that mangrove production was decreased by increasing air temperature and increasing humidity. These two parameters have considerable effects on leaf temperature, transpiration and net photosynthesis (figure 80).

Although factors such as light, humidity and temperature undoubtedly affect the seasonal patterns of primary production and

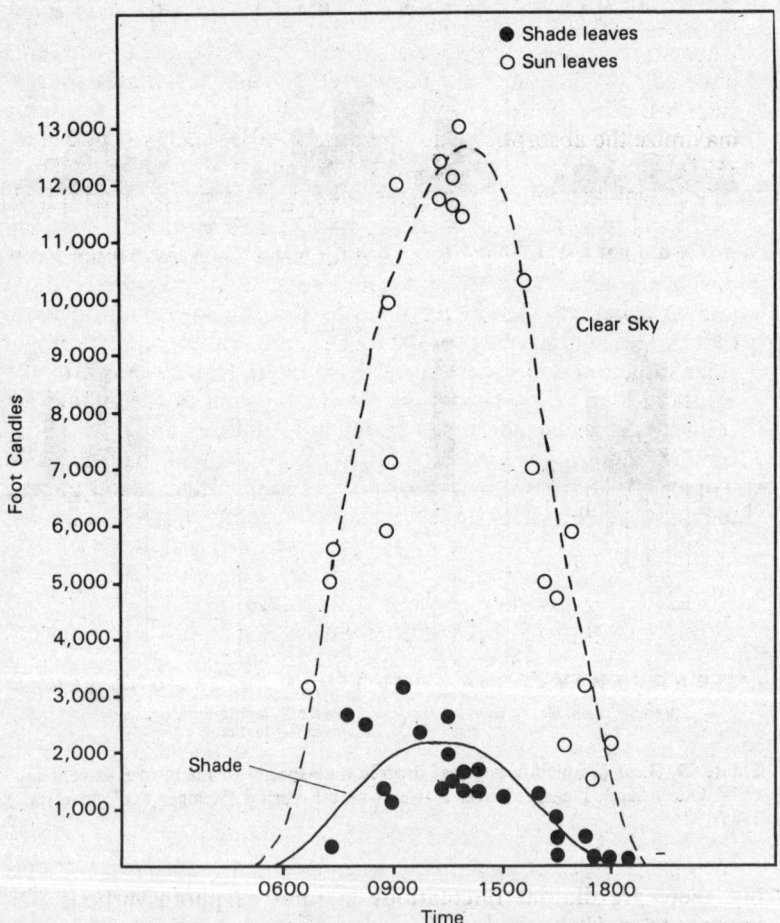

Figure 80 The diurnal sequence of light intensity in the top of the forest (sun leaves) and under the forest canopy (shade leaves) in May in a Puerto Rican red mangrove forest (after Golley, Odum and Wilson 1962).

leaf formation (see chapter 3), some internal factors are also involved. For example, growth rings in the wood of *Avicennia* appear to be under endogenous control (Gill 1971).

Species vary in their productivity and in their response to environmental factors. A unit leaf surface area of *Rhizophora mangle* exhibited a higher net primary productivity than one of *Avicennia germinans* where the two species were interspersed. *Avicennia* exhibited comparable rates with other species only when growing in conditions conducive to its dominance (Lugo and Snedaker 1974). This suggests that when a species is out of its usual zone its primary productivity is low in comparison to that when it is in the zone characteristic of it. Perhaps the zonation of mangrove species also involves zonation of their photosynthesis, respiration and transpiration rates. The species of each zone and their symbionts have adaptations that take advantage of auxiliary energy sources such as tidal flushing and nutrient run-off and in so doing maximize the absorption and retention of solar energy (Lugo et al. 1975).

Seasonal differences in gas exchange characteristics (Moore et al. 1973) and marked seasonal variations in leaf growth and litter fall imply that rates of net primary production may vary appreciably during the year. Also, these rates may vary latitudinally. Clough and Attiwill (1982) attempted to account for this seasonal variation by relating photosynthesis and respiration to seasonal trends in solar radiation and temperature. In Westernport Bay towards the southern limit of *Avicennia marina*, 67 per cent of the annual net primary production occurs in spring and summer, and only 15 per cent in winter. Very different patterns of growth may occur in tropical Australia where seasonal variation in solar radiation and temperature is less pronounced, but where there is a distinct wet season.

Tidal Control of Primary Production

The tidal regime of an estuary determines the rate of transport of oxygen to the root system. Tidal flushing affects the rate of sediment deposition or erosion within a given stand, and is responsible for the physical exchange of soil water with the overlying water mass. Such flushing may remove toxic sulphides and may reduce the total salt content of the soil water. The vertical movement of the ground watertable during a tidal cycle may transport nutrients regenerated by detrital food chains into the root zone of the mangroves.

There have been no measurements of the changes in rates of photosynthesis during a tidal cycle. If it varies, then the rate may vary across a particular stand of mangroves. The rate may also vary from neap to spring tides. Along the eastern Australian coast considerable differences occur in tidal amplitude between neap and spring tides. Within a range of salinities (8–30°/oo), the gross primary productivity of mangroves increases as fresh water becomes available. However, respiration-nutrient rates along the same gradient also increase. The increase in respiration is a reflection of the amount of physiological work associated with the problems of higher salinity. Scholander et al. (1962, 1965) showed that rates of water loss are related to the salinity regime of the plant and its environment. Hence, plants living in high salinities tend to transpire less than those growing in less saline environments. They also discussed some of the physiological costs of these adaptations and indicated that metabolic energy was involved in the process of transpiration. As the supply of metabolic energy available for translocation is finite, a limit is set on the amount of water that a plant can effectively take up and transport against an osmotic gradient within its environment. At very high salinities, a decrease in the net productivity of mangroves should occur. Carter et al. (1973), using these data and their own results, suggested that a linear function of increasing salinity and increasing productivity beyond those salinities tested (18–30°/oo) could not be expected to occur, for respiration would exceed gross production at higher salinities and the species would be eliminated. Instead they suggested a U-shaped relationship of mangrove metabolic dynamics along tidal and water chemistry gradients. At the two extremes of the U, the energetic costs of survival are high and most of the production is used in respiration (self-maintenance). The two extremes represent either high nutrients and low amplitude tides, or low nutrients and high amplitude tides. Between these two extremes, or in the middle of the U, nutrients and tidal amplitude are of the correct proportions and net productivity is maximized. Some preliminary data by Lugo and Snedaker (1974) support this concept.

The osmotic pressure gradient between the soil water and the plant vascular system will affect transpiration rates of the leaves (see chapter 3). This gradient will be determined largely by the salt concentration in the soil water, which in turn is affected by tidal flushing and freshwater run-on. This run-on may also contain essential nutrients for the mangroves and the fauna. Preliminary work from Hinchinbrook strongly supports this idea.

Latitude and geographic location will determine the frequency and duration of such freshwater run-on. However, in several areas in Australia such as northwestern Western Australia and South Australia, freshwater input is minimal. The effects of this have not been investigated. A high macro-nutrient content of the soil solution has been suggested as a factor causing the high productivity in mangrove ecosystems despite the low transpiration rates caused by high salt concentrations in sea water.

Role of Nutrient Supply of Primary Production

As mentioned in the preceding section and in chapter 3, the transport of nutrients regenerated by the detrital food chains into the mangrove root zone by tidal flushing may be important in determining productivity rate. Boto (1982) summarized the sources of inorganic nutrients in the mangrove ecosystem and suggested how they can enter the system and perhaps be crucial to the mangroves. In summary they are:
1. Rainfall
2. Freshwater run-on from surrounding terrestrial forests
3. Nitrogen fixation
4. Mineralization (decomposition, heterotrophic conversion of organic N, P to an inorganic form)
5. Tidal-borne dissolved or particulate-bound nutrients
6. Chemical release from fixed states in soil by changes in soil Eh and pH
7. Human influence, e.g., agriculture land drainage, sewage, clearing of mangrove areas.

Subsequently, Boto and Wellington (1983) described the nutritional status of a northern Australian mangrove forest in terms of phosphorus and nitrogen. They found that mangrove growth in this locality is generally nitrogen-limited with phosphorus limitation also a factor at the more elevated sites. It is likely that complex interactions between soil nutrient status, salinity and redox potential are involved in the control of mangrove growth.

Leaf Production

Clough and Attiwill (1982), using Christensen's (1978) data on total leaf production (556 g m^{-2} yr^{-1}) in southern Thailand, and his estimate of an average ash-free dry weight of 0.69 g per leaf, calculated annual gross leaf production to be 383 g ash-free dry weight m^{-2} or, in terms of total dry weight, about 437 g m^{-2} yr^{-1} (ash-free and excluding bud scales), indicating a substantial net loss of leaf biomass from the community. In contrast, Clough and Attiwill (unpublished, but quoted in Clough and Attiwill 1982), working with *Avicennia* in Westernport Bay, estimated gross leaf production to be 324 g m^{-2} yr^{-1} of which 50 per cent was lost as litter.

Wood Production

Clough and Attiwill (1982) summarized data on wood production. There is some variation: 307 g m^{-2} yr^{-1} for *Rhizophora* in Puerto Rico (Golley, Odum and Wilson 1962); 2,000 g m^{-2} yr^{-1} for *Rhizophora* in Thailand (Christensen 1978); 14 g C m^{-2} day$^-$ $\simeq$ 10 kg dry matter m^{-2} yr^{-1} for *Rhizophora* in Malaysia (Noakes, cited by Walsh 1974); 356 g m^{-2} yr^{-1} for *Avicennia* in Westernport Bay (Clough and Attiwill 1982). More measurements are needed to confirm whether these variations are characteristics of particular mangrove forests.

Gill (1971) and Duke, Birch and Williams (1981) have shown from growth ring studies that seasonal changes in wood formation occur. In *Avicennia* this appears to be under endogenous control whereas in *Diospyros* it appears to be regulated by rainfall.

There appear to be no estimates of root production in the literature, reflecting the difficulty of sampling and distinguishing between living and dead roots. Clough and Attiwill (1982) suggested that mangrove roots may have considerable potential for cycling and recycling organic and inorganic material; reliable methods of estimating growth and turnover of roots are urgently needed.

Mathematical Models

Miller (1972) attempted to develop a model for predicting rates of photosynthesis for *Rhizophora* in southern Florida. The model suggests that canopies with steeply inclined leaves will have high rates of photosynthesis, both because of reduction in leaf temperature and because of more efficient light interception.

Lugo, Sell and Snedaker (1976) developed a model of energy flow for mangroves in southern Florida. The validity of their model was tested by comparing some of the simulated results with values reported in the literature for other stands. The model suggests that mangrove forests take about ten years to reach a steady state and predicts that most mangroves reach maturity at about twenty to twenty-five years. Interestingly, this is the mean time interval for major cyclones in southern Florida (Lugo, Sell and Snedaker 1976), and suggests that the whole ecosystem has adapted to this externally imposed periodicity.

Before attempting to construct budgets of primary production, the widely used practice of extrapolating figures of respiration and converting them to primary productivity figures should be validated. Estimates of primary production are obtained over a short period of time, and often for only parts of the system. As already stressed, considerable variation may occur in both space and time. Therefore, the validity of such estimates is in question, as are comparisons between mangrove systems in various parts of the world.

Primary Production Budgets

Relatively few attempts have been made to construct overall budgets for primary production in which net gains and losses of biomass in different compartments (for example, leaves, wood, roots and propagules) are balanced against net production. Net production will occur only when photosynthesis exceeds respiration. Golley, Odum and Wilson (1962) were the first to attempt to study this. They measured the total photosynthesis, leaf respiration, and the amount of air exchanged through the prop roots. Estimates of leaf fall, trunk growth, tidal export of particulate matter, underwater respiration of the soil, and soil respiration in air were made for a stand of *Rhizophora* in Puerto Rico. However, there appear to be some inadequacies in the techniques used (Clough and Attiwill 1982) so that interpretation of their results is difficult. They calculated that the forest had a total gross production and respiration exceeding 8 g C m^{-2} day^{-1} or about 16 g organic matter m^{-2} day^{-1}. These figures are based on measurements taken over a few weeks, so no account has been taken of seasonal variation in production which is marked in mangrove communities. They attempted to produce a balance sheet for the mangrove community. Gross photosynthesis from sun leaves, shade leaves, seedlings and algae on the soil surface was 8.23 g C m^{-2} day^{-1}. Respiration of the community, including the organisms in the soil, was 7.79 g C m^{-2} day^{-1}, with 1.37 g C m^{-2} day^{-1} exported out of the system. Thus, during the study period, losses owing to respiration and export exceeded the amounts produced by photosynthesis by 0.93 g C m^{-2} day^{-1}. This may reflect problems of measuring accurately all the components or may represent the real situation at that time. It is known that seasonal variations occur with fluctuations in the balance between photosynthesis and respiration. Lugo and Snedaker (1975) and Lugo, Sell and Snedaker (1976) published an overall model of carbon flow in the mangroves of southern Florida, and calculated an overall primary production budget.

Clough and Attiwill (1982) attempted to devise an overall budget for a temperate stand of *Avicennia marina* in Westernport Bay (table 45). They calculated net primary production by integrating over a full year the daily rates of net photosynthesis and respiration calculated from light-response curves, daily solar radiation and temperature. Leaf fall was measured over a period of two years and leaf production was estimated from changes in leaf area and leaf

Table 45 Preliminary annual production budget for *Avicennia* in Westernport Bay. Data are expressed as g dry matter m^{-2} yr^{-1}

Component	Total production	Losses	Net growth
Leaves	324	162	162
Branchwood	294	22	272
Trunk	84	0	84
Roots*	138	?	?
Net primary production	840 (= 2.358 g dry matter m^{-2} day^{-1})		

* Production of roots is estimated as the difference between net primary production and total production of above ground biomass.
Source: After Clough and Attiwill (1982).

number. Finally, wood growth was estimated from changes in branch diameter and allometric relationships between diameter and biomass.

Net primary productivity exceeded the total production of above-ground biomass, this small surplus presumably being used for root production. Clough and Attiwill (1982) suggested that there are certain anomalies in these data. Although the estimates of production and loss are reasonable, the estimate for annual root production seems too low considering the biomass of the roots, suggesting either that there is little net growth or that turnover of roots is very low. Neither is likely. Another anomaly is that net leaf production, a reliable measurement, suggested that leaf biomass doubles in about four years; other more general observations suggest a very much slower rate of increase in leaf biomass.

Finally, in Westernport Bay, fruit production is not annual. Propagules are produced only every second or third year, indicating cycles of production over a two- or three-year period rather than an annual one. In more tropical areas, fruit production is an annual event. Fruit production did not occur during the year that Clough and Attiwill (1982) were measuring leaf and wood growth. This suggests that in some years productivity is higher than they measured, thereby allowing fruit production in temperate mangroves. A similar phenomenon is found in tropical mangroves. Periodically, tropical mangroves are subjected to cyclonic winds which may cause considerable damage. Subsequently, the damaged trees grow rapidly and regain steady-state conditions (with respect to biomass).

The question which appears not to have been asked and yet is one of the most interesting is why mangrove forests are among the most productive ecosystems (Odum 1959; Deevey 1960).

Mangrove communities are among the highest in production of the marine plant communities listed in table 46 and rank with terrestrial cultivated crops. The leaf-litter production of several species of mangroves in Queensland has been found to be as high as 2,200 g (dry weight) m^{-2} yr^{-1} (Bunt 1982). Moving southwards,

Table 46 Experimentally determined rates of primary production of selected terrestrial and marine plant communities

Community	Comments	g C m^{-2} day^{-1}	t ha^{-1} yr^{-1}	Reference
Terrestrial				
Napier grass	Above-ground	9.1	88	Boardman and Larkum 1974
Sugarcane	Above-ground	6.8	66	Boardman and Larkum 1974
Tropical reed swamp	Above-ground	6.1	59	Boardman and Larkum 1974
Annual crops	Above-ground	2.3	22	Boardman and Larkum 1974
Evergreen crops	Above-ground	2.2	21	Boardman and Larkum 1974
Marine				
Spartina marsh (subtropical)	Above-ground	2.06	20	Turner 1976
Mangroves, Thailand	Litter fall	2.57	25	Christensen 1978
Mangroves, North Qld	Litter fall	2.0	19	Bunt 1979
Thalassia testudinum (tropical seagrass)	Above-ground	0.8–1.9	9–18	Zieman 1975b
Zostera marina (temperate seagrass)	Above-ground	0.9	8.6	Sand-Jensen 1975
Laminaria longicruris	Minimum estimate	1.65	16	Mann 1973
Laminaria hyperborea	Minimum estimate	1.07	11	Kain 1977
Phytoplankton, low nutrient levels		0.13	1.3	Ryther 1969
Phytoplankton, coastal waters		0.27	2.6	Ryther 1969
Phytoplankton, eutrophic waters		0.8	7.9	Ryther 1969

there seems to be a declining gradient of production with a value of about 162 g m^{-2} yr^{-1} in Victoria (Clough and Attiwill 1979; Goulter and Allaway 1977). Litter production is only part of the production of mangroves, and Christensen (1978) suggested that wood production is four to five times that of litter. If this is correct, then production of tropical mangroves would rank as one of the highest recorded in any natural environment. What unique features do they possess? Is it because they are at the interface of two environments and receive nutrients from terrestrial and marine sources as well as being daily inundated by sea water which replenishes the oxygen supply and nutrients and removes toxic substances? This question is a vital one.

Biomass Available for Export or Reuse

Mangrove systems appear to be very productive. Rates vary according to season and presumably will vary according to the geographic location of the stands and hence species composition and other physical characteristics of that region such as rainfall and freshwater run-off. What happens to this primary productivity? Its fate is varied. Some of it is stored in trunks and roots for years. That in leaves is cycled more quickly as leaf litter which is broken down *in situ*. Goulter and Allaway (1979) calculated that *A. marina* leaf litter in the Sydney region had a half-life of about eight weeks. This is similar to rates found by Albright (1975) and Woodroffe (1982) in New Zealand (lat. 36°S), but much faster than rates determined for the tougher leaves of *Rhizophora mangle* (Heald 1971; Lugo and Snedaker 1975). The loss of leaves through export out of the system in Sydney was not determined, nor was the loss of particulate organic matter.

In contrast Boto and Bunt (1981) suggested that the majority of export of particulate C and N from Hinchinbrook mangroves is via the flushing of mangrove litter. Little litter is evident in the mangroves themselves and there was little evidence of grazing. Mangrove litter was found considerable distances from the mangroves and little returned on the incoming tide. Whether this is characteristic of tropical mangroves with good tidal flushing or whether it is more common for the litter to be retained in the mangroves where it is broken down by resident bacteria, fungi and fauna is unclear. It seems likely that the composition of the net export from mangroves may be determined by many variable factors, including:
1. Seasonal fluctuations in leaf-litter production
2. Seasonal activity of biodegrading organisms
3. Tidal cycles and amplitudes and variations during a neap–spring tidal cycle
4. Freshwater run-on
5. Storms
6. Size and structure of the mangroves
7. Topography of the estuarine floor immediately adjacent to the mangroves (which will determine flushing rates)
8. The presence or absence of seagrass beds immediately in front of the mangroves which may trap some of the mangrove litter.

Recently, Woodroffe (1985a, 1985b) analysed the flux of organic and inorganic particulate matter out of a mangrove basin in New Zealand.

Many of these factors need to be considered and it may be that some mangrove communities contribute more particulate matter to estuarine ecosystems than do others. Whatever the actual values, it is clear that the mangrove system is not a closed one, but rather one that exports materials to estuaries and is influenced by adjacent marine and terrestrial communities.

8. The Role of Mangroves and Other Wetlands in Estuarine Ecosystems

Mangroves are intimately associated with terrestrial forests or salt flats on the landward side and with mudflats, eelgrass beds or other marine or estuarine communities on the seaward side. Fresh water flows through to seaward and tidal flows surge into and out of mangroves from the sea or estuaries. Such water movement transports materials and organisms from one community to another, and hence a mangrove cannot be considered in isolation from contiguous or, in some cases, even rather distant communities.

Some of the interactions of mangroves with other communities have been mentioned in previous chapters and will not be treated further here. In this chapter, mangroves are considered in the wider context of tidal wetlands and estuaries, and their role in this larger system is assessed. Mangroves are compared with other tidal and estuarine communities.

Primary Productivity

Primary production is the basic driving force of the estuarine ecosystem. Much of it is contributed by estuarine wetlands, salt marshes, mangroves and seagrass beds. The rest of it is provided by phytoplankton and benthic diatoms within the estuarine and nearshore waters. The proportions contributed by each of these components is largely unknown and will vary according to the estuary (figure 81). Until recently it was thought that the wetlands were the major contributors, but a recent article by Haines (1979) suggested that the algal contribution may be much higher than was previously thought, at least in estuaries along the Georgian coast of the United States. In Australia there are few figures on the productivity of algal communities other than the value of 400 g m^{-2} yr^{-1} for *Hormosira* in Botany Bay (King 1981). Consequently, in the ensuing discussion on estuarine ecosystems, the algal communities, of necessity, must be ignored. However, it seems likely that in the next few years measurements of the productivity of algae will be made which perhaps will modify current views of estuarine ecosystems in Australia considerably. The reader should also bear in mind the

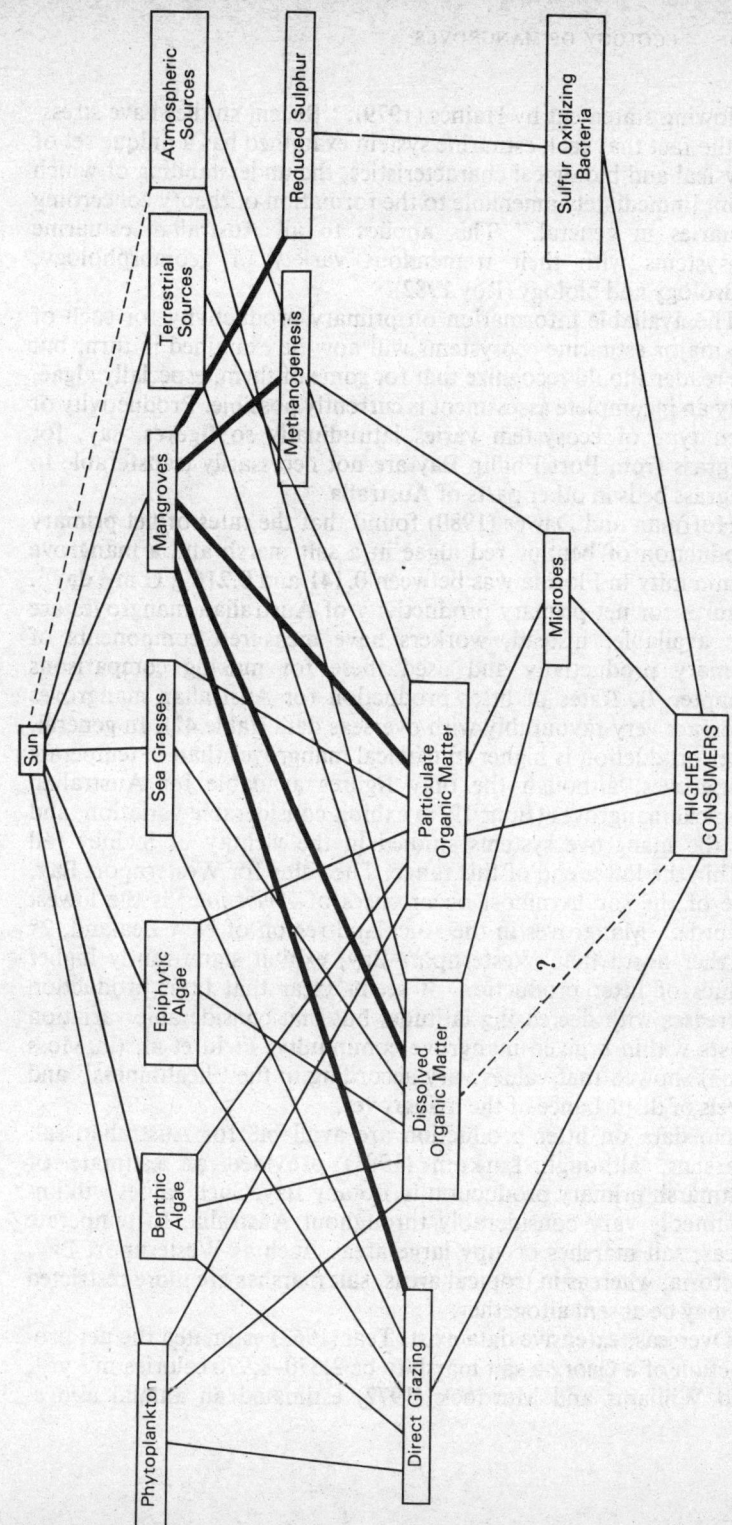

Figure 81 Potential pathways of energy flow in mangrove ecosystems — not all possible pathways have been drawn (after Odum, McIvor and Smith 1982).

following statement by Haines (1979): "Recent studies have stressed the fact that each estuarine system examined has a unique set of physical and biological characteristics, the understanding of which is not immediately amenable to the formation of theory concerning estuaries in general." This applies to all Australian estuarine ecosystems with their tremendous variety of geomorphology, hydrology and biology (Roy 1982).

The available information on primary productivity for each of the major estuarine ecosystems will now be examined in turn, but the reader should recognize that for some of them, especially algae, only an incomplete assessment is currently possible. Productivity of each type of ecosystem varies latitudinally so figures, say, for seagrass from Port Phillip Bay are not necessarily transferable to seagrass beds in other parts of Australia.

Hoffman and Dawes (1980) found that the rates of net primary production of benthic red algae in a salt marsh and a mangrove community in Florida was between 0.141 and 0.216 g C m^{-2} day^{-1}. Figures for net primary productivity of Australian mangroves are not available; instead, workers have measured components of primary productivity and used these for making comparisons (chapter 7). Rates of litter production for Australian mangroves compare very favourably with overseas data (table 42). In general, litter production is higher in tropical mangroves than in temperate mangroves, although the only figures available for Australian tropical mangroves (Bunt 1978) exhibit considerable variation, and all the mangrove systems studied in the vicinity of Sydney fall within the lower end of this range. The value for Westernport Bay, one of the southernmost occurrences of *Avicennia*, is the lowest recorded. Mangroves in the Auckland region of New Zealand, 2° further north than Westernport Bay, exhibit significantly higher values of litter production. It seems clear that litter production increases with decreasing latitude, but that considerable variation exists within a given mangrove community. Field et al. (in Moss 1983) showed that values vary according to the "healthiness" and levels of disturbance of the mangroves.

No data on litter production are available for Australian salt marshes, although Larkum (1981) provided an estimate of saltmarsh primary production in Botany Bay. Such values will undoubtedly vary considerably throughout Australia. In temperate areas, salt marshes occupy large areas, such as Westernport Bay, Victoria, whereas in tropical areas, salt marshes are more restricted or may be absent altogether.

Overseas, extensive data exist. Teal (1962) estimated the net production of a Georgia salt marsh to be 2,570–8,970 calories m^{-2} yr^{-1}, and Williams and Murdock (1972) estimated an annual above-ground production of 754 g dry weight m^{-2} yr^{-1} ($\simeq$230 g carbon) of black needle rush (*Juncus roemerianus*) in North Carolina.

Compared with mangroves and salt marshes, some rather extensive data on the productivity of Australian seagrasses are available, but to date only from temperate areas. *Posidonia australis* has a leaf productivity of between 0.2 and 1.7 g cm^{-2} day^{-1} in New South Wales and South Australia (West and Larkum 1979). *Zostera capricorni* has a mean annual above-ground productivity of 512.7 ± 51.3 t yr^{-1}, that is, 1.66 ± 0.17 t ha^{-1} yr^{-1} in Sydney waters (Larkum, Collett and Williams 1984). Using these figures, Larkum (1981) estimated primary productivity for each of the plant communities in Botany Bay, New South Wales (table 38).

Mangrove communities occupy only 5.78 per cent of the total area of Botany Bay, yet contribute 30.4 per cent to the primary productivity of the bay, and together with the salt marsh and seagrasses contribute 61.6 per cent of the total productivity. These figures stress the importance of wetlands in this particular temperate estuarine system.

The estimates for the production by seagrasses and mangroves are based on growth rates of seagrasses or on litter production (Goulter and Allaway 1979; Larkum, Collett and Williams 1984; West and Larkum 1979) and are the most reliable. The other estimates are based upon calculations of data collected elsewhere: epiphytic algae (McRoy and McMillan 1977), salt marsh (Jefferies 1972), benthic algae (Pomeroy 1959) and phytoplankton (Revelante and Gilmartin 1978).

Larkum (1981) did not comment on the possible contributions of phytoplankton from near-shore coastal waters, but it may be a reasonable component in an area like Botany Bay which is well flushed. Data from the United States and Australia (Jeffrey 1981) suggest that high levels of phytoplankton production may occur (Haines 1979).

Types of Food Chains

In the sea, as on land, the primary producers are plants which convert the energy from the sun into organic matter by photosynthesis. Three main groups of plants occur in the sea, phytoplankton (single-celled plants), marine angiosperms or seagrasses, and the multicellular algae or seaweeds. These plants form the basis of the two types of food chains occurring within the marine environment: (1) those based directly on living plants and (2) those based on decaying or detrital plant material.

The phytoplankton food chain dominates the ocean where animals either feed on the living phytoplankton or on their dying remains as they slowly sink to the ocean floor. In shallow coastal and estuarine environments, the type of food chain is determined by whether the estuary is dominated by the benthos or by plankton (Odum and Heald 1975a, 1975b). Benthos-dominated estuaries (those characterized mainly by bottom-dwelling organisms) have the higher primary productivity, produced by the seagrass, mangrove and saltmarsh communities. The relative contribution of the detritus and phytoplankton to the food chains will be largely determined by the depth profiles of a particular estuary. Shallow estuaries will have a greater development of wetlands (salt marsh, mangroves and seagrasses) than deep estuaries with steeply shelving shores.

In the detritus-based food chain, the plant matter produced by the wetlands in the form of fallen leaves, seagrass blades and so on is broken down into fine particulate matter by a series of steps; most of the benthic organisms subsequently utilize this particulate organic matter.

Botany Bay is an estuary dominated by detritus and benthic organisms (table 38). Such a pattern is probably characteristic of all Australian estuaries where wetland communities occur, that is, the majority of estuaries on the eastern coast of Australia.

Until recently it was assumed that, in estuaries with large areas of wetlands, the food web was mainly based on non-living plant material and that animals in the second trophic level were, for the most part, detritivores rather than herbivores (Teal 1962; Odum and Heald 1975a, 1975b). Haines (1979) suggested that this concept has caused misunderstanding because of the confusion over whether animals can assimilate dead plants directly, or indirectly after digestion by microbes living on the plant matter (Adams and Angelovic 1970), and because in the past estuarine food chains have been oversimplified.

In the strictest sense the only "detritivores" are the bacteria, fungi and perhaps polychaete worms, which can assimilate the dead plant material. Associated with these organisms are protozoans, nematodes and other meiofaunal consumers of the detritus degraders. Benthic algae in the sediment and phytoplankton in the water column are closely associated with the microbial and meiofaunal community, so some of the animals which are often considered to be detritivores (many types of crabs, gastropods and some shrimps) should be referred to as opportunistic omnivores, feeding on a mixture of microbes, small multicellular animals and algae (Haines 1979). Recently, a technique has been developed which should allow additional checks to be made on the original sources of detritus upon which estuarine animals feed. Phytoplankton and seagrasses appear to lay down their carbon in different ratios of C^{12} and C^{13}, and by determining these ratios the source can be ascertained. Using these differences in the carbon isotopes, the source of the detritus within the guts of common estuarine animals such as prawns can be identified (Fry and Parker 1979; McConnaughey and McRoy 1979; Thayer et al. 1978). In these studies, estuarine prawns were feeding on detritus produced from seagrasses rather than from phytoplankton.

However, isotopic composition reveals nothing about the number of trophic steps between plants and animals since there is virtually no change in isotopic composition from the live plant through microbial decomposition and animal assimilation (Haines and Montague 1979). Caution needs to be exercised in interpreting these data. In Australia there are virtually no detailed data on the various links in the food chain, but they are probably complex and variable, and it seems likely that many of the animals are highly opportunistic.

Fate of Primary Production

The plant matter produced as a result of primary production probably can be used by only a few organisms directly. Fell and Master (1980) investigated the role of fungi in mangrove detrital systems and found that in the presence of fungi there was an increase in the amount of organic carbon leached from the leaves, and an increase in nitrogen as a result of the decomposition of cellulose and the infiltration of fungal mycelia. There was also an increase in the carbon levels of the leachate that flocculated, that is, precipitated as either semi-transparent flakes or small globules (Lush and Hynes 1973). The ratio of carbon to nitrogen (>12) of flocculated material suggests a high food value available to the animals associated with litter decomposition (Fell and Masters 1980).

Fell and Masters (1980), working in southern Florida on the decomposition of *Rhizophora mangle* leaves, found a sequence of fungal populations during different stages of decay, and Cragg and Swift (1980) in New Guinea showed the importance of fungi in the breakdown of wood by facilitating the penetration of boring organisms (see also chapter 4). Very little is known about other types of detritivores, bacteria or polychaetes and no work has been done on these groups in Australian wetlands.

Van der Valk and Attiwill (1984) studied decomposition of *Avicennia marina* material at Westernport Bay, and Robertson, Mills and Zieman (1982) monitored the breakdown of dried leaves of the seagrasses *Thalassia* and *Syringodium*. The latter found that the dissolved organic carbon was rapidly converted to bacterial aggregates of a size that could be eaten by macroconsumers. Large populations of ciliates and flagellates also developed; these were presumably feeding on the unaggregated bacteria.

The detritus produced by different types of wetlands undergoes various types of degradation when it enters the aquatic environment. Much of the initial breakdown of the leaves occurs on the surface of the sediment and it is not until small particles are produced that they are transported out of the wetlands into the tidal body of the estuary. Initially, soluble compounds are rapidly leached from fresh detritus (Harrison and Mann 1975; Fell et al. 1975; Rice and Tenore 1981). Epiphytic growth on the detritus will enhance the nutritional value of the detritus and facilitate its further breakdown. Because of the complex structure of vascular plants and their relative resistance to microbial decay, breakdown of this material is slow (Gosselink and Kirby 1974; de la Cruz 1975; Harrison and Mann 1975). The decay-resistant portion of these substrates is available to macroconsumers only after long periods (for example, months) of "ageing" and related microbial activity (Tenore and Hanson 1980). For example, detritus derived from seaweeds was utilized rapidly and incorporated by the polychaete *Capitella capitata*, whereas more decay-resistant detritus derived from seagrasses or saltmarsh plants took a longer period of ageing before it could be utilized (Tenore 1975, 1977; Tenore and Hanson 1980).

The populations of micro-organisms on the surface of the detritus increase with age (Odum and de la Cruz 1967; Fenchel 1977; Heald 1971). Fenchel (1970) found an average of 5×10^4 ciliates, 3×10^9 bacteria, 5×10^7 small zooflagellates and 2×10^4 diatoms per gram dry weight of detritus derived from the seagrass *Thalassia*. The numbers were highest in detrital samples with particles of small to average diameter and lowest on new, large-sized particles. Briggs, Tenore and Hanson (1979) showed that more detritus is incorporated in the presence of ciliates than when they are lacking. The increase in net incorporation probably resulted from a combination of stimulation of bacteria by grazing, and from fragmentation and ingestion of detritus by ciliates. Briggs, Tenore and Hanson (1979) stressed the importance of the microfauna in the transfer of energy from detritus to the macro-detritivores, and the role of macrofauna in enhancing microbial activity in the sediment through bioturbation.

When these animals die, they also contribute to the supply of detritus, as do their faecal pellets. This also applies to other estuarine animals. Many animals ingest large quantities of plant matter, and their role in the system is the mechanical break up of plant matter either by their jaws or by grinding plates in their guts. The agents responsible for this breakdown are numerous. Amphipods and isopods have been identified as major ones (Robertson and Mann 1980; Goulter and Allaway 1979). They break down leaves into smaller particles which are then attacked by bacteria and fungi. Nutrient enrichment of these particles facilitates further shredding by other herbivorous animals. Some animals obtain much of their nutrition from the bacteria, fungi and algae growing on the surface of the plant fragments rather than from the plants themselves. Plant fragments are also broken down by physical abrasion of the particles and bioturbation by meio- and macro-fauna (Tenore and Rice 1980). The deposit feeders, which are mainly infaunal species, disturb the sediment (Rhoads and Young 1971), putting into suspension fine particulate matter which is then collected by the filter-feeding species such as bivalves and some polychaetes.

In laboratory-controlled experiments, Tenore and Dunstan (1973) and Tenore, Goldman and Clarner (1973) found that suspension-feeding bivalves filter approximately 8 g carbon/g dry weight of animal per day at food concentrations typical of coastal environments. Of this filtered food, 20–30 per cent is deposited as faeces and pseudofaeces.

The distribution and density of the bivalve fauna is regulated by the amount of organic matter in the sediment and its state of decomposition. As organic matter increases in the sediment, density of bivalves increases until the point that bacterial decomposition of organic matter results in deoxygenation, and there is a predominance of lignin. Then the density of bivalves decreases again.

The macroconsumers, apart from physically breaking up the plant matter, may facilitate its further breakdown in other ways. Some early work suggested that microbes colonize faecal pellets and are used for food by benthos; the faecal pellets *per se* pass through the gut undigested (Newell 1965; Fenchel 1970, 1972). As micro-organisms recolonize fresh faecal pellets the nitrogen content of the pellets increases. If these faecal pellets are reingested, the nitrogen level falls again, presumably because the macroconsumers have removed the microbes and their contained nitrogen from the pellets. Recently Tenore et al. (1982) suggested that this concept may need modification. Changes in nitrogen level may or may not reflect a real increase in available nitrogen content, depending on a combination of differential C–N mineralizations and accumulation by microbes of nitrogen from extraneous sources (de la Cruz 1976; Haines and Hanson 1979). Also, phenolic complexation of nitrogenous compounds may lead to an increased nitrogen content, but the extra nitrogen may not be available to the consumer (Rice 1979). More work is needed to clarify the role of faecal pellets in the detrital food chain.

Many studies have stressed the importance of detritus in the estuarine ecosystem and some have indicated the importance of the sources of that detritus (Tenore et al. 1982). Tenore and Gopalan (1974) found significant differences in feeding rates and growth efficiencies of the polychaete *Nereis virens* fed on detritus derived from different materials. Such information would be invaluable for management and assessing the minimum area of wetlands necessary for maintaining an estuarine ecosystem. Whether detritus ever becomes a limiting factor or regulates the size of benthic communities is unknown.

Seasonal fluctuations in the estuarine benthos have been documented by Hutchings and Recher (1974), Saenger, Stephenson and Moverley (1980) and Poiner (1980). However, Collett et al. (1984) were unable to detect seasonal changes in the benthos of *Posidonia* seagrass beds in Port Hacking, New South Wales. These infaunal communities are dominated by deposit feeders, primarily polychaetes. This is particularly marked in estuarine sites of *Posidonia* compared with those in more open water. Whether these deposit feeders are obtaining nutrients from the detritus or only from the micro-organisms on the surface of the particles is unclear. The presence of cellulases has not been documented in these animals.

Tenore (1977) reviewed the food chains in detritus-feeding benthic communities. He stressed the need for information on the availability and nutritive value of detritus to the benthos and on the inter-relations that regulate energy flow in the detrital food chain. Also important are the amounts of litter produced (discussed in chapter 7) and the rate of its breakdown and hence the time taken to make detritus available to consumers. Each of the major sources of detritus will now be considered.

Mangroves

Goulter and Allaway (1979) measured the rate of breakdown of leaves of *Avicennia marina* in Sydney (lat. 34°S) and they reported that leaf dry-matter had a half-life of about eight weeks (figures 82 and 83) during May. The amount of leaf fall varies during the year with the greatest concentration of leaf litter available for breakdown during the summer months. Similar but slightly faster rates were found by Albrecht (1975) working on the same species in New Zealand (lat. 36°S). More recently, Woodroffe (1982), also working in New Zealand, found a slightly higher rate than Goulter and Allaway (figures 84 and 85). Much higher rates were recorded from Thailand (Boomruang 1978). Decomposition of mangrove leaves appears to be seasonal (Hesse 1961), with more rapid degradation in the summer than in the winter. With increasing decomposition time, the percentage of nitrogen decreases, leaves of

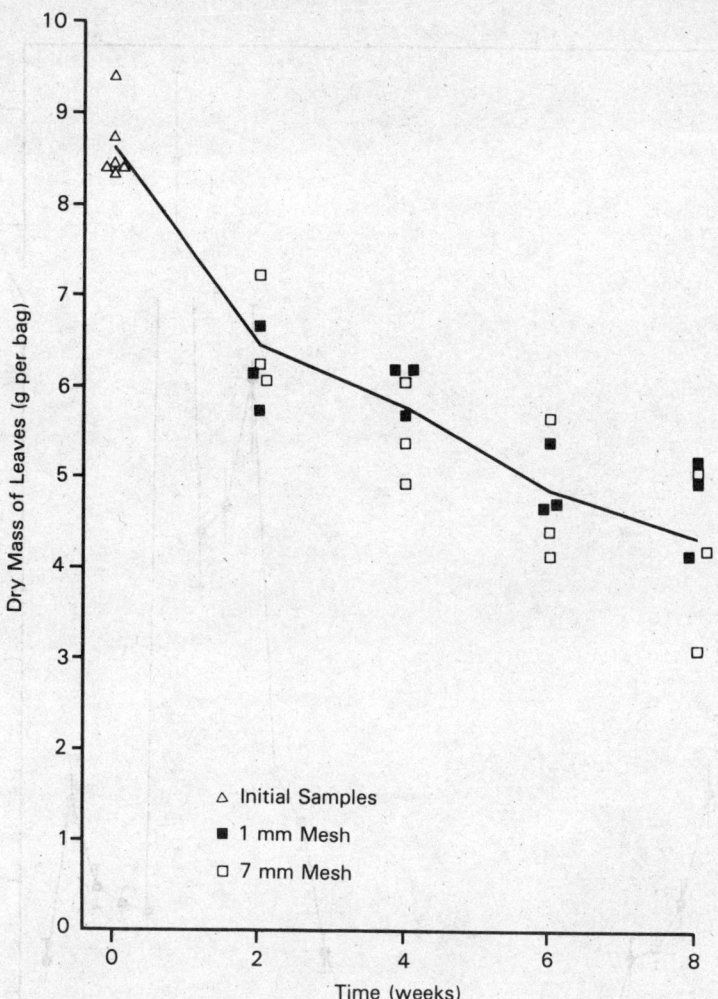

Figure 82 Decomposition of *Avicennia marina* leaves in mesh bags pegged out on the mud at Roseville, Sydney. Each point represents dry mass of leaves in one bag. All bags were set out on 8 May 1977 (after Goulter and Allaway 1979).

Avicennia having higher levels than those of *Bruguiera* (figure 86). Similarly, there is a decrease in mass of nitrogen per litter bag during decomposition (figure 87) (Steinke, Naidoo and Charles 1983). Factors important in decomposition may vary with distance from tidal creeks (Heald 1969; Heald, Odum and Tabb 1974), and the distribution of fungi important in decomposition may vary within mangroves (Ulken 1981).

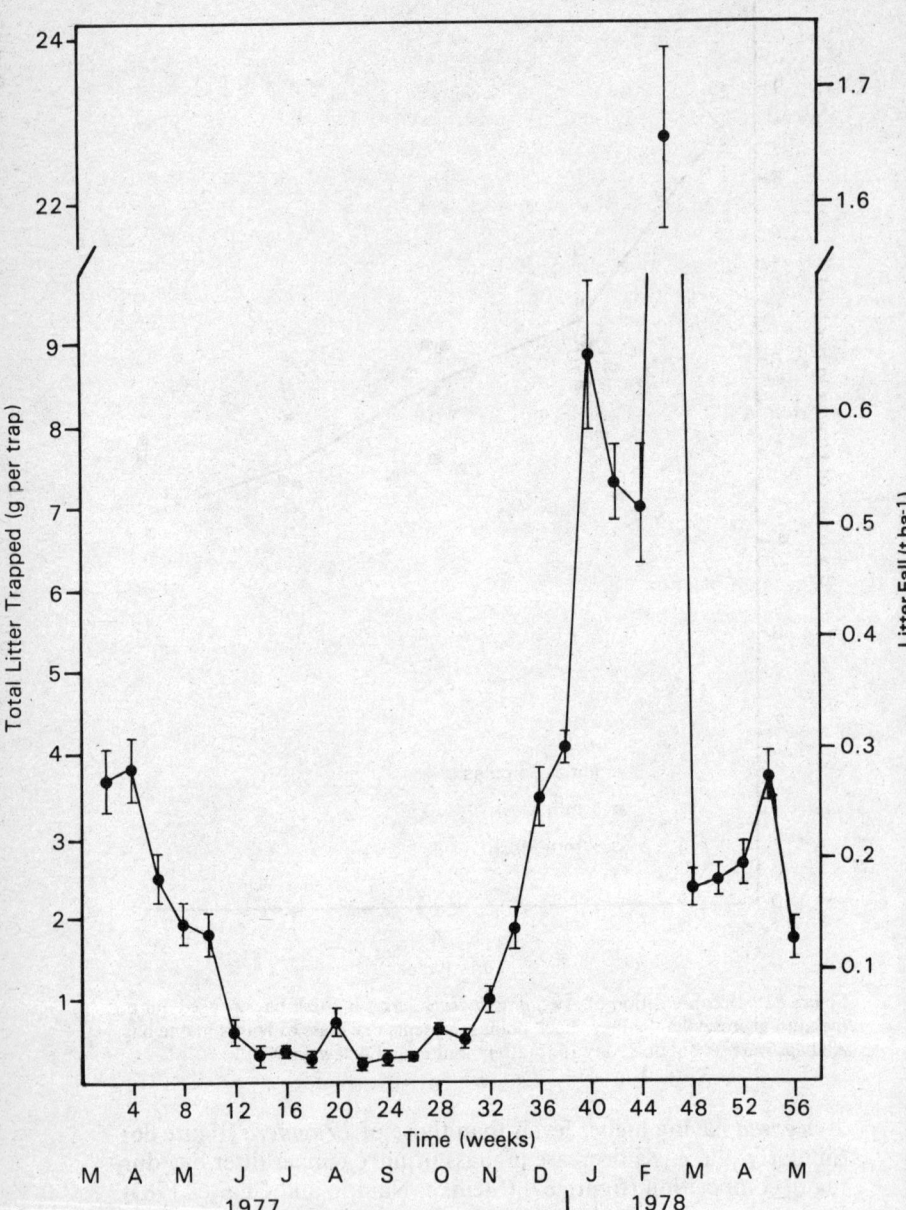

Figure 83 *Avicennia marina* litter fall at Roseville, Sydney. Data points are means of amounts of dry matter collected from twenty-four traps (after Goulter and Allaway 1970).

Salt Marshes

No work has been done in Australia on the productivity of saltmarsh plants and the decomposition of this material. Rates available from the United States are not directly comparable because of the different species involved.

Teal (1962) showed that only 5 per cent of cord grass (*Juncus*) is grazed on by herbivores; about 45 per cent of the total annual production is exported as detritus to the adjacent estuary. Estimates in the literature (Williams and Murdock 1972) of the rate of decay of marsh grasses based on loss of weight of bagged material are: 0.27 g m^{-2} yr^{-1} (Latter and Craig 1967), 0.36 g m^{-2} yr^{-1} (Heald 1971) and 0.49 g m^{-2} yr^{-1} (Waits 1967). Haines (1979) suggested that the major export of saltmarsh production of the east coast of the United States may not be as particulate detritus but as living organisms. Vascular plants decay *in situ* and the algae living on the surface of the mud are eaten by a large number of animals which invade the salt marsh at high tide to feed. Thus, salt marshes may be the true nursery grounds of the estuary (Haines 1979).

No work has been carried out on the pathways of breakdown of Australian saltmarsh plants, but probably similar pathways of breakdown and decomposition occur as in mangroves. Overseas, considerable work has been carried out mainly on beds of *Spartina* in Georgia (for review see Haines 1979). Ribelin and Collier (1979) suggested that in the Gulf Coast salt marshes of the United States, contrary to previously accepted ideas, the vascular plant tissue is decomposed beneath the layer of benthic algae and is retained in the marsh. More than 98 per cent of the detrital material exported from these marshes is made up of amorphous aggregates. These detrital aggregates are produced by the benthic microflora of the marsh rather than by microbial decomposition of the dominant vascular plant, *Juncus roemerianus*.

Seagrasses

Detailed information on the rates of breakdown of plant material in temperate Australian seagrass beds are available. Kirkman and Reid (1979) estimated the following components of the carbon budget:
1. Seasonal variations in biomass of *Posidonia*
2. Loss of living seagrass blades by grazing fish or other herbivores
3. Amount of floating detritus (detached seagrass leaves) on a weekly basis over a one-year period
4. Rate of accumulation of old leaves and attached biota on the substrate

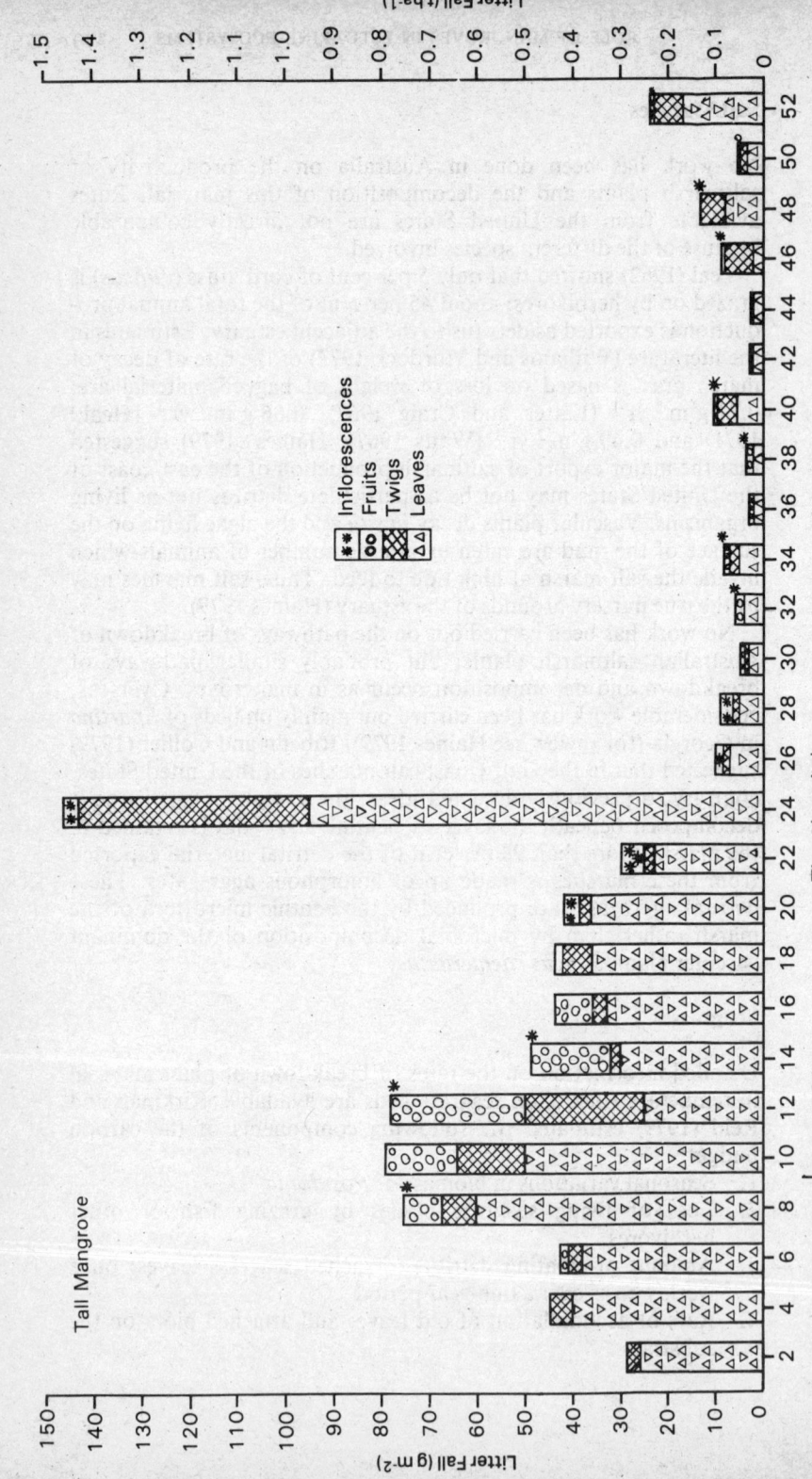

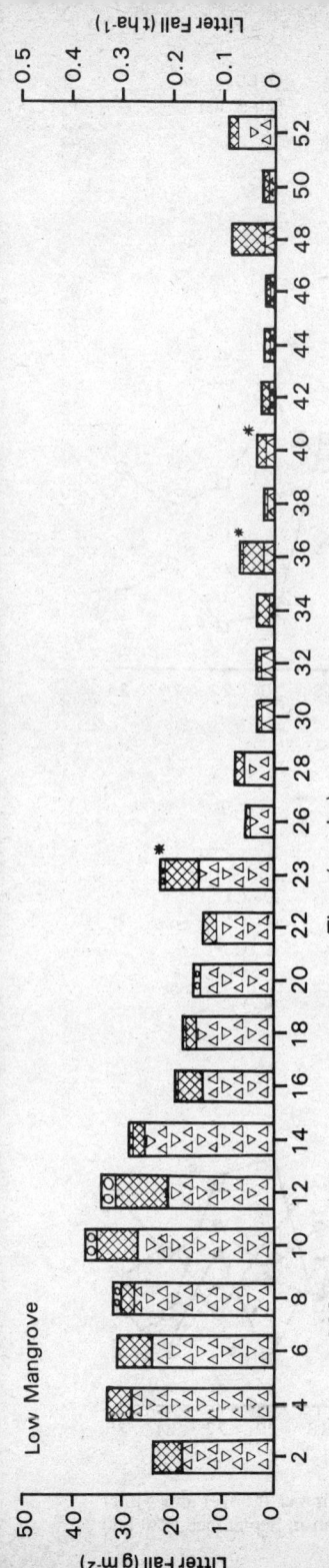

Figure 84 Litter fall beneath tall *Avicennia* (above) and low *Avicennia* (below) in Tuff Crater, Auckland, New Zealand, November 1980–November 1981 (after Woodroffe 1982).

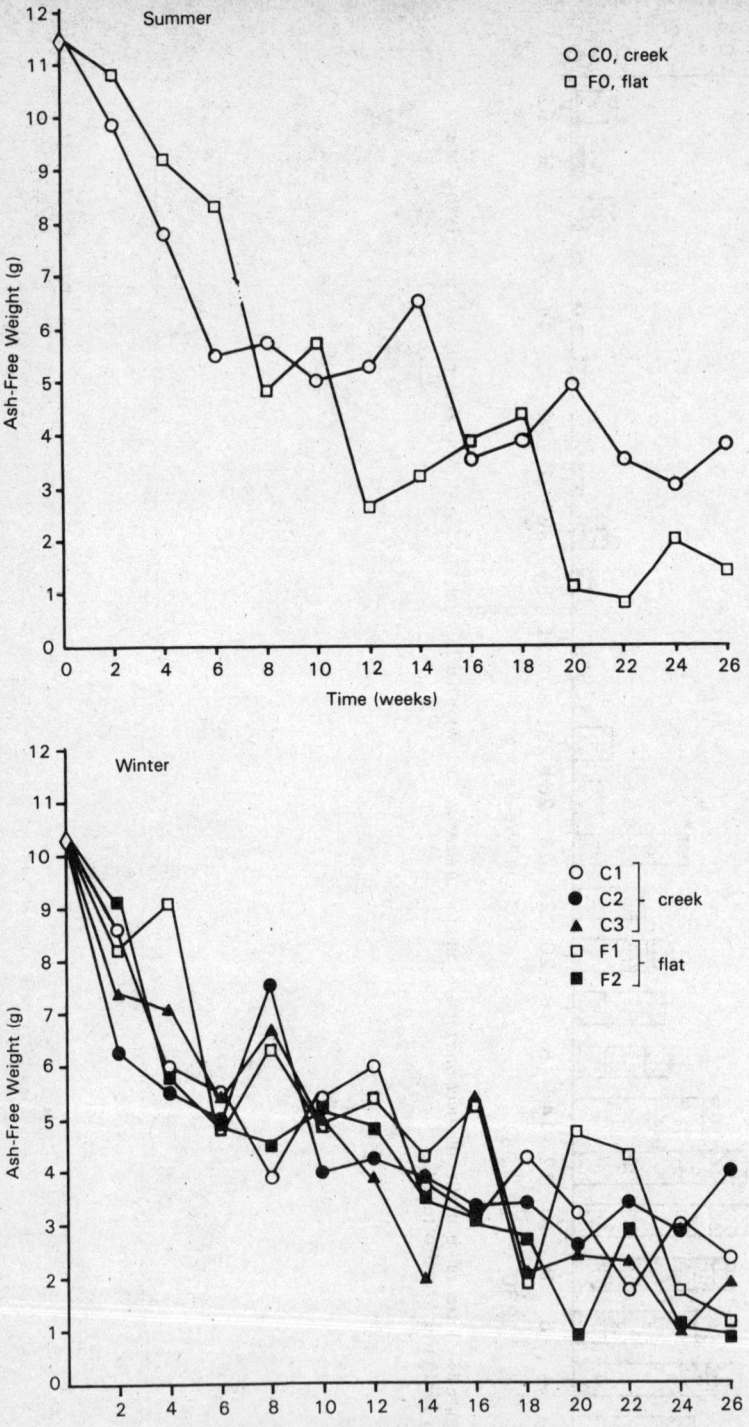

Figure 85 Decomposition of *Avicennia* leaf litter in summer (above) and winter (below) in Tuff Crater, New Zealand. Experiments began in September 1980 and March 1981 respectively (after Woodroffe 1982).

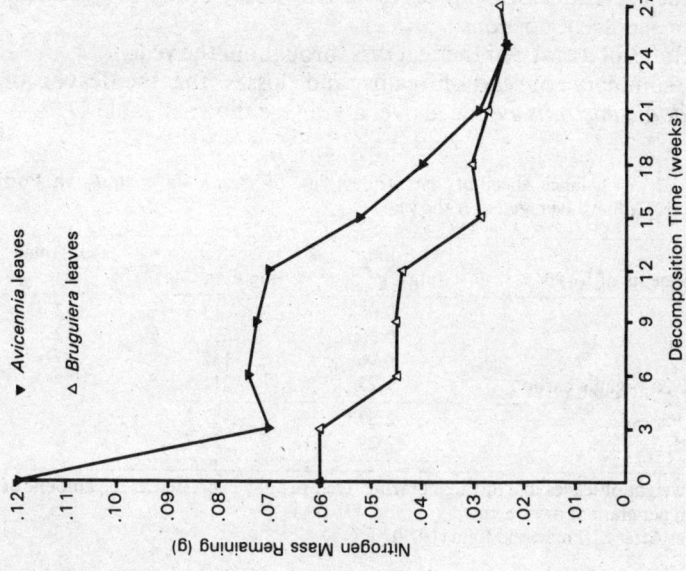

Figure 87 Decrease in mass of nitrogen per litter bag during decomposition (after Steinke, Naidoo and Charles 1983).

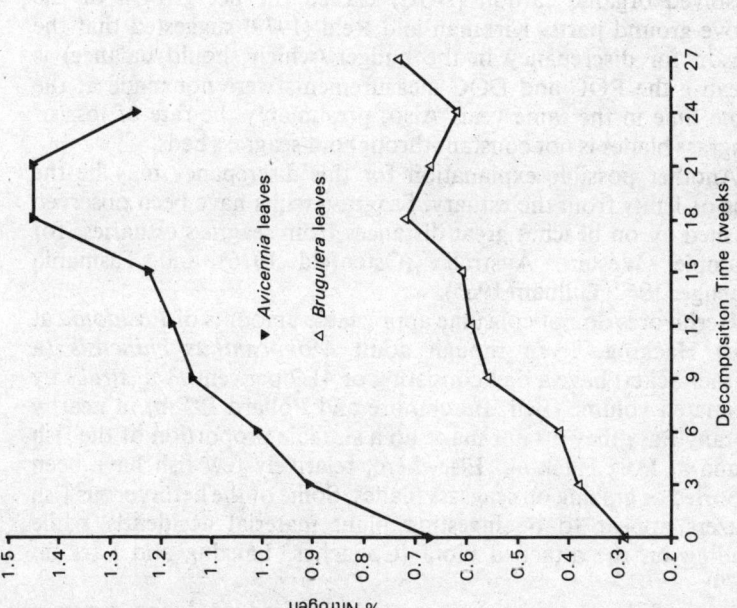

Figure 86 Decrease in nitrogen percentages of *Avicennia* and *Bruguiera* leaves during decomposition (after Steinke, Naidoo and Charles 1983).

5. Dissolved organic carbon in the water column, contributed by the excretions of living leaves and by decay of the non-growing or senescent portions
6. Rates of growth of the seagrass throughout the year.

A summary of carbon gains and losses for the leaves of *Posidonia australis* averaged over a year are shown in table 47.

Table 47 A balance sheet of the productivity of *Posidonia australis* in Port Hacking, Sydney, averaged over the year

Components of losses	Rate (mg C g^{-1} day^{-1}*)	Rate (mg C·m^{-2} day^{-1})	% of total losses
Grazing	0.08	13.5	3
Floating	0.30	50.6	12
Sinking	0.96	162.3	37
Dissolved organic carbon	1.25	211.3	48
Total losses	2.59	437.7	
Growth	2.29	388.7	

* dry weight of leaves lost through grazing, etc., per day expressed as the amount of carbon per gram of dry weight.
Source: After Kirkman and Reid (1979).

Losses exceed gains by about 13 per cent in this budget, that is, losses from the leaves of particulate organic carbon (POC) and dissolved organic carbon (DOC) exceed the net growth of the above-ground parts. Kirkman and Reid (1979) suggested that the reason for discrepancy in the budget (which should balance) is because the POC and DOC measurements were not made at the same time in the same year. Also, presumably the rate of loss of seagrass blades is not constant throughout seagrass beds.

Another possible explanation for this discrepancy may be the loss of fruits from the estuary. Seagrass fruits have been observed washed up on beaches great distances from seagrass estuaries, for example, Western Australia (Ostenfeld 1916) and Tasmania (Saenger 1967; Gillham 1965).

Herbivores do not consume appreciable amounts of *Posidonia* at Port Hacking. Even though adult *Monocanthus chinensis* (a leatherjacket) have a diet consisting of 41.7 per cent *P. australis* by estimated volume (Bell, Burchmore and Pollard 1978b) in nearby Botany Bay, they do not make up a sizeable proportion of the fish fauna at Port Hacking. Elsewhere, relatively few fish have been reported as grazing on seagrass blades. Some of the herbivorous fish grazers appear to be ingesting plant material accidently while feeding on the attached biota (Conacher, Lanzing and Larkum 1979).

Waterbirds such as swans and ducks feed on living seagrass

blades (Ferguson-Wood 1959); such feeding is often markedly seasonal (Hodgkin 1978). In tropical beds, dugongs and some turtles may remove large quantities of seagrass (Rebel 1974; Jones 1967; Heinsohn et al. 1977).

Seagrass leaves plus the organic detrital content of the sediments make up the major part of the particulate organic carbon (POC) lost from the plants. These leaves are skeletons of live plants in which cell walls and lacunae have broken down and are covered by epibiota. Apart from adding detritus to the sediment they offer surfaces upon which micro-organisms, plants and animals can grow which in turn break down to add more dissolved organic carbon to the overlying water. Smith and Penhale (1980) demonstrated uptake of ^{14}C-labelled organic compounds by the seagrass *Zostera marina* and its epiphytes. The heterotrophic component of the epiphytic community appears to be an active part of the epiphytic microflora and capable of utilizing substantial amounts of dissolved organic matter and converting it into particulate matter. It seems likely that eelgrass has the potential to transport organic materials through its vascular tissue and release it into the estuary in a manner similar to the transport of carbon and phosphate (Wetzel and Penhale 1979; Penhale and Thayer 1980). Seagrass blades free of epiphytes release almost twice as much dissolved organic carbon than do encrusted, colonized blades (Penhale and Smith 1977). If this excreted material is converted into particulate carbon at the assimilation rate found by Smith and Penhale (1980), then heterotrophic production is a sizeable portion of the autotrophic fixation. This role of epiphytic heterotrophs may be very important in providing food for grazing organisms. It may provide a significant source of new particulate matter in estuarine food chains.

Detrital Export

Some of the particulate matter is exported out of the estuary. Some seagrass blades have been reported in deep-sea sediments (Menzies, Zanevald and Pratz 1967), and as already mentioned the fruits of seagrasses and mangroves may be washed up on beaches great distances from their origin. Zieman et al. (1979) discussed the export of seagrass from a bay in the Virgin Islands. Approximately 60–100 per cent of the *Syringodium* production is carried out of the system, whereas only about 1 per cent of *Thalassia* production is exported. The net amount exported depends upon the season (primary production is seasonal), tidal flushing and rainfall. In temperate systems, leaf litter accumulates around the bases of the plants and breakdown occurs *in situ*. During storms this litter may be washed up on to beaches or out of the estuary.

In tropical estuaries in northern Queensland where the tidal range is of the order of 3 metres, litter rarely accumulates. Boto and Bunt (1981) suggested that most of it is exported out of the estuary still as litter. They calculate that 10 kg C ha^{-1} day^{-1} is exported in this form by tidal action. However, some of the primary productivity is broken down *in situ* and then exported. This occurs in backwaters or in upper regions of the mangroves where tidal flushing is reduced. The net export rate attributable to tidal-borne particles in their study area, Hinchinbrook Island, is approximately 1.5 kg C ha^{-1} day^{-1}. Combining these figures gives an export of 11.5 kg C ha^{-1} day^{-1}, a value much higher than for Florida mangroves, where Lugo and Snedaker (1973) calculated only 2.5 kg C ha^{-1} day^{-1} to be exported. The amount of POC exported from Australian estuaries seems very high, and Boto and Bunt (1981) suggested it may be because of the unusually long time it stays within the system before tidal flushing occurs.

Presumably these rates may be even higher during the wet season when terrestrial litter is added to the wetland litter. In temperate estuaries, similar terrestrial inputs occur during flood conditions. Evidence of plant litter of terrestrial origin has been commonly observed in mangroves. Heavy rains after bushfires may also carry other organic matter of terrestrial origin into the wetlands. Estuaries in southern and western Australia situated in arid regions generally have little terrestrial input.

Terrestrial pulses of organic matter have been observed overseas. In Florida, salt marshes are replaced by extensive sawgrass (*Cladium jamaicense*) flats, and it has been shown that, after heavy rains, organic matter flows out of the sawgrass into mangroves and other shallow coastal systems. Between floods, the organic matter builds up and accumulates in the soil as partially decomposed sawgrass plants. Perhaps a similar process occurs in Australian beds of sedges (*Juncus*).

The impact of the variation of detritus export caused either by seasonal primary productivity or by terrestrial pulses of organic matter into the estuarine ecosystems is unclear. There is some evidence that catches of prawns in the Gulf of Carpentaria are related to good wet seasons (Staples, 1980a, 1980b).

Feeding Strategies in an Estuarine Community

Odum and Heald (1975b) analysed the feeding strategies of an estuarine community in Florida. This community is characterized by a large production by vascular plants and a low algal production. Most of the mangrove leaf and stem production are not eaten directly but become part of the decaying matter in the forest floor.

The mangrove detritus forms the basic component in the primary marine consumers' diet. Among the invertebrate herbivores and herbivorous fish species recognized by Odum and Heald, 20 per cent of the material they ate consisted of vascular plant detritus; they concluded that mangrove leaf detritus is the most important single element in the food web. Of the 120 species examined, approximately one-third could be classified as detritus consumers. These animals had guts which contained at least 20 per cent of vascular plant detritus by volume on an annual basis. They included herbivores and omnivorous species of crustaceans, molluscs, insect larvae, nematodes, polychaetes and a few fishes. Probably similar chains occur in Australian estuaries where mangroves dominate. However, Redfield (1983) suggested that the food chains described by Odum and Heald (1975b) overemphasized the role of bacteria. Based on studies carried out in Australian mangroves, Redfield suggested that more emphasis should be placed on the primary consumption of mangrove leaves by insects (Onuf, Teal and Valiela 1977) and mud-dwelling herbivores (Malley 1978), and less emphasis on importance of the route through bacteria and fungi. Insect attack, which may occur while the leaves are still on the trees, may facilitate breakdown once the leaves fall. Indications of insect grazing on otherwise healthy leaves are indeed common, particularly within dense and shaded mangrove foliage.

Estuarine Fish Communities

Because of the potential commercial importance of estuaries for fish populations, some information on the importance of wetlands for fish communities is available. Many species of fish move from the deeper parts of the estuary into the shallow wetlands to feed and seek shelter during high tides. Such migrations of larval and juvenile stages in turn encourages predators to come into the wetlands to feed (Littlejohn, Watson and Robertson 1974).

Beumer (1978) studied the diets of several fishes in mangrove creeks in northern Queensland. Black bream (*Acanthopagrus berda*), and hair-finned goby (*Ctenogobius criniger*) are predominantly carnivorous, bony bream (*Andontostoma chacunda*) are detritivorous and the milk-spotted toadfish (*Chelonodon patoca*) is omnivorous. These species utilize the creek as a nursery and feeding area. Similarly, Blaber (1980) suggested that if Trinity Bay is representative of areas of open water adjacent to tropical estuaries, the significantly lower numbers of piscivores in the estuary may increase the value of estuaries as sanctuaries for juvenile fish.

Recently Bell et al. (1984) confirmed that mangrove habitats in temperate Australia are important nurseries for fish inhabiting adjacent estuarine and inshore marine habitats as adults. The gut contents of abundant species were examined and they all had very different diets; some species were selective whereas others were omnivorous, feeding on both plant and animal matter.

Moriarty and Pollard (1981) showed a high bacterial biomass and high productivity in seagrass sediment, suggesting that a large proportion of the primary productivity is cycled through bacteria to the higher trophic levels. Some species such as mullet (*Mugil cephalus*) (figure 88) and prawn (*Metapenaeus bennettae*) make direct use of the bacteria present in muddy estuarine sediments. It comprises between 15 and 35 per cent of the organic carbon in their gut contents (Moriarty 1976).

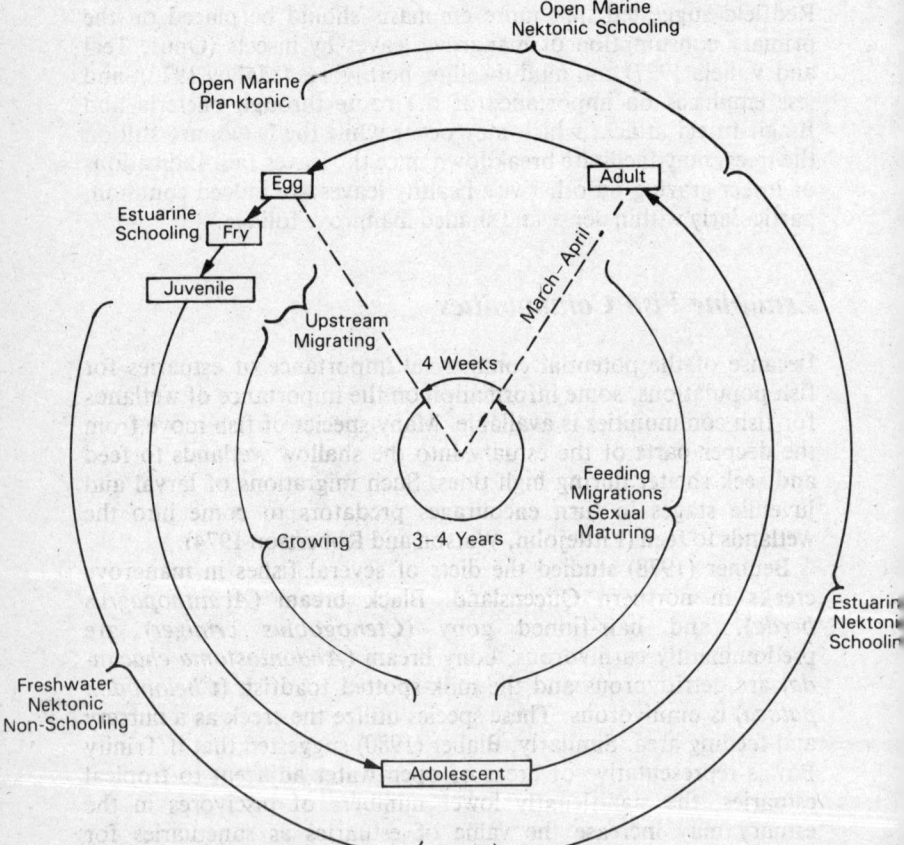

Figure 88 Schematic diagram of the life cycle of the sea mullet, *Mugil cephalus*.

Other fish in the seagrass beds are carnivorous, feeding on the zooplankton (Robertson and Howard 1978). As the zooplankton undertake vertical migration over a twenty-four-hour period, the fish must either cease to feed at certain times or change their diet to bottom-living amphipods.

In the Caribbean, Ogden and Zieman (1977) documented the feeding migration of fish from seagrass beds to nearby coral reefs. Such events also may occur in the tropical regions of Australia where seagrasses and coral reefs are contiguous.

A detailed analysis of the feeding strategies of the fish community in a *Rhizophora mangle* forest in Florida was carried out by Odum and Heald (1972). They analysed the gut contents of many species of fish. These included a few species which fed primarily on vascular plant detritus, but the majority were carnivorous, either feeding on pelagic or benthic organisms.

The trophic relationships of seagrass communities were studied in detail by Brook (1977) for *Thalassia* beds in southeastern Florida. Using gut contents, the principal interaction between the primary consumers and the higher trophic level predators was via the polychaetes and the peracaridean crustaceans.

Exit Links

As already mentioned, some detritus, litter and POC is exported directly out of the estuary. These components of primary production are not easy to measure but may be important for understanding offshore communities like coral reefs, and the dynamics of offshore commercial fishing (Redfield 1983).

Some of the estuarine primary productivity is exported out of the estuary by migration of animals into open water. Commercially important migrations are undertaken by several species of prawns, crabs and fish. Banana prawns (*Penaeus merguiensis*) (figure 89) are caught in Edgecumbe Bay, Bowen (an offshore marine community), after they have been flushed out of the estuaries in February. The prawns carry with them much of the energy assimilated in the proximity of the mangrove community. Moriarty and Barclay (1981), working on four species of prawns from the Gulf of Carpentaria, found the main component of their diet to be meiofauna. The low bacterial density and high protein content of their food indicates that most of these prawns are not feeding on detritus, but rather on living animals (Moriarty 1977). However, as already discussed, this abundant meiofauna is probably dependent on the detritus in the sediments in the Gulf which have been washed out of the surrounding wetlands. Moriarty and Barclay (1981) also suggested that bacteria are less important in the food of prawns

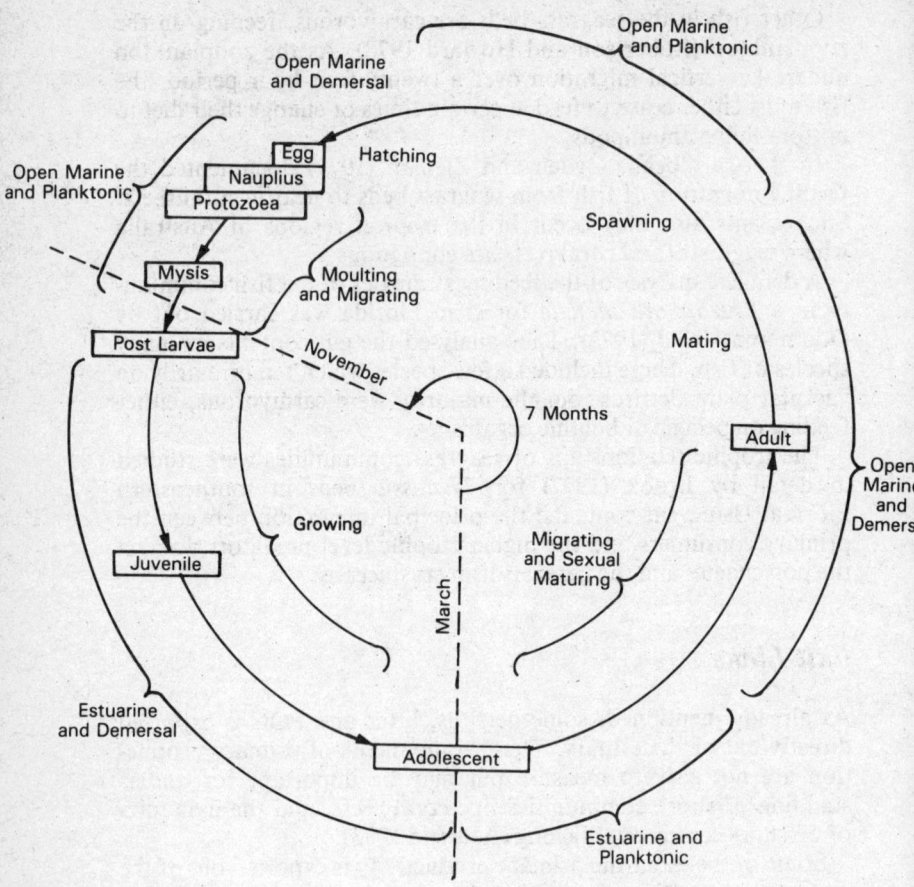

Figure 89 Schematic diagram of the life cycle of the banana prawn, *Penaeus merguiensis*.

than previously reported. Bacteria constituted less than 2 per cent of the organic matter in the adults of all species, but in many juvenile *Penaeus merguiensis* bacteria were more important, constituting up to 14 per cent of the total organic matter. Thus, the link from wetlands to prawns may not be direct, but prawns and other migratory animals, such as fish, clearly remove much of the production of a mangrove ecosystem (Redfield 1983). The life cycles of banana prawns (*P. merguiensis*), sea mullet (*Mugil cephalus*) and barramundi (*Lates calcarifer*), all commercially important species, are shown in figures 88 to 90, and indicate the importance of estuaries to these species both as habitat and probably as sources of

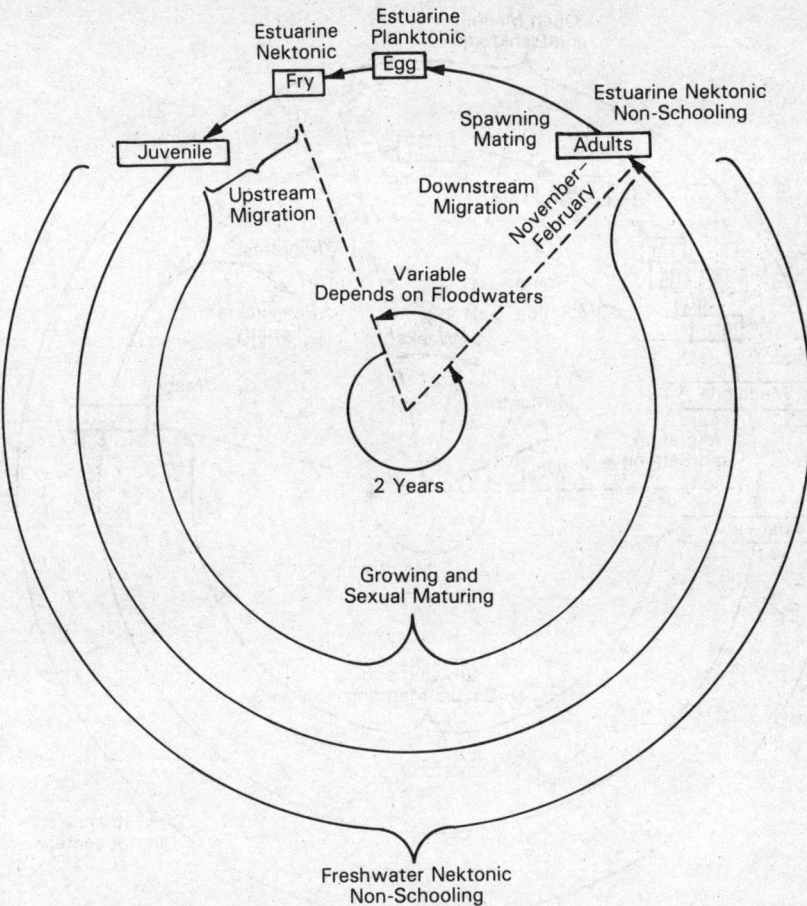

Figure 90 Schematic diagram of the life cycle of the barramundi, *Lates calcarifer*.

food. Pollard (1981) suggested that about 70 per cent of commercially important fish species in New South Wales are dependent on estuaries.

Another example of a species moving out of the wetlands into offshore marine zones for spawning are female mud crabs, *Scylla serrata* (Hill 1979; pers. comm.). The male remains in the estuary at this time (figure 91).

Other exit links from the cycle are provided by wading birds feeding on the benthos and fish during low tide. Juvenile crocodiles in tropical areas feed exclusively on crabs and small fish on the mudflats among the mangroves (Taylor 1979).

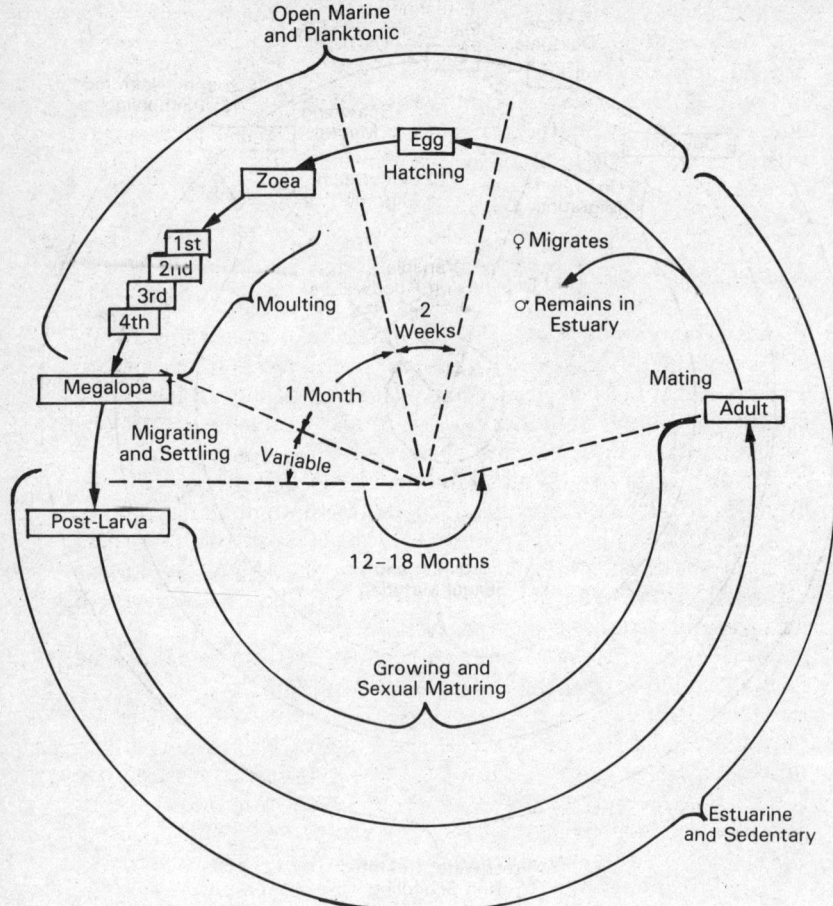

Figure 91 Schematic diagram of the life cycle of the mud crab, *Scylla serrata*.

Generalized Detrital Cycle

A basic detrital cycle is shown in figure 81. Probably it is actually more complex and the number of exits and returns far more numerous than shown. Superimposed upon this detrital cycle are seasonal and annual variations in the amount of primary productivity, the supply of terrestrial inputs, variations in recruitment success of animals, and fishing pressures on the fish and prawns. Managed properly, this detrital cycle is self-sustaining, but removal of primary productivity components (such as reclamation or degradation of wetlands) will rapidly degrade the entire system. This degradation also will have impact on offshore ecosystems dependent upon a supply of detritus and POC from the estuaries. In figure 92, the substantial contribution played by wetlands in production of organic matter is shown throughout the world, and contrasted with production figures for other types of vegetation (Deevey 1960).

Exploitation at Higher Trophic Levels

In Australia, virtually the only exploitation by humans of wetlands is for catching of bait, oyster farming, prawning, and commercial and sport fishing. Oyster farming and the fishing industry are economically very important, and proper management of the wetlands will ensure that they remain so. In table 48, figures for oyster, fish and prawn production in New South Wales for selected estuaries are given to illustrate this point (West, unpubl. data).

In the past, some mangrove trees were cut and used as wharf piles, as the wood is resistant to teredo attack. Also, tannin was extracted from mangrove trees and formed the basis of small localized industries, but these are now defunct.

Overseas, apart from the fisheries associated with wetlands, the main exploitation of mangroves by humans is the use of the trees as firewood. In Indonesia and India especially, mangroves may be the only source of firewood available for domestic use. As a result of this extensive culling in parts of India, no primary mangrove forests remain, only secondary forests. Perhaps the most serious and totally destructive exploitation of mangroves is that of using the trees for wood chips and paper pulp production, as is occurring in parts of southeastern Asia. These processes, which remove the primary producers and hence the supply of "fuel" to the detrital cycle, will rapidly lead to the destruction of the estuarine ecosystem and consequently the loss of the fishing industry.

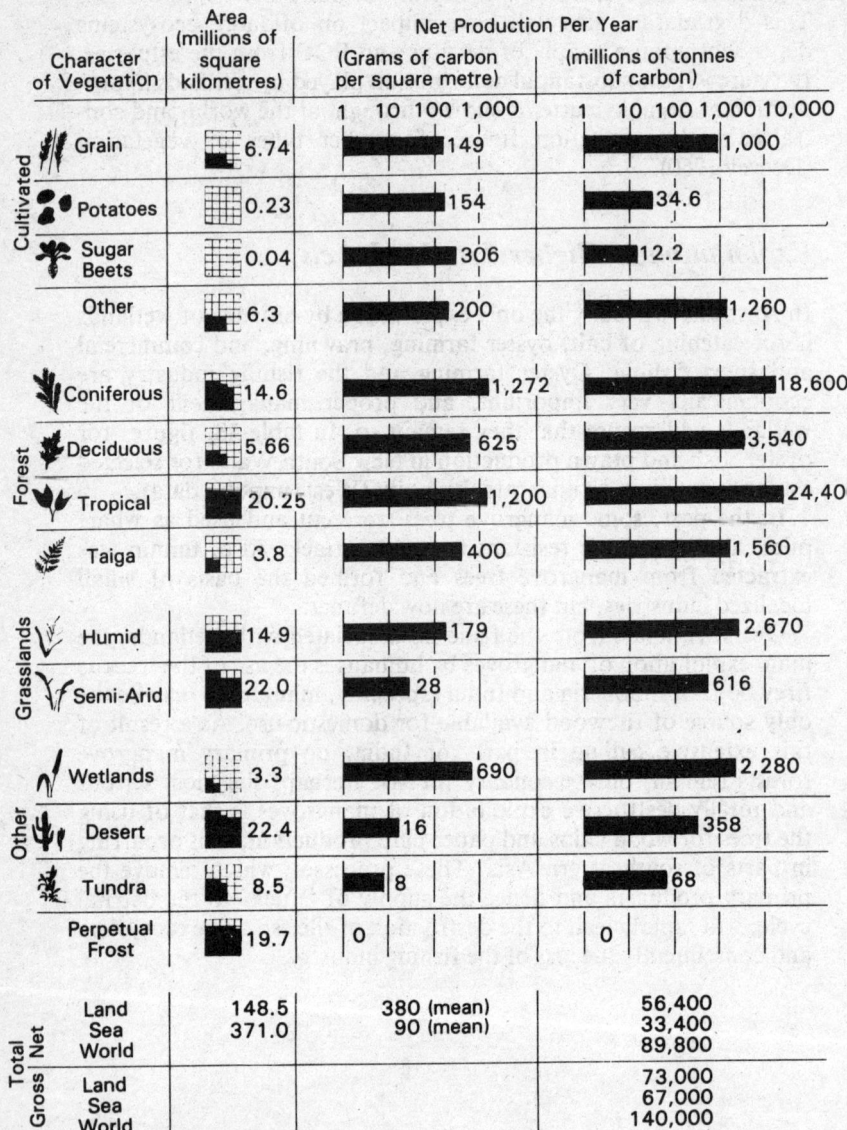

Figure 92 Production of organic matter per year by the land vegetation of the world — and thus its ultimate food-producing capacity — is shown in terms of carbon content. Cultivated vegetation (top left) is less efficient than forest and wetlands vegetation, as indicated by the uptake of carbon per square metre (third column), and it yields a smaller overall output than do forest, humid grasslands and wetlands vegetation (fourth column). Difference between net production and gross production is accounted for by the consumption of carbon in plant respiration (after Deevey 1960).

Table 48 A Fisheries-orientated inventory of selected New South Wales estuaries

Statistic	Northern Region	Hunter Region	Gosford–Wyong Region	Sydney Region	Illawara Region	Southern Region	Totals
Nos. of estuaries	39	8	7	9	26	44	133
Mapped water area (sq. km)	252	480	100	142	221	181	1376
Mangrove area (sq. km)	29.425	45.025	12.289	7.456	5.840	6.690	106.725
Seagrass area (sq. km)	29.824	57.057	18.031	8.266	28.255	13.573	155.006
Salt marsh area (sq. km)	18.618	24.093	2.051	2.097	4.584	6.607	58.051
Total mean fish prod. 1972–82 (kg/yr)	1,586,090	1,258,881	413,584	274,916	407,846	418,390	4,359,671
Total mean prawn prod. 1972–82 (kg/yr)	338,699	174,478	160,153	70,009	65,053	12,136	820,528
Total mean prod. of other crustaceans 1972–82 (kg/yr)	47,076	121,692	11,203	14,428	5,439	2,509	202,447
Total mean oyster prod. 1972–82 (kg/yr)	1,252,632	3,405,342	1,333,702	1,762,904	168,773	859,007	8,782,360
Total mean prod. of other molluscs 1972–82 (kg/yr)	455	8,776	95,777	71,890	173,869	747	351,514

Source: Fisheries-orientated inventory consisting of tabular information and an atlas of estuarine wetlands covering 133 estuaries and embayments along the New South Wales coastline by R.J. West, et al., 1985.

9. Conservation and Management of Mangrove Communities

Australia is the driest of all the continents; consequently its estuarine resources are limited. As a result of past and present human activities, these resources have been depleted. Keen competition for use of the remaining estuarine resources is apparent, and it is likely to further intensify with an increasing population, most of which is concentrated along the eastern seaboard. The considerable range of competing uses presents the decision-makers with confusing options. Their task is made all the more difficult by (1) the strong emphasis on economic evaluation of alternatives in an ecosystem whose biological values have rarely been quantified, (2) the need to resolve conflicts in resource use in the interests of a community which, by and large, fails to appreciate the value of the resource, (3) the absence of a realistic ecological basis on which to evaluate and manage the various forms of resource utilization (AMSA 1977) and (4) differences in the framework of engineers and planners on the one hand and biologists on the other (because of the seasonal and annual fluctuations in biological systems).

The decision-maker can manage the mangrove ecosystem as a renewable resource producing fuel, oysters, fish, fence material, dyes, tannins, honey and, not least, providing fisheries products and a range of scientific, educational and recreational opportunities. Alternatively, the mangrove ecosystem can be considered to be non-renewable and exploited for the space it occupies, for agriculture, buildings, residential wharves, airports, marinas and roads. Somewhere between these two extremes lie additional alternative uses for this ecosystem: mari-culture, waste disposal, and wood chips and other forestry products.

In theory then, the ideal decision-maker will manage the ecosystem so as to leave as many resource-use options open in the long term as possible. In practice, unfortunately, decisions are usually taken in the short-term interests of expediency, mostly dominated by a desire for economic and political gains. Making decisions purely by economic and political yardsticks is acceptable provided that the economic costs are truly inclusive of all the elements involved (including economic losses owing to permanent loss of resources for alternative uses), and that the community is sufficiently well informed so as to translate the misuse of a valuable resource into a political weapon. In relation to the management of Australia's mangrove resource, both the decision-makers and the electorate at large have failed miserably, often because of a lack of awareness and because of inadequate public education.

The situation in Australia today is that mangrove ecosystems are not recognized as a valuable national asset by most decision-makers nor by the community generally. Mangroves are not managed by any single authority as a national resource as are terrestrial forests, national parks, mining and fishing. In the absence of a coherent attitude towards this resource, management decisions in relation to mangroves are taken in a piecemeal fashion. The development area is seen in isolation and the regional context or the catchment area of the particular site is conveniently ignored. The result is a constant gnawing away of the resource without taking into account the full implications of the impact it makes (Odum 1982). Not only is this a wasteful process in relation to the resource, but it may, in fact, give rise to problems that with proper consideration would not have arisen.

Are Mangroves Endangered?

If a resource is not endangered, a management policy is unnecessary. However, in today's world of burgeoning human populations, few resources, even those that may be considered as "inexhaustible" such as air, soil and water, can be looked upon as unlimited. Few, if any, forest communities can be ignored from the standpoint of conservation. This is equally true of mangroves.

In many parts of the world mangroves are being destroyed at very rapid rates (Saenger, Hegerl and Davie 1983). For example, the island of Puerto Rico originally had a maximum area under mangroves of 26,300 ha. Rate of destruction was low in the early history of the island (for example, only about 1.7 per cent per decade between 1930 and 1960), but the overall destruction from the maximum extent was 28 per cent by the mid 1960s. The following decade, especially the five-year period between 1965 and 1970, saw massive destruction, with the largest single mangrove area (325 ha) being 88 per cent destroyed and showing little regeneration. Other large swamps were similarly severely damaged and many of the smaller ones were completely destroyed (Heatwole 1985). The causes of this severe depletion of the mangroves were many fold and included petrochemical pollution, cutting for charcoal, construction of marinas, and altering drainage patterns from swamps by road construction and sand mining.

This is not a unique case. In many parts of the world the destruction of mangroves is proceeding on a large scale, but has seldom been documented in detail. Places in which massive destruction is now occurring are Indonesia, the Philippines, Gambia, Ecuador and Benin (Saenger, Hegerl and Davie 1983).

Australia has not seen as great a devastation as many countries have, and had rather extensive tracts initially. Nevertheless, it is clear that clearing of mangroves is increasing markedly without sufficient monitoring, concern or thought.

Are Mangroves Worth Managing?

This question must underlie the entire management or usage concept, for unless it is recognized that the mangrove resource is valuable, it is unlikely that sound management procedures will ever be developed (Greeson, Clark and Clark 1978). Estuarine wetlands are among the most productive natural systems in the world (Clark 1974; Gore 1977); for example, the amount of organic material produced annually by certain temperate salt marshes in North America exceeds that produced by the world's best strains of wheat, corn or sugarcane (Odum 1973). Because of their high productivity, estuaries function as nursery and feeding grounds for a very large percentage of coastal fish taken by commercial and amateur fishermen (Newell and Barber 1975; Haedrich and Hall 1976; Pollard 1976; Bell et al. 1984; Blaber, Young and Dunning 1985). Prawn and oyster production is almost entirely estuarine-dependent in New South Wales (Malcolm 1971; Ruello 1973). The value of the commercial and amateur fishery in Australia is now over $100 million annually (Newell and Barber 1975).

In addition, many species of local and migratory waterbirds breed, roost and feed in estuarine areas (Holmes 1970; Fox 1973; Braithwaite 1975; Briggs 1977b). Finally, coastal wetlands store nutrients and regulate their passage into the estuary and near-shore region (Axelrad, Moore and Bender 1976). Wetlands also have the ability to remove contaminants (various hydrocarbons and heavy metals) and suspended sediments from estuarine waters (Valiela and Vince 1976; Boto and Patrick 1978; Harbison 1981).

The biological importance of estuaries and estuarine wetlands necessitates their conservation and management, especially as society is now looking to the shallow coastal seas and the estuaries to augment the world's supply of protein (Ryther 1975).

Careful planning and sound management are essential to the proper use of any resource, and this principle also must be applied to the use of coastal and estuarine resources, including mangroves. The need for conservation and management of estuaries near the larger cities is particularly urgent as these estuaries are subjected to the greatest stresses (Shepherd 1970; Dunstan 1973; Odum 1973;

Gilmour 1974; Hodgkin 1974; Shapiro 1975; Hutchings and Recher 1977).

The main benefits from planning and sound management of estuarine areas include:
1. Maintenance of attractive and readily accessible areas of high scenic and aesthetic value, suitable for both passive and active recreational pursuits by all members of the community
2. Conservation of important wetland and estuarine habitats and of the breeding and nursery grounds of many marine organisms and waterbirds
3. Conservation of feeding areas for migratory birds
4. Retention of a "drought refuge" habitat which can be used by inland waterbirds in dry years
5. The continuing profitability of shellfish cultivation and of the inshore and estuarine prawning and fishing industries
6. Maintenance of a range of natural ecosystems which are suitable for teaching and research purposes
7. Reduction of the problems of sedimentation and erosion and consequently the avoidance of expensive corrective engineering works (AMSA 1977).

Many other direct but lesser benefits from proper management of estuarine resources could be added to this list (Cocks 1975; Lugo and Brinsen 1978; Hegerl 1982).

In view of these benefits, it seems that the answer to the original question should be in the affirmative. Having established that mangroves do indeed warrant management, the specific question "Management for what purpose?" must be approached. From the benefits described above as emanating from proper management of the resource, it is apparent that the aim should be to maintain its use as a renewable resource, providing fisheries products and possessing an inherent amenity value based on its geomorphological, recreational and scientific characteristics. Only the most pressing and essential community demand should be considered to justify treatment of the mangrove resource as non-renewable (Saenger, Hegerl and Davie 1983). Canal estates, garbage tips, industrial land, playing fields or other uses requiring reclamation would need to be justified in the light of the fact that a valuable renewable resource was being permanently destroyed.

Management — Whose Responsibility?

Having decided that the mangrove resource is worthy of careful management, it is essential to determine who should be responsible for policy. Since nearly all of the mangroves in Australia occur on Crown land, it is reasonable to expect that responsibility should reside with the State. Accordingly, in all mainland states, the State has assumed legislative control over mangroves, generally under the respective Fisheries Acts, and has afforded a minimal measure of protection. However, the situation is not as simple as it appears, for the State covers a multitude of governmental instrumentalities, often with overlapping areas of interest, and always with a finely tuned sense of interdepartmental rivalry and hierarchy.

The situation as it exists in Queensland illuminates the difficulties in defining who is responsible for mangrove management (Baines 1975; Saenger 1981). Early legislation in Queensland (Fish and Oyster Act of 1914) was reasonably enlightened and it set aside certain areas as "protected"; in these areas the cutting of oyster stakes was controlled by permits issued by Fisheries authorities. The Fisheries Act of 1957 extended this protection to all mangrove areas of the Queensland coastline with the exception of specific areas excluded by Order-in-Council. The Fisheries Act of 1976 protects all marine plants and permits are required for their chopping, burning, collecting or other use.

These Fisheries Acts were clearly useful in preventing direct damage to the mangrove and, through the permit system, in controlling their exploitation. However, in relation to the mangrove system as a whole, these Acts proved to be inadequate. In addition to the general concern felt by Fisheries authorities about disruption of mangrove ecosystems, the activities of commercial diggers of polychaete bait worms and the yabbie *Callianassa australiensis* from the intertidal mudflats had become increasingly destructive, interfering with the traditional seining grounds of professional net fishermen. Much to their dismay, the Fisheries authorities found they were unable to prosecute to protect the ecosystem owing to a legal interpretation of the Fisheries Act. The Court held that worms did not constitute "fish" or a "marine product" under the meaning of the Act, and were therefore not covered by the Act, despite the Act's intent. These definitions were enlarged in the revised 1976 Fisheries Act.

To prevent sediment disturbance in the mangrove ecosystem, legislation enabling the setting up of Fisheries Habitat Reserves was passed in 1968 (The Fisheries Habitat Reserve Regulations of 1968), and similar regulations have been continued under the Fisheries Act of 1976. The status of Fisheries Habitat Reserve is granted by an Order-in-Council of the Queensland government and this can be altered only by the same authority. Executive Council is composed of the governor and the cabinet, assuring the Fisheries Habitat Reserves of the strongest authority. Further, the authority is vested in fifteen men, not (as so often is the case with natural areas) in a single minister. This spread of responsibility among a number of senior decision-makers reduces, though it does not eliminate, the risks of removal of "reserve" status through the political and economic pressures which can be focused so effectively on a single minister.

At present, then, the legislative framework seems to exist for the proper decision-making on mangrove ecosystem utilization. However, the present structures and procedures exhibit certain weaknesses which it is instructive to discuss (Saenger 1981).

Authority for the development of mangrove ecosystems is vested in two separate ministries (minister for Lands and minister for Harbours and Marine), and the Coastal Zone Management Study was authorized by the Premier's Department. An attempt is made to overcome this split responsibility by referrals to the various interested sections of government. However, since authority to grant or refuse an application for exploitation is not vested solely in one section (for example, Lands Administration for permission to reclaim mangrove areas, Fisheries Branch for permission to cut mangroves, and Harbours and Marine for permission to develop canals in mangrove areas), advice tendered may, if it conflicts with others' views, still be ignored. This diffusion of responsibility for the mangrove ecosystem is further compounded by the responsibility for waste discharges to mangrove areas: the discharge of liquid wastes is controlled by the Water Quality Council, gaseous wastes by the Clean Air Council, solid wastes by the Department of Local Government and radiological wastes by the Health Department. Each of these have varying permit requirements and are able to make final decisions without necessarily referring to the other instrumentalities which may have an interest, or to the Queensland Fisheries Branch, which has the primary responsibility for the mangroves themselves.

It is doubtful whether meaningful policies of natural resource utilization can be devised when the administrative responsibility is so fragmented. A more balanced assessment of alternatives for the utilization of tidal ecosystems could be provided by a decision-making unit which had sole responsibility for them. However, administrations are rarely structured on the basis of ecosystems. Mangrove and other tidal ecosystems may suffer more from this division of authority than do most others as their fate is complicated by a sea-level line (high water springs) which separates the ecosystem into two parts to which different legislation applies and over which different instrumentalities hold authority. This is a problem for other coastal ecosystems as well, although in Queensland the Beach Protection Act of 1968 has enabled an environmentally more realistic administrative approach to beach areas which extend into both marine and terrestrial environments.

Similarly complex and fragmented control over mangroves exists in other states as well.

Management — On What Basis?

To develop effective management plans for mangrove resources, it is necessary to relate them to management problems of the adjoining tidal lands and estuarine waters (AMSA 1977). Mangroves must be viewed as a part of a complex estuarine system of inter-related habitat and dependent biota which, in turn, is maintained by natural drainage patterns and rates of freshwater discharge from the catchment on the one hand and the natural tidal and salinity regimes on the other. It is the natural movement of water that provides the essential linkage of the terrestrial and aquatic elements in these coastal ecosystems (Clark 1974). Thus, in planning the management of estuaries it is important to recognize that some activities in the catchment can have far-reaching effects on associated near-shore regions through their effects on the quality of the water in the catchment streams (Shapiro 1975; Atkinson et al. 1981). Clearly, then, the catchment of an estuary should be considered as part of the estuarine ecosystem and land use in the catchment must be coordinated with the overall aims of estuarine planning.

Management within the physical boundaries established above must proceed primarily on ecological data. Estuaries and mangrove systems cannot be assessed in "the twinkling of an eye"; a static vegetation map resulting from a cursory reconnaissance does not constitute an adequate assessment, nor is the simple application of forestry principles, largely developed to manage temperate deciduous trees, an adequate foundation for the good management of mangroves. In other words, management planning requires sufficient field data for each specific mangrove system to enable recognition of those processes, qualities and organisms which are in need of protection and specifically how these may be vulnerable to human activities. Although different estuaries share similar geomorphological and hydrological characters and have similar sorts of wetlands, individual estuaries do possess unique characteristics which further complicate assessment. Social characteristics and their significance must be evaluated. It is only after the ecological factors have been adequately assessed that other factors, be they economic, social or political, should be brought into focus. The managers of ecosystems of any kind must work primarily within an ecological framework, just as a designer of aircraft must work primarily within the constraints imposed by aeronautical and safety principles, and not base decisions solely on the laws of short-term economics. Any attempt to work outside an ecological framework must ultimately meet with difficulties and involve remedial expense or irreversible losses.

Management of mangroves must be based on a philosophy of conservation which, as a first step, seeks to prevent further destruction of existing mangrove ecosystems. Most importantly, it should recognize the need to devise management practices which optimize the conservation of mangrove resources in such a way as to provide for traditional and contemporary human needs, while ensuring adequate provision of reserves suitable for protection of the diversity of plant and animal life within them. Being a renewable resource, mangrove ecosystems must be managed on a sustainable-use basis (Saenger, Hegerl and Davie 1983). The concept of sustainable use involves sustainable harvest and economic benefit, and sustainable economic returns, at the same time maintaining the ecosystem as close to its natural or original state as possible. This tends to be difficult to attain except in a few cases such as use for tourism. Consequently, sustainable use often does not mean the original natural system in its pristine condition; a compromise may be reached which allows sustainable yield and reasonable resemblance to an undisturbed or non-harvested system. However, preservation or maintenance in a completely undisturbed or unexploited state may be a desirable management policy for certain localities or for some parts of extensive mangroves. Such unexploited areas may serve as a refuge for fauna and flora and as a resource for restoring areas in which management policies have failed or accidents have occurred. Preservation of some proportion of a mangrove area can buffer the area generally and can be an advantageous part of an overall sustained-use management plan.

The potential for implementation of conservation and management strategies differs within existing patterns of legislation and governmental organizations (Saenger, Hegerl and Davie 1983). A higher potential is discernible in those administrative systems in which mangroves are regarded as an integral component of coastal regions and not a botanical curiosity. Decisions concerning the use of mangroves can then be made in the proper context of its dependency on land use in the adjacent water catchment and on its important inter-relationships with estuaries, lagoons and coral reefs.

Some Specific Management Problems

Many coastal urban centres already have had detrimental effects on nearby estuaries and their mangroves. Management plans which inventory existing wetlands, seek to eliminate or reduce the stresses already imposed by urban centres, and prevent similar conditions developing in other areas should be developed by appropriate authorities (AMSA 1977). Such plans should involve the drawing up of specific management schemes for each of the following stress-producing factors:

1. discharge of wastes (thermal effluents, sewage and industrial wastes, urban and street run-off)
2. foreshore development for harbour installations, residential and commercial purposes, and garbage and other waste tips
3. flood mitigation and swamp draining works
4. reclamation and dredging and
5. bund walls.

These are five major means whereby estuaries and associated mangroves are often needlessly destroyed, and it is valuable to examine some of the specific ecological effects that can result.

Discharges of Wastes

The utilization of estuaries and mangrove areas as sinks for the discharge of liquid wastes is a well-established and still-growing phenomenon despite the passage of various Clean Water Acts by several state governments. In most instances, these Acts merely license the discharge and endeavour to keep the discharge level to an amount which it is assumed can be absorbed by the water body without any permanent deleterious change.

Little information is available in Australia on the effects of liquid wastes on the mangrove fauna and flora, although there has been some monitoring of sewage effluents (Poore and Kudenov 1978), thermal effluents (Saenger, Stephenson and Moverley 1982) and heavy metals (Bebbington et al. 1977; Ellis and Kanamori 1977; de Forest, Murphy and Pettis 1978; Arnott and Ahsanullah 1979).

The work of Nedwell (1974a) in Fiji showed that mangroves appear to have the capacity to absorb high levels of nutrients, particularly those contained in sewage. Nedwell suggested that a suitable tertiary treatment for sewage may well be attained by simply discharging secondary effluent into shallow retaining ponds in mangrove areas and allowing the overflow to discharge into the mangroves. This may be feasible only where no industrial wastes are included in the sewage because, as Nedwell (1974b) pointed out subsequently, when there is a danger of toxic bioaccumulation because of industrial contamination of domestic sewage the suggested use of mangrove areas must be discounted. More recently, it has been shown that heavy metals contained in sewage sludge are indeed released and are accumulated by species in the mangrove and seagrass ecosystems to which it is discharged (Montgomery and Price 1979). The seedlings of *Avicennia alba*, *Rhizophora mucronata* and *R. mangle* appear not to be adversely affected by zinc, lead, cadmium and mercury (Walsh, Ainsworth and Rigby 1979; Thomas and Ong 1984) but such materials, once accumulated in their tissues, can be passed along the food chain, ultimately with a potential of attaining unacceptably high levels in species consumed by humans.

Numerous studies of the effects of thermal wastes on estuaries and mangroves have been reported from overseas (Roessler 1971; Bader, Roessler and Thorhaug 1972; Thorhaug, Segar and Roessler 1973; Kolehmainen, Marten and Schroeder 1974). In all of these studies, marked reductions in the invertebrate fauna was reported, at least in the zone where mixing of the heated water occurred. Seagrasses also were affected adversely, although the mangroves themselves appeared to be able to cope with the increased temperatures. What these changes mean to the ecosystem in the long term is not known, nor have investigations been undertaken to determine the reversibility of such induced changes once a power station ceases operation.

Foreshore Development

At present, 80 per cent of Australia's population lives within 160 km of the coast and 60 per cent lives directly at the mouth of estuaries (Rooney, Talbot and Clark 1978). It is not surprising that the most widespread destruction of both mangroves and tidal marshes has resulted from filling the wetlands to create dry land for industrial purposes, airports, port facilities and places for people to live and play.

Some social justification exists for the use of estuaries and mangroves for such essential purposes as harbours and loading facilities, but these need to be restricted to particular ones. There needs to be a coordinated management plan for large geographical regions (perhaps whole states). Destruction of mangroves appears less justified in the case of airports, and totally unjustified for housing and playing fields. The practice of dredging artificial canals in estuarine wetlands and using dredge spoils for land-fill to create canal or key-type residential subdivisions is accompanied by many insidious effects (Odum 1970; Lindall 1974; Lindall and Trent 1975; Saenger and McIvor 1975; Westman 1975).

In southern Queensland, deficiencies in siting and in engineering design have led to flooding and erosion problems, as well as reduced water quality because of poor circulation in canals. These problems, coupled with public concern at the increasing loss of a public resource and the discovery that canals were creating a biting midge problem (Reye 1982), prompted the Queensland government to commission the "Coastal Management Investigation" in March 1974. That study summarized existing knowledge of the environmental problems associated with canal-estate housing projects. It recommended engineering improvements for any future canals but, more importantly, recommended the protection of most of the remaining mangroves, salt marshes and seagrass beds in southern Queensland (Gutteridge, Haskins and Davey 1975). Despite this, many local councils continue to press for canal developments and other engineering projects which are, in a regional context, a disastrous use of coastal resources (Hegerl 1982).

Flood Mitigation and Swamp Draining Works

One of the reasons for the decline in the extent of wetlands is the practice of draining them under the pretext of flood mitigation (Goodrick 1970; Pressey and Middleton 1982). The high fertility of floodplain soils encourages land developments and investments which are often at risk because of the flood-prone nature of these areas. Future town-planning practice should prevent inappropriate developments (especially residential development) in flood-prone areas.

Although it is recognized that some flood mitigation works may be necessary for the protection of existing urban areas, the need for such schemes should not be used merely as a justification for draining of swamps for agricultural land (Briggs 1977). Future flood mitigation practice should not isolate wetlands from the estuary. Where wetlands are already cut off from tidal influence by floodgates, a revised schedule of flood-gate operations should be introduced to allow for tidal flushing of wetlands. As the natural pattern of drainage provides for the optimum function of the ecosystem, land-use planners and civil engineers should seek to retain the natural drainage pattern of the land, especially at the wetland's boundary with the land (AMSA 1977; Pressey and Middleton 1982).

Reclamation and Dredging

The sediments in estuaries (particularly sand) are important to society either as a resource or as an accumulation that hinders navigation or causes river flooding. Dredging in estuaries, whether for resource recovery or removal of "spoils", is a common activity which unfortunately has many adverse effects on estuarine ecosystems (Saila, Pratt and Polgar 1972; Clark 1974).

Interference with natural vegetation in the catchment through forestry, agriculture, mining or urban development often leads to an accelerated rate of sedimentation in the estuary. The accumulated sediments can cause problems which sometimes make it necessary to remove them. Disposal of the unwanted dredge "spoil", either in deeper parts of the estuary or on estuarine wetlands, should not be allowed. Preferably, planners should set aside non-tidal areas (not wetlands) near estuaries and rivers for the disposal of "spoils" from such operations. Once disposed of in this way, steps should be taken to ensure minimal return to the estuary by soil erosion.

Washing of the sand to remove the fines and shellgrit is a common activity associated with commercial dredging for sand or gravel. This activity, unless properly controlled, can cause serious problems in estuaries by decreasing photosynthesis of aquatic plants (Wolanski and Collis 1976), by smothering benthic organisms (Kaplan, Walker and Kraus 1974), by reducing the level of dissolved oxygen and by releasing toxic substances which have accumulated in the sediments. If the sediment is discharged over the roots of mangroves, it leads to their death.

Dredging at the mouth of a river or lagoon is often required when cutting or maintaining navigation channels or in training its opening. When such dredging is commissioned without adequate knowledge of the dynamics of sand movement in the area, the resulting instability of nearby channels, sand-bars or beaches can have undesirable effects which may be costly to rectify. Such dredging programmes should always be preceded by an analysis of the estuary's sand budget (Stephens 1973; Posford et al. 1976).

It must also be remembered that the provision of a navigable channel will open an area which was previously less accessible. The detrimental effect of excessive human activity can be lessened by controlling or restricting access to certain areas. When the estuarine inventories have been compiled, priorities for wetland and waterway use can be determined, and these priorities implemented by controlling the types and extent of access available.

Large-scale alteration of the floor of the estuary resulting from dredging can produce far-reaching effects on natural flow patterns and circulation and thus on the estuary's salinity regime (Nichols 1972; Simmons and Herrmann 1972). Alterations of this type can induce changes in the kinds, abundance and distribution of estuarine organisms and thus in the fisheries. A study of the hydraulic consequences of large-scale dredging should precede any major dredging works (Stephens 1973).

Similarly, the construction of sea walls and revetments may cause changes to the shoreline far distant from the actual structure owing to reflection of wave energy. The gentle slopes found on most beaches lead to the dissipation of almost all the energy of waves incident upon them. The steep sides of a sea wall or revetment can easily produce an increase of an order of magnitude in the energy of the reflected waves. Thus, areas which previously were not subjected to wave activity may now be, and shoreline erosion will inevitably follow. There is evidence of this process along the southern shores of Botany Bay.

Bund Walls

Bund walls are sometimes constructed around mangrove and

saltmarsh areas to prevent tidal inundation, or may be incidentally built by the construction of roadways. With waterlogging, the mangroves within the bund walls generally are killed within approximately six weeks, and the clearing of dead mangroves usually follows. These reclaimed areas can then be put to some other use — to provide effluent ponds for industry, to be filled for playing fields, and so on. The bund walls in the majority of cases, however, are vulnerable to breaching by floods or are permeable to the effluents they contain, or may be designed to overflow after a period of retention. The red mud ponds of central coastal Queensland contain the toxic wastes of an aluminium refinery. Once filled, they are topdressed and lawns are planted. Nevertheless, subsurface drainage of these effluents is continuing, and these effluents are killing the adjacent mangroves of South Trees Inlet.

Bund walls also have been used in central coastal Queensland to create areas of lowered salinity for cattle grazing, as on Curtis Island, near Gladstone, or alternatively to provide evaporation ponds for solar salt production, as at Port Alma. Around Cairns, mangroves and salt marshes have been bunded in an attempt to grow sugarcane on intertidal soils (Hegerl 1982). Apart from problems with acid sulphates in the soil, this project does not appear to be economically feasible.

In parts of Asia, bund walls have been constructed in mangrove areas for aquaculture projects (Macnae 1968; Ong 1982). Pollard (1973) concluded that "alteration of estuarine areas by construction of levee banks or by other means should be avoided in order that culture practices supplement and not endanger production from natural resources". It is hoped that the practice of bunding for aquaculture will not become widespread in Australia.

The Future of Mangroves and Salt Marshes in Australia

The Australian coastline supports approximately 1.1×10^6 ha of mangroves (Galloway 1982) and probably a similar area of salt marshes and saltflats, and these intertidal communities span thirty-four degrees of latitude and forty-one degrees of longitude. As a result of this geographic range, these communities extend over tropical and cool temperate regions, and possess a concomitant diversity of species and patterns.

Threats to Australian mangroves and salt marshes emanate not from the immediate pressures of subsisting, but from failure of the general community and of decision-makers specifically to appreciate the value of these intertidal communities. Failure to recognize their ecological value in terms of direct products, indirect products and the amenities or "free services" they provide has resulted in an attitude that these communities are wastelands, which, in turn, has led to their exploitation as a non-renewable resource. This tendency has been compounded by the failure both of developers and of legislators to recognize the extended boundaries of these communities, with the consequent but nevertheless frequent unintended destruction of other valuable resources. The failure of biologists to undertake appropriate research and then communicate the results to the public and the decision-makers must also receive its share of the blame.

Despite these failures, however, Australia apparently possesses some of the largest tracts of relatively undisturbed mangroves in the world (table 49), and this is likely to continue into the foreseeable future, although the exploitation and conversion, particularly of mangroves, continues elsewhere with undiminished intensity, largely because of economic and population pressures (Hamilton and Snedaker 1984). Because of the economic and population characteristics of Australia, the Australian mangrove

Table 49 Population density and economic status of countries with the world's major mangrove areas

	Area of mangroves (ha)	Population density** (N km^{-2})	Per capita GNP** (US$)
Brazil	2,500,000	14.2	2,220
Indonesia	2,176,271	77.9	530
Australia	1,161,700	1.9	11,080
Nigeria	973,000	94.8	870
Venezuela	673,600	16.9	4,220
Mexico	660,000	36.1	2,250
Malaysia	652,219	43.0	1,840
Burma	517,077	50.4	190
Senegal	500,000	30.1	430
Panama	486,000	24.7	1,910
Colombia	440,000	23.2	1,380
Bangladesh	417,013	629.9	140
Papua New Guinea	411,600	6.7	462
India	356,500	209.9	260
Malagasy	320,700	15.3	330
Vietnam	286,400	168.8	—
Gabon	250,000	2.8	2,420
Pakistan	249,489	105.1	350
Philippines	246,699	165.3	790
Ecuador	215,852	30.3	1,180
United States of America	205,000	24.5	12,820
Cameroon	200,000	18.3	880

* Saenger, Hegerl and Davie (1983).
** 1981 data from the 1983 World Development Report, World Bank, USA, Washington, D.C.

and saltmarsh communities are in a unique position in the world (table 49); with forethought and husbandry, these communities could ultimately serve as a global genetic resource. Viewed in this light, the need for more research and greater protection of these Australian communities cannot be overemphasized. Detailed management studies and enlightened decisions as, for example, on Towra Point, Kakadu National Park, Careel Bay, Riley's Island and the St Kilda foreshore in Adelaide, give some sustenance to the hope that Australian mangrove ecosystems will receive adequate protection and be used on a proper rational basis in the future.

Glossary

Accreting. The laying down of additional layers on top of each other, e.g. in sedimentation.
Adventitious roots. Roots arising from plant parts other than the (main) root system, e.g. from the stem or from a leaf cutting.
Aerenchyma/aerenchymatous tissue. A tissue of unthickened cells surrounding large air spaces.
Albinism. Lack of pigment.
Allometric. A change in the state of factors in which the relationship between the factors is maintained proportionally, i.e. allometric growth as diameter increases biomass increases proportionally.
Anoxia. A habitat devoid of oxygen.
Apical decollation. The dropping off of the upper whorls of a gastropod shell, when the animal has ceased to occupy them.
Axil. The angle between leaf or branch and axis from which it springs.
Berm. The large deposits of dry loose sediment above the high tide line on a beach.
Bioturbation. The mixing of a sediment by the burrowing, feeding or other activity of living organisms.
Bole. That part of a tree trunk from the ground up to the first branch.
Boundary layer. That layer of air surrounding a leaf lamina.
Buccal cavity. "Mouth" chamber prior to pharynx.
Calyx. A collective term for the sepals.
Capillarity. The movement of fluid due to surface tensions of the tube geometry. A phenomenon associated with surface tension, which occurs in fine bore tubes or channels.
Carapace. "Shield of exoskeleton" covering part of the body of some Arthropoda, e.g. crabs.
Cay. Shoal or island on a coral reef.
Chelae. Pincer or "nipper" claw of crustacean, e.g. crab.
Chenier plains. Alluvial plains interspersed with storm-deposited beach ridges.
Ciliate. Possessing short fine hair-like projections; can be used for locomotion and feeding.
Cladodes. A stem which takes on the function of a leaf and which bears scale leaves.

Coelom. Main body cavity of many three layered animals in which gut is suspended.

Crenulate. Margin indented regularly forming shallow lobes.

Crespuscular. Active during twilight hours, i.e. dusk and dawn.

Cryptic. For example, colouration/shape which conceals an organism by blending in with surroundings.

Cytokinin. A class of plant hormones important in the regulation of nucleic acid and protein metabolism, in cell division, delaying senescence, and organ initiation.

Detritus. Fragmented particulate organic matter derived from the decomposition of plant and animal remains; organic debris.

Diaspore. Any part of an organism produced either sexually or asexually that is capable of giving rise to a new individual — propagule.

Dorsiventral leaves. Having differing upper and lower surfaces.

Edaphic factor. The physical, chemical and biological properties of the soil or substratum, which influence the associated biota.

Epicormic shoots. Shoots arising from dormant buds under the bark — stimulated to growth after the event of fire or other damage.

Flocculate. An aggregate of fine particles in a liquid; such aggregates are formed when clay particles are mixed with salt water.

Foliose. Having leaf-like lobes in lichens.

Guano. An accumulation of seabird droppings rich in phosphates and nitrates.

Gypseous. Containing gypsum (calcium sulphate $CaSO_4 \cdot 2H_2O$).

Halophyte. A plant which is living in and is tolerant of salty conditions.

Hepatopancreas. Large glands in some crustaceans, especially in malacostracans, which produce digestive enzymes.

Heterocystous. Bearing heterocysts, i.e. clear cells occurring at intervals on filaments of some blue-green algae.

Holdfast. Part of an algal plant which anchors the thallus to the substrate.

Hydrostatic pressure potential. Pressure exerted on one side of a semi-permeable membrane against that pressure resulting from osmosis from the other side of the membrane.

Incident radiation. Incoming solar radiation on any surface, e.g. on leaf.

Insolation. The energy emitted by the sun which reaches the surface of the earth.

Instar. Stage of development of an insect.

Interstitial. Pertaining to, or occurring within, the spaces between sediment particles.

Isobilateral leaf. A leaf having the upper and lower surfaces essentially similar.
Lacunae. Spaces between cells.
Lignin. An organic substance impregnating the cellulose framework of certain plant cell walls.
Littoral. The intertidal zone of the coast line.
Mantle. Of mollusca — surface layer of visceral hump. Secreting shell. Outer soft fold of integument next to shell of molluscs and brachiopods.
Maxilliped. Mouth part appendage in decapod crustaceans.
Meiofauna. The small interstitial animals that pass through a 1 mm mesh sieve but are retained by 0.1 mm mesh.
Meropodites. Fourth segment of thoracic appendage in crustaceans.
Mesic. Pertaining to conditions of moderate moisture or water supply; used of organisms occupying moist habitats.
Mycelium. Collective term for mass of hyphae that constitutes vegetative part of a fungus.
Neap tide. The tide of minimum range occurring at the time of first and third quarters of the moon, when the gravitational attraction of the sun and moon act at right angles to each other during quadrature.
Nephridium. Excretory organ present in some invertebrate groups. Helps control water content in body.
Obligate. Necessary to survival.
Oocyte. (Animal) cell which undergoes meiosis and thereby forms ovum.
Osmolarity. An ability to reduce the osmotic pressure of a fluid.
Osmotic potential. The pressure exerted against a semi-permeable membrane due to potential osmotic flow from that side to the other.
Oviposition. The laying of eggs in insects.
Oxygen Tension. The amount of oxygen available measured in mm of Hg (mercury).
Pericarp. The wall of a fruit, developed from the ovary wall after fertilization.
Petiole. The stem of a leaf — between the leaf lamina and branch.
Physiognomy/-ic features. The characteristic features or appearance of a plant community/or vegetation.
Phytotoxic. Poisonous to plants.
Primordia — of leaf. The small mass of tissue from which a leaf starts its development.
Propagule. A structure derived from a parent organism (by asexual or sexual means) which is capable of developing into a new individual.

Proteolytic. Relating to the metabolic breakdown of proteins.

Prothallus/i. The sexually reproducing form (in ferns) that alternates with the sporophyte, spore bearing, form (which is the familiar fern plant).

Proximal. Regarding position — closest to (cf distal).

Pseudofaeces. End product of digestion not passed through but rather regurgitated.

Radicle. The rudimentary root in the embryo of a plant.

Resistance — of leaf. The degree of inhibition of water loss through stomata by transpiration.

Saponin. Steroid vegetable glycosides that act as emulsifiers of oils. They dissolve red corpuscles, irritate the eyes and organs of taste and are toxic to lower animals.

Saprophyte. A plant which obtains nutrients from dead or decaying material.

Sclereids. Type of cells within plant sclerenchyma tissue (thick walled supporting tissue). Sclereids (or stone cells) are usually not much longer than wide. Common in fruits and seed coats.

Setae. Bristles of invertebrates — produced by epidermis.

Sublingual. Beneath the tongue.

Sympatric. Describes overlapping species distributions; (cf allopatric, i.e. species with different distributions — not overlapping).

Terrigenous. Derived from the land.

Thallus. A simple plant body which is not differentiated into stem, branches and roots, e.g. as in ferns and algae.

Trophic levels. The sequence of steps in a food chain or pyramid from producer to primary, secondary or tertiary consumer.

Turgor potential. The pressure within the cell resulting from the absorption of water into the vacuole and the imbibition of water by the protoplasm.

Upwelling. An upward movement within an ocean of cold water bringing nutrients to the surface.

Xeromorphic. Regarding plants — bearing characters which apparently enable it to survive dry conditions (adaptation).

Zooflagellate. Protozoans possessing flagella as adult locomotor organelles — not containing chromoplasts (cf phytoflagellate).

Bibliography

Adam, P., and B.M. Wiecek. 1983. The salt glands of *Samolus repens*. *Wetlands* 3: 2-11.

Adams, S.M., and J.W. Angelovic. 1970. Assimilation of detritus and its associated bacteria by three species of estuarine animals. *Chesapeake Science* 11: 249-54.

Ahmadjian, V. 1967. A guide to the algae occurring as lichen symbionts: Isolation, culture, cultural physiology and identification. *Phycologia* 6: 127-60.

Albright, L.J. 1976. In situ degradation of mangrove tissues. *NZ J. Mar. Freshw. Res.* 10: 385-89.

Allen, O.N., and E.K. Allen. 1981. *The Leguminosae. A source book of characteristics, uses and nodulation*. Madison: Univ. Wisconsin Press.

Almodovar, L.R., and R. Biebl. 1962. Osmotic resistance of mangrove algae around La Parguera, Puerto Rico. *Rev. Algol.* 6: 203-8.

AMSA. 1977. Guidelines for the protection and management of estuaries and estuarine wetlands. Sydney: Australian Marine Sciences Association.

Andriesse, J.P., N. van Breeman and W.A. Blokhuis. 1973. The influence of mudlobsters (*Thalassina anomala*) on the development of acid sulphate soils in mangrove swamps in Sarawak (East Malaysia). In *Acid sulphate soils*, vol. 2, ed. H. Dost, pp. 11-32. Wageningen, Holland: Intern. Inst. Land Reclamation and Improvement.

Arnott, G.H., and M. Ahsanullah. 1979. Acute toxicity of copper, cadmium and zinc to three species of marine copepod. *Aust. J. Mar. Freshw. Res.* 30: 63-71.

Atherton, G., and G. Dyne. 1975. Survey of the algal flora of the Serpentine Creek area. In *Brisbane Airport development*. Vol. 4, *Marine study factor reports*, pp. 83-109. Canberra: AGPS.

Atkinson, G., P. Hutchings, M. Johnson, W.D. Johnson and M.D. Melville. 1981. An ecological investigation of the Myall Lakes region. *Aust. J. Ecol.* 6: 299-328.

Atkinson, M.R., G.P. Findlay, A.B. Hope, M.G. Pitman, H.D.W. Saddler and K.R. West. 1967. Salt regulation in the mangroves *Rhizophora mucronata* Lam. and *Aegialitis annulata* R. Br. *Aust. J. Biol. Sci.* 20: 589-99.

Attiwill, P.M., and B.F. Clough. 1980. Carbon dioxide and water vapour exchange in the white mangrove. *Photosynthetica* 14: 40-47.

Austin H.M. 1971. A survey of the ichthyofauna of the mangroves of western Puerto Rico during December 1967-August 1968. *Caribb. J. Sci.* 11: 27-39.

Australian Littoral Society. 1977. *An investigation of management options for Towra Point, Botany Bay*. Report prepared for the Aust. Nat. Parks and Wildl. Serv.

Australian Water Resources Council. 1976. *Review of Australia's water resources 1975*. Canberra: AGPS.

Axelrad, D.M., K.A. Moore and M.E. Bender. 1976. Nitrogen, phosphorus and carbon fluxes in Chesapeake Bay marshes. *Bull. Wat. Res. Centre* 79: 1-182.

Bader, R.G., M.A. Roessler and A. Thorhaug. 1972. Thermal pollution in a tropical marine estuary. In *Marine pollution and sea life*, ed. M. Ruivo. London: Fishing News (Books).

Baijnath, H., and L.M. Charles. 1980. Leaf surface structures in mangroves. I. The genus *Rhizophora* L. *Proc. Electr. Microscopy Soc. S. Afr.* 10: 37-38.

Baines, G.B.K. 1975. Patterns of exploitation of mangrove ecosystems. In *Proceedings of the International Symposium on Biology and Management of Mangroves*, eds G.E. Walsh, S.C. Snedaker and H.J. Teas, vol. 2, pp. 742-52. Gainesville: Univ. Florida.

Baker, R.T. 1915. The Australian "grey mangrove" (*Avicennia officinalis*, Linn.). *J. Proc. Roy. Soc. NSW* 49: 257-88.

Baldwin, G.F., and L.B. Kirschner. 1976. Sodium and chloride regulation in *Uca* adapted to 75% seawater. *Physiol. Zool.* 49: 158-76.

Ball, M.C., and C. Critchley. 1982. Photosynthetic responses to irradiance by the grey mangrove, *Avicennia marina*, grown under different light regimes. *Plant Physiol.* 70: 1101-6.

Baltzer, F. 1969. Les formations végétales associées au delta de la Dumbéa. Cah. ORSTOM, *Sér. Géol.* 1: 59-84.

Baltzer, F., and L.R. Lafond. 1971. Marais maritimes tropicaux. *Rev. Géogr. Phys. Géol. Dynamique* 13: 173-96.

Banerji, J. 1958. The mangrove forests of the Andamans. *Trop. Silviculture* 20: 319-24.

Barlow, B.A. 1966. A revision of the Loranthaceae of Australia and New Zealand. *Aust. J. Bot.* 14: 421-99.

⸺. 1967. Parasitic flowering plants. *Aust. Nat. Hist.* 15: 365-68.

⸺. 1981. The loranthaceous mistletoes in Australia. In

Ecological biogeography of Australia, ed. A. Keast, pp. 555-74. The Hague: Dr W. Junk.

———. 1984. Loranthaceae. In *Flora of Australia*, vol. 22, ed. A.S. George, pp. 68-131. Canberra: AGPS.

Barlow, B.A., and D. Wiens. 1977. Host-parasite resemblance in Australian mistletoes: The case of cryptic mimicry. *Evolution* 31: 69-84.

Barnes, R.S.K. 1976. The osmotic behaviour of a number of Grapsoid crabs with respect to their differential penetration of an estuarine system. *J. Exper. Biol.* 47: 535-51.

Barth, H. 1981. The biogeography of mangroves. In *Contributions to the ecology of halophytes*, eds D.N. Sen and K.S. Rajpurohit. Vol. 2, *Tasks for vegetation science*, pp. 66-131. The Hague: Dr W. Junk.

Batista, A.C., H. ca Silva Maia and A.F. Vital. 1955. Ascomycetidae aliquot novarum (Some new ascomycetes). *Amer. Soc. Biol. Pernambuco* 13: 72-86.

Bavor, J. 1978. *Microbiological studies in Westernport Bay. I. Seagrass biodegradation*. Report to Westernport Bay Environmental Study, Ministry for Conservation, Melbourne.

Bavor, J., and N.F. Millis. 1976. *Bacteriological studies in Westernport Bay*. Report to Westernport Bay Environmental Study, Ministry for Conservation, Melbourne.

Baylis, G.T.S. 1940. Leaf anatomy of the New Zealand mangrove. *Trans. Roy. Soc. NZ* 70: 164-70.

Beadle, N.C.W. 1954. Soil phosphate and the delimitation of plant communities in eastern Australia. *Ecology* 35: 370-75.

Beanland, W.R., and W.J. Woelkerling. 1982. Studies on Australian mangrove algae. II. Composition and geographic distribution of communities in Spencer Gulf, South Australia. *Proc. Roy. Soc. Vic.* 94: 89-106.

———. 1983. *Avicennia* canopy effects on mangrove algal communities in Spencer Gulf, South Australia. *Aquat. Bot.* 17: 309-13.

Beard, J.S. 1967. An inland occurrence of mangroves. *West. Aust. Nat.* 10: 112-15.

Bebbington, G.N., N.J. Mackay, R. Chvojka, R.J. Williams, A. Dunn and E.A. Auty. 1977. Heavy metals, selenium and arsenic in nine species of Australian commercial fish. *Aust. J. Mar. Freshw. Res. 28: 277-86*.

Beever, J.W., D. Simberloff and L.L. King. 1979. Herbivory and predation by the mangrove crab, *Aratus pisonii*. *Oecologia* 43: 317-28.

Bell, J.D. 1980. Aspects of the ecology of fourteen economically important fish species in Botany Bay, New South Wales, with special emphasis on habitat utilisation and a discussion of the effects of man-induced habitat changes. M.Sc. thesis, Macquarie Univ., Sydney.

Bell, J.D., J.J. Burchmore and D.A. Pollard. 1978a. Feeding ecology of a scorpaenid fish, the fortescue *Centropogon australis* from a *Posidonia* seagrass habitat in New South Wales. *Aust. J. Mar. Freshw. Res.* 29: 175-85.

———. 1978b. Feeding ecology of three sympatric species of leatherjacket (Pisces: Monacanthidae) from a *Posidonia* seagrass habitat in New South Wales. *Aust. J. Mar. Freshw. Res.* 29: 631-43.

Bell, J.D., D.A. Pollard, J.J. Burchmore, B.C. Pease and M.J. Middleton. 1984. Structure of a fish community in a temperate tidal mangrove creek in Botany Bay, New South Wales. *Aust. J. Mar. Freshw. Res.* 35: 33-46.

Bennett, I. 1968. The mudlobster. *Aust. Nat. Hist.* 16: 22-25.

Benson, A.A., and M.R. Atkinson. 1967. Choline sulphate and phosphate in salt excreting plants. *Fed. Proc.* 26: 394.

Berjak, P., G.K. Campbell, B.I. Huckett and N.W. Pammenter. 1977. In *The mangroves of southern Africa*, pp. 42-63. Natal Branch of the Wildlife Society of Southern Africa.

Berry, A.J. 1958. Fluctuations in the reproductive condition of *Cassidula aurisfelis* a Malayan mangrove ellobiid snail (Pulmonata: Gastropoda). *J. Zool.* (Lond.) 154: 377-90.

———. 1963. Faunal zonation in mangrove swamps. *Bull. Nat. Mus.* 32: 91-98.

———. 1972. The natural history of west Malaysian mangrove faunas. *Malayan Nature J.* 25: 135-62.

———. 1975. Molluscs colonizing mangrove trees with observations on *Enigmonia rosea* (Anomiidae). *Proc. Malac. Soc.* (Lond.) 41: 589-600.

Beumer, J.P. 1978. Feeding ecology of four fishes from a mangrove creek in north Queensland, Australia. *J. Fish. Biol.* 12: 475-90.

Bhosale, L.J., and L.S. Shinde. 1983. Significance of cryptovivipary in *Aegiceras corniculatum* (L.) Blasco. In *Tasks for vegetation science*, ed. H.J. Teas. Vol. 8, pp. 123-29. The Hague: Dr W. Junk.

Biebl, R., and H. Kinzel. 1965. Blattbau und Salzhaushalt von *Laguncularia racemosa* (L.) Gaertn. f. und anderer Mangrovenbaume auf Puerto Rico. *Osterreichische Botanische Zeitung* 112: 56-93.

Bird, E.C.F. 1970. Coastal evolution in the Cairns district. *Aust. Geogr.* 11: 327-35.

———. 1971. Mangroves as land-builders. *Vic. Nat.* 88: 189-97.

———. 1972. Mangroves and coastal morphology in Cairns Bay, north Queensland. *J. Trop. Geogr.* 35: 11-16.

Bjorkman, O. 1970. Characteristics of the photosynthetic apparatus as revealed by laboratory measurements. In *Prediction and measurements of photosynthetic productivity*, pp. 267-81. Wageningen: Pudoc.

Blaber, S.J.M. 1980. Fish of the Trinity Inlet systems of north Queensland with notes on the ecology of fish faunas of tropical Indo-Pacific estuaries. *Aust. J. Mar. Freshw. Res.* 31: 137-46.

Blaber, S.J.M., J.W. Young and M.C. Dunning. 1985. Community structure and zoogeographic affinities of the coastal fishes of the Dampier Region on northwestern Australia. *Aust. J. Mar. Freshw. Res.* 36: 247-66.

Blackmore, A.V. 1976. Salt sieving within clay soil aggregates. *Aust. J. Soil. Res.* 14: 149-58.

Blake, J.A., and J.D. Kudenov. 1978. The Spionidae (Polychaeta) from southeastern Australia and adjacent areas with a revision of the genera. *Mem. Nat. Mus. Vic.* 39: 170-280.

Blake, S.T., and C. Roff. 1972. *The honey flora of Queensland*. Brisbane: Govt. Printer.

Blakers, M., S.J.J.F. Davies and P.N. Reilly. 1984. *The atlas of Australian birds*. Royal Australasian Ornithologist Union. Melbourne University Press.

Bond, G. 1956. A feature of root nodules of *Casuarina*. *Nature* 177: 192.

———. 1963. The root nodules of non-leguminous angiosperms. *Symp. Soc. Gen. Microbiol.* 13: 315-47.

Boonruang, P. 1978. The degradation rates of mangrove leaves of *Rhizophora apiculata* (Bl.) and *Avicennia marina* (Forsk.). Vierh at Phuket Island, Thailand. *Phuket Mar. Biol. Center Res. Bull.* 26: 1-7.

Bostrom, T.E., and C.D. Field. 1973. Electrical potentials in the salt gland of *Aegiceras*. In *Ion transport in plants*, ed. W.P. Anderson, pp. 385-92. London: Academic Press.

Boto, K.G. 1982. Nutrient and organic fluxes in mangroves. In *Mangrove ecosystems in Australia. Structure, function and management*, ed. B.F. Clough, pp. 239-59. Proceedings of the Australian National Mangrove Workshop AIMS. Published by AIMS with ANU Press.

———. 1983. Nutrient status and other soil factors affecting mangrove productivity in northeast Australia. *Wetlands* 3(1): 45-49.

Boto, K.G., and J.S. Bunt. 1981. Tidal export of particulate organic matter from a northern Australian mangrove system. *Estuar. coastal Shelf Sci.* 13(3): 247-57.

Boto, K.G., and W.H. Patrick. 1978. Role of wetlands in the removal of suspended sediments. In *Wetland functions and* values. *The state of our understanding*, eds P.E. Greeson, J.R. Clark and J.E. Clark, pp. 479-89. Amer. Wat. Res. Assoc., Technical Report. No. TPS 79-2.

Boto, K.G., and J.T. Wellington. 1983. Phosphorus and nitrogen nutritional status of a northern Australian mangrove forest. *Mar. Ecol. Progr. Ser.* 11: 63-69.

Boucher, D.H., S. James and K.H. Keeler. 1982. The ecology of mutualism. *Ann. Rev. Ecol. Syst.* 13:315-47.

Bower, C.A., and L.V. Wilcox. 1965. Soluble salts. In *Methods of soil analysis*. Part 2, Agronomy 9, ed. C.A. Black, pp. 933-51. Madison: Amer. Soc. Agron.

Bowman, H.H.M. 1917. Ecology and physiology of the red mangrove. *Proc. Amer. Phil. Soc.* 56: 589-672.

――――. 1921. Histological variations in *Rhizophora mangle* L. Pap. *Mich. Acad. Sci.* 22: 129-34.

Braithwaite, L.W. 1975. Managing waterfowl. *Proc. Ecol. Soc. Aust.* 8: 107-28.

Branch, G.M., and M.L. Branch. 1980. Competition in *Bembicium auratum* (Gastropoda) and its effect on microalgal standing stock in mangrove muds. *Oecologia* (Berl.) 46: 106-15.

Breen, C.M., and B.J. Hill. 1969. A mass mortality of mangroves in the Kosi Estuary. *Trans. Roy. Soc. S. Afr.* 38(3): 285-303.

Breteler, W.C.M.K. 1975. Oxygen consumption and respiratory levels of juvenile shore crabs *Carcinus maenas* in relation to weight and temperature. *Neth. J. Sea Res.* 9: 243-54.

Bridgewater, P.B. 1975. Peripheral vegetation at Westernport Bay. *Proc. Roy. Soc. Vic.* 87: 69-78.

Briggs, K.B., K.R. Tenore and R.B. Hanson. 1979. The role of microfauna in detrital utilisation by the polychaete *Nereis succinea* (Frey and Leuckart). *J. Exp. Mar. Biol. Ecol.* 36: 225-35.

Briggs, S.V. 1977a. Estimates of biomass in a temperate mangrove community. *Aust. J. Ecol.* 2:369-73.

Briggs, S.V. 1977b. Flood mitigation. *Nat. Parks J.* 21(2):5-9.

Bright, D.B. 1977. Burrowing Central American mangrove land crabs and their burrow associates. *Mar. Res. Indonesia* 18: 87-99.

Brillet, C. 1970. Relations entre comportement sexuel territoire et agresivité chez les Périophthalmes. *C.R. Acad. Sc. Paris Sér. D.* 270: 1507-10.

――――. 1975. Relations entre territoire et compartment agressif chez *Periphthalmus sobrinus* (Eggert) (Pisces, Periophthalmidae) au laboratoire et en milieu naturel. *Z. Tierpsychol.* 39: 283-331.

Brook, I.M. 1977. Trophic relationships in a seagrass community (*Thalassia testudinum*) in Card Sound, Florida. Fish diets in

relation to macrobenthic and cryptic faunal abundance. *Trans. Amer. Fish. Soc.* 106(3): 219-29.
Brown, D.S. 1971. Ecology of Gastropoda in a South African mangrove swamp. *Proc. Malac. Soc.* (Lond.) 39: 263-79.
Brunnich, J.C., and F. Smith. 1911. Some Queensland mangrove barks and other tanning materials. *Qld Agric. J.* 27: 86-94.
Bunt, J.S. 1978. The mangroves of the eastern coast of Cape York Peninsula, north of Cooktown. Great Barrier Reef Marine Park Authority Workshop on Northern Sector, Townsville.
_____. 1982. Studies of mangrove litter fall in tropical Australia. In *Mangrove ecosystems in Australia — structure, function and management*, ed. B.F. Clough, pp. 223-39. Published by AIMS with ANU Press.
Bunt, J.S., K.G. Boto and G. Boto. 1979. A survey method for estimating potential levels of mangrove forest primary production. *Mar. Biol. 52(2): 123-29.*
Bunt, J.S., and W.T. Williams. 1980. Studies in the analysis of data from Australian tidal forests ("mangroves"). I. Vegetational sequences and their graphic representation. *Aust. J. Ecol.* 5: 385-90.
_____. 1981. Vegetational relationships in the mangroves of tropical Australia. *Mar. Ecol. Progr. Ser.* 4: 349-59.
Bunt, J.S., W.T. Williams and H.J. Clay. 1982. River water salinity and the distribution of mangrove species along several rivers in North Queensland. *Aust. J. Bot.* 30(4): 401-12.
Bunt, J.S., W.T. Williams and N.C. Duke. 1982. Mangrove distributions in northeast Australia. *J. Biogeogr.* 9(2): 111-20.
Bunt, J.S., and E. Wolanski. 1980. Hydraulics and sediment transport in a creek — mangrove swamp system. *Proceedings 7th Australasian Hydraulics and Fluid Mechanics Conference* Brisbane, pp. 492-95.
Burbidge, N.T. 1960. The phytogeography of the Australian region. *Aust. J. Bot.* 8: 75-211.
Burchmore, J.J., D.A. Pollard and J.D. Bell. 1984. Community structure and trophic relationships of the fish fauna of an estuarine *Posidonia australis* seagrass habitat in Port Hacking, New South Wales. *Aquat. Bot.* 18: 71-87.
Butler, A.J., A.M. Depers, S.C. McKillup and D.P. Thomas. 1977a. Distribution and sediments of mangrove forests in South Australia. *Trans. Roy. Soc. Sth Aust.* 101: 35-44.
_____. 1977b. A survey of mangrove forests in South Australia. *Sth Aust. Naturalist* 51: 34-49.
Butler, V., and T.D. Steinke. 1976. Ultrastructural studies on *Avicennia marina* propagules. *Proc. Electr. Microscopy Soc. S. Afr.* 6: 67-68.

Cameron, A.M. 1966. Some aspects of the behaviour of the soldier crab, *Mictyris longicarpus*. *Pac. Sci.* 20: 224–34.

Camilleri, J.C., and G. Ribi. 1983. Leaf thickness of mangroves (*Rhizophora mangle*) growing in different salinities. *Biotropica* 15(2): 139–41.

Campbell, N., and W.W. Thomson. 1976. The ultrastructure of *Frankenia* salt glands. *Annals of Bot.* 40: 681–86.

Cann, J. 1978. *Tortoises of Australia*. Sydney: Angus & Robertson.

Canoy, M.J. 1975. Diversity and stability in a Puerto Rican *Rhizophora mangle* L. forest. In *Proceedings of the International Symposium on Biology and Management of Mangroves*, eds G.E. Walsh, S.C. Snedaker and H.J. Teas, vol. 1, pp. 344–56. Gainesville: Univ. Florida.

Cardale, S., and C.D. Field. 1971. The structure of the salt gland of *Aegiceras corniculatum*. *Planta* 99: 183–91.

_____. 1975. Ion transport in the salt gland of *Aegiceras*. In *Proceedings of the International Symposium on Biology and Management of Mangroves*, eds G.E. Walsh, S.C. Snedaker and H.J. Teas, vol. 2, pp. 608–14. Gainesville: Univ. Florida.

Carey, G. 1934. Further investigations on the embryology of viviparous seeds. *Proc. Linn. Soc. NSW* 59: 392–410.

Carey, G., and L. Fraser. 1932. The embryology and seedling development of *Aegiceras majus* Gaertn. *Proc. Linn. Soc. NSW* 57: 341–60.

Carlquist, S. 1975. *Ecological strategies of xylem evolution*. Berkeley: Univ. Calif. Press.

Carter, M.R., L.A. Burns, T.R. Cavinder, K.R. Dugger, P.L. Fore, D.B. Hicks, H.L. Revells, T.W. Schmidt and R. Farley. 1973. *Ecosystems analysis of the big cypress swamp and estuaries*. USEPA South Florida Ecological Study, Atlanta, Georgia.

Chanda, S. 1977. An eco-floristic survey of the mangrove of Sundarbans, West Bengal, India. *Trans. Bose Res. Inst.* (Calcutta) 40(1): 5–14.

Chandrashekar, M., and M. Ball. 1980. Leaf blight of grey mangrove in Australia caused by *Alternaria alternata*. *Trans. Br. Mycol. Soc.* 75: 413–18.

Chapman, V.J. 1938. Studies of salt marsh ecology, sections I–III. *J. Ecol.* 26: 144–79.

_____. 1944. The 1939 Cambridge University expedition to Jamaica. III. The morphology of *Avicennia nitida* Jacq. and the function of its pneumatophores. *J. Linn. Soc.* (Lond.) (*Bot. Ser.*) 52: 487–533.

_____. 1947. Secondary thickening and lenticels in *Avicennia nitida* Jacq. *Proc. Linn. Soc.* (Lond.) 158: 2–6.

_____. 1960. *Salt marshes and salt deserts of the world*. London: Hill.
_____. 1975. Mangrove biogeography. In *Proceedings of the International Symposium on Biology and Management of Mangroves*, vol. 1, eds G.E. Walsh, S.C. Snedaker and H.J. Teas, pp. 3–22. Gainesville: Univ. Florida.
_____. 1976. *Mangrove vegetation*. Vaduz: Cramer.
_____. 1977. Introduction. In *Ecosystems of the world. I. Wet coastal ecosystems*, pp. 1–29. Amsterdam: Elsevier.
Chapman, V.J., and D.J. Chapman. 1973. *The algae*. 2nd ed. London: Macmillan.
Chapman, V.J., and J.W. Ronaldson. 1958. The mangrove and salt-marsh flats of the Auckland isthmus. *NZ Dept. Sci. Indust. Res. Bull.* 125: 1–79.
Charmantier, G. 1973. Resistance à la desiccation chez *Sphaeroma serratum* (Isopoda; Flabellifera). *Arch. Zool. Exp. Gen.* 114: 513–24.
Charmantier, G., and J.P. Trilles. 1973. Physiologie des invertébrés. La pression osmotique de l'hémolymphe de *Sphaeroma serratum* (Crustacea, Isopoda): variations en fonction de la salinité et de la senescence. *C.R. Acad. Sci. Paris* 276: 69–72.
Chou, L.M., S.H. Ho, H.W. Khoo, T.J. Lam, D.H. Murphy and W.H. Tan. 1980. Report on the impact of pollution on mangrove ecosystems in Singapore and related national research programs. Unpubl. report, Singapore Univ.
Christensen, B. 1978. Biomass and primary production of *Rhizophora apiculata* Bl. in a mangrove in southern Thailand. *Aquat. Bot.* 4: 43–52.
Christensen, B., and S. Wium-Anderson. 1977. Seasonal growth of mangrove trees in southern Thailand. I. The phenology of *Rhizophora apiculata* Bl. *Aquat. Bot.* 3: 281–86.
Churchill, D.M. 1973. The ecological significance of tropical mangroves in the early tertiary floras of southern Australia. *Geol. Soc. Aust. Spec. Publ.* 4: 79–86.
Cintron, G., A.E. Lugo, D.J. Pool and G. Morris. 1978. Mangroves of arid environments in Puerto Rico and adjacent islands. *Biotropica* 10: 110–21.
Clark, J. 1974. *Coastal ecosystems: Ecological considerations for management of the coastal zone*. Washington, DC: The Conservation Foundation.
Clarke, L.D., and N.J. Hannon. 1967. The mangrove swamp and saltmarsh communities of the Sydney district. I. Vegetation, soils and climate. *J. Ecol.* 55: 753–71.
_____. 1969. The mangrove swamp and saltmarsh communities of the Sydney district. II. The holocoenotic complex with particular reference to physiography. *J. Ecol.* 57: 213–34.

_____. 1970. The mangrove swamp and saltmarsh communities of the Sydney district. III. Plant growth in relation to salinity and waterlogging. *J. Ecol.* 58: 351-69.

_____. 1971. The mangrove swamp and saltmarsh communities of the Sydney district. IV. The significance of species interaction. *J. Ecol.* 59: 535-53.

Clarke, S.M., and H.B.S. Womersley. 1981. Cross-fertilization and hybrid development of forms of the brown alga *Hormosira banksii* (Turner) Decaisne. *Aust. J. Bot.* 29: 497-505.

Clements, A.N. 1963. *The physiology of mosquitoes.* Oxford: Pergamon Press.

Clifford, H.T., and R.L. Specht. 1979. *The vegetation of North Stradbroke Island, Queensland (with notes on the fauna of mangrove and marine meadow ecosystems* by M.M. Specht). St Lucia: Univ. Qld Press.

Clough, B.F., and P.M. Attiwill. 1975. Nutrient cycling in a community of *Avicennia marina* in a temperate region of Australia. In *Proceedings of the International Symposium on Biology and Management of Mangroves*, eds G.E. Walsh, S.C. Snedaker and H.J. Teas, vol. 1, pp. 137-46. Gainesville: Univ. Florida.

_____. 1982. Primary productivity of mangroves. In *Mangrove ecosystems in Australia — structure, function and management*, ed. B.F. Clough, pp. 213-23. Published by AIMS with ANU Press.

Clough, B.F., T.J. Andrews and I.R. Cowan. 1982. Physiological processes in mangroves. In *Mangrove ecosystems in Australia — structure, function and management*, ed. B.F. Clough, pp. 193-210. Published by AIMS with ANU Press.

Cockcroft, V.G., and A.T. Forbes. 1981. Growth, mortality and longevity of *Cerithidea decollata* (Linnaeus) (Gastropoda: Prosobranchia) from bayhead mangroves, Durban Bay, South Africa. *Veliger* 23(4): 300-307.

Cocks, K.D. 1975. The social functions of aquatic ecosystems. *Proc. Ecol. Soc. Aust.* 8: 167-73.

Cogger, H.G. 1979. *Reptiles and amphibians of Australia.* Sydney: Angus & Robertson.

Coleman, J.M., and L.D. Wright. 1975. Modern river deltas: Variability of processes and sand bodies. In *Deltas, models for exploration*, ed. M.L. Broussard, pp. 99-150. Houston: Houston Geological Society.

Collett, L.C., P.A. Hutchings, P.J. Gibbs and A.J. Collins. 1984. The macrobenthic fauna of *Posidonia australis* meadows in New South Wales, Australia. *Aquat. Bot.* 18: 111-34.

Collette, B.B. 1983. Mangrove fishes of New Guinea. In *Tasks for vegetation science*, vol. 8, ed. H.J. Teas, pp. 1-102. The Hague: Dr W. Junk.

Collins, M.I. 1921. On the mangrove and saltmarsh vegetation near Sydney, NSW, with special reference to Cabbage Creek, Port Hacking. *Proc. Linn. Soc. NSW* 46: 376-92.

Common, I.W.B. 1970. Lepidoptera. In *The insects of Australia*, ed. I.M. Mackerras, pp. 765-866. Melbourne: Melb. Univ. Press.

Common, I.W.B., and D.F. Waterhouse. 1972. *Butterflies of Australia*. Sydney: Angus & Robertson.

Conacher, M.J., J.R. Lanzing and A.W.D. Larkum. 1979. Ecology of Botany Bay 11. Aspects of the feeding ecology of the fanbellied leatherjacket, *Monacanthus chinensis* (Pisces: Monacanthidae) in *Posidonia australis* seagrass beds in Quibray Bay, Botany Bay, New South Wales. *Aust. J. Mar. Freshw. Res.* 30: 387-400.

Conaghan, P.J. 1966. Sediments and sedimentary processes in Gladstone harbour, Queensland. *Univ. Qld. Pap., Dept. Geol.* 6: 1-52.

Congdon, R.A. 1981. Zonation in the marsh vegetation of the Blackwood River estuary in southwestern Australia. *Aust. J. Ecol.* 6: 267-78.

Connor, D.J. 1969. Growth of grey mangrove (*Avicennia marina*) in nutrient culture. *Biotropica* 1: 36-40.

Cook, M.T. 1907. The embryology of *Rhizophora mangle*. *Bull. Torrey Botanical Club* 34: 271-77.

Corpe, W.A. 1974. Periphytic marine bacteria and the formation of microbial films on solid surfaces. In *Effect of the ocean environment on microbial activities*, eds R.R. Colwell and R.Y. Morita, pp. 397-417. Baltimore: Univ. Park Press.

Cousens, N.B.F. 1974. Some physiological and behavioural adaptations to anoxic tolerance in the burrowing shrimp *Upogebia pugettensis* (Dana). *Abstracts of symposia and contributed papers of Western Society of Naturalists, 55th Annual Meeting, 27-30 Dec. 1974, Vancouver*.

Cowan, I.R. 1978. Stomatal responses in mangroves. Paper presented at the 19th General Meeting Australian Society Plant Physiologists, Sydney, Australia.

Cragg, S.M., and M.J. Swift. 1980. The contribution of fungi and marine borers to wood decay of some mangrove communities of Papua New Guinea. *Second International Symposium Biology and Management of Mangroves and Tropical Shallow Water Communities* (abstract only), p. 22.

Crane, J. 1975. Fiddler crabs of the world (Ocypodidae: genus *Uca*). New Jersey: Princeton Univ. Press.

Crawford, R.M.M. 1978. Metabolic adaptations to anoxia. In *Plant life in anaerobic environments*, eds D.D. Hook and R.M.M. Crawford, pp. 119-36. Michigan: Ann Arbor Science.

Crawford, R.M.M., and P.D. Tyler. 1969. Organic acid metabolism in relation to flooding tolerance in roots. *J. Ecol.* 57: 235-44.

Creager, D.B. 1962. A new *Cercospora* on *Rhizophora mangle*. *Mycologia* 54: 536-39.

Cribb, A.B. 1979. Algae associated with mangroves in Moreton Bay, Queensland. In *Proceedings Northern Moreton Bay symposium*, eds A. Bailey and N.C. Stevens, pp. 63-69. Brisbane: Roy. Soc. Qld.

Cribb, A.B., and J.W. Cribb. 1975. Marine fungi from Queensland. *I. Pap. Dept. Bot. Univ. Qld.* 3: 77-81.

Critchley, C. 1982. Stimulation of photosynthetic electron transport in a salt-tolerant plant by high chloride concentrations. *Nature* 298: 483-85.

———. 1983. Further studies on the role of chloride in photosynthetic O_2 evolution in higher plants. *Biochim. Biophys. Acta* 724: 1-5.

Critchley, C., I.C. Baianu, Govindjee and H.S. Gutowsky. 1982. The role of chloride in O_2 evolution by thylakoids from salt-tolerant higher plants. *Biochim. Biophys. Acta* 682: 436-45.

Curtis, W.M., and J. Sommerville. 1947. Boomer Marsh — a preliminary botanical and historical survey. *Proc. Roy. Soc. Tas.* 1947: 151-57.

Curtiss, A.H. 1888. How the mangrove forms islands. *Garden and Forest* 1: 100.

Dacraemes, W., and A. Coomans. 1978. Scientific report on the Belgian expedition to the Great Barrier Reef in 1967. Nematodes XII. Ecological notes on the nematode fauna in and around mangroves on Lizard Island. *Aust. J. Mar. Freshw. Res.* 29: 497-508.

Davey, A., and W.J. Woelkerling. 1980. Studies on Australian mangrove algae. I. Victorian communities: composition and geographic distribution. *Proc. Roy. Soc. Vic.* 91: 53-66.

Davey, J.E. 1975. Note on the mechanism of pollen release in *Bruguiera gymnorhiza*. *J. S. Afr. Bot.* 41: 269-72.

Davie, J.D. 1983. Pattern and process in mangrove ecosystems in Moreton Bay, Queensland. Ph.D. thesis, Queensland Univ.

Davie, P. 1982. List of mangrove crabs in Australia. *Operculum* 5(4): 204-7.

Davis, J.H. 1940. The ecology and geological role of mangroves in Florida. In *Carnegie Inst. Washington Publ. No. 517, Papers from the Tortugas Lab*, 32: 303-412.

Debenham, M.L. 1978. *An annotated checklist and bibliography of Australasian region Ceratopogonidae (Diptera: Nematocera)*. Monograph Series, Entomology Monograph No. 1. Canberra: AGPS.

Deevey, E.S. 1960. The human population. *Scient. Am.* 203: 195–204.
De Forest, A., S.P. Murphy and R.W. Pettis. 1978. Heavy metals in sediments from the central New South Wales coastal region. *Aust. J. Mar. Freshw. Res.* 29: 777–85.
De Fraine, E. 1912. The anatomy of the genus *Salicornia*. *J. Linn. Soc. Bot.* 41: 317–46.
De La Cruz, A.A. 1975. Proximate nutritive value changes during decomposition of saltmarsh plants. *Hydrobiologia* 47: 475–80.
De Witt, C.T. 1965. Photosynthesis of leaf canopies. *Agr. Res. Rep. 663. Inst. Biol. Chem. Res. on Field Crops and Herbage*, Wageningen.
Diamond, J.M. 1975. Assembly of species communities. In *Ecology and evolution of communities*, eds M. Cody and J. Diamond. Cambridge, Mass: Harvard Univ. Press.
_____. 1976. Preliminary results of an ornithological exploration of the islands of Vitiaz and Dampier Straits, Papua New Guinea. *Emu* 76: 1–7.
Dickinson, C.H. 1965. The mycoflora associated with *Halimione portulacoides*. III. Fungi on green and moribund leaves. *Trans. Br. Mycol. Soc.* 48: 603–10.
Ding Hou. 1958. Rhizophoraceae. *Flora Malesiana* (Ser. I) 5: 429–93.
_____. 1972. Rhizophoraceae. *Flora Malesiana* (Ser. I) 6: 965–67.
Doskotch, P.W., H.Y. Cheng, T.M. Odell and L. Girard. 1980. Nerolidol: An antifeeding sesquiterpene alcohol for gypsy moth larvae from *Melaleuca leucodendron*. *J. Chem. Ecol.* 6(4): 845–51.
Dov Por, F., and I. Dor. 1984. Hydrobiology of the mangal. The ecosystem of the mangrove forests. *Developments in Hydrobiology* 20. The Hague: Dr W. Junk.
Downton, W.J.S. 1982. Growth and osmotic relations of the mangrove *Avicennia marina*, as influenced by salinity. *Aust. J. Plant Physiol.* 9: 519–28.
Ducker, S.C., and R.B. Knox. 1976. Submarine pollination in seagrasses. *Nature* (Lond.) 263: 705–6.
Duke, N.C., W.R. Birch and W.T. Williams. 1981. Growth rings and rainfall correlations in a mangrove tree of the genus *Diospyros* (Ebenaceae). *Aust. J. Bot.* 29: 135–42.
Duke, N.C., J.S. Bunt and W.T. Williams. 1981. Mangrove litter fall in northeastern Australia. I. Annual totals by component in selected species. *Aust. J. Bot.* 29: 547–53.
_____. 1984. Observations on the floral and vegetative phenologies of northeastern Australian mangroves. *Aust. J. Bot.* 32: 87–99.

Dunson, W.A. 1970. Some aspects of electrolyte and water balance in three estuarine reptiles, the diamond back terrapin, American and "salt water" crocodiles. *Comp. Biochem. Physiol.* 32: 161-74.

———. 1974. Salt gland secretion in a mangrove monitor lizard. *Comp. Biochem. Physiol.* 47(A): 1245-55.

———. 1978. Role of the skin in sodium and water exchange of aquatic snakes placed in seawater. *Amer. J. Physiol.* 235: R 151-R 159.

———. 1980a. The relation of sodium and water balance to survival in sea water of estuarine and freshwater races of the snakes *Nerodia fasciata*, *N. sipedon* and *N. valida*. *Copeia* 1980: 268-80.

———. 1980b. Adaptations of nymphs of a marine dragonfly, *Erythrodiplax berenice*, to wide variations in salinity. *Physiol. Zool.* 53: 445-52.

Dunson, W.A., and M.K. Dunson. 1973. Convergent evolution of sublingual salt glands in the marine file snake and the true sea snakes. *J. Comp. Physiol.* 86: 193-208.

———. 1979. A possible new salt gland in a marine homalopsid snake (*Cerberus rhynchops*). *Copeia* 1979: 661-72.

Dunson, W.A., and J.D. Lazell, Jr. 1982. Urinary concentrating capacity of *Rattus rattus* and other mammals from the Lower Florida keys. *Comp. Biochem. Physiol.* 71A: 17-21.

Dunstan, D.J. 1973. Estuaries, their value, vulnerability and their future. *Fisherman* 4: 7-11.

Dwyer, P.D., M. Hockings and J. Willmer. 1979. Mammals of Cooloola and Beerwah. *Proc. Roy. Soc. Qld* 90: 65-84.

Dye, A.H. 1983. Composition and seasonal fluctuations of meiofauna in a Southern Africa mangrove estuary. *Mar. Biol.* 73: 165-70.

Edmonds, E. 1935. The relations between the internal fluid of marine invertebrates and the water of the environment with special reference to Australian Crustacea. *Proc. Linn. Soc. NSW* 60: 233-47.

Edmonds, S.J. 1980. A revision of the systematics of Australian Sipunculans (Sipuncula). *Rec. Sth Aust. Mus.* 18(1): 1-74.

Eggert, B. 1929. Bestimmungstabelle und Beschreibung der Arten Familie *Periophthalmus*. *Z. Wiss. Zool.* 133(1-2): 398-410.

Eiseley, L. 1971. *The night country*. New York: Scribners.

Ellis, J., and S. Kanamori. 1977. Water pollution studies on Lake Illawarra. III. Distribution of heavy metals in sediments. *Aust. J. Mar. Freshw. Res.* 28: 485-96.

Ellway, C.P. 1974. An ecological study of Corio Bay, Central Queensland. Habitat. Environmental survey prepared for Capricorn Coast Protection Council.

Elsol, J.A., and P. Saenger. 1983. A general account of the mangroves of Princess Charlotte Bay with particular reference to zonation of the open shoreline. In *Tasks for vegetation science*, ed H.J. Teas, vol. 8, pp. 37-46. The Hague: Dr W. Junk.

Enright, J. 1973. Mangroves shores in Westernport Bay. *Vic. Res.* 15: 12-15.

Ericksen, N.J. 1970. Measurement of tide-induced change to watertable profiles in coarse and fine sand beaches along Pegasus Bay, Canterbury. *Earth Sci. J.* 4: 24-31.

Essig, F.B. 1973. Pollination in some New Guinea palms. *Principes* 17: 75-83.

Everist, S.L. 1974. *Poisonous plants of Australia*. Sydney: Angus & Robertson.

Faegri, K., and L. Van der Pijl. 1971. *The principles of pollination ecology*. New York: Pergamon.

Fahn, A. 1963. The fleshy cortex of articulated Chenopodiaceae. *J. Indian Bot. Soc.* 42: 39-45.

Farrell, M.J. 1973. Studies on the ecology of Victorian mangrove and saltmarsh communities. B.Sc. (Hons.) thesis, Univ. of Melbourne.

Fell, J.W., and I.M. Masters. 1973. Fungi associated with the degradation of mangrove (*Rhizophora mangle* L.) leaves in south Florida. In *Estuarine microbial ecology*, eds H.L. Stevenson and R.B. Colwell, pp. 455-66. Columbia: South Carolina Press.

_____. 1980. The association and potential role of fungi in mangrove detrital systems. *Botanica Mar.* 23: 257-64.

Fell, J.W., R.C. Cefalu, I.M. Masters and A.S. Tallman. 1975. Microbial activities in the mangrove (*Rhizophora mangle* L.) leaf detrital systems. In *Proceedings of the International Symposium on Biology and Management of Mangroves*, eds. G.E. Walsh, S.C. Snedaker and H.J. Teas, vol. 2, pp. 661-79. Gainesville: Univ. Florida.

Fenchel, T. 1970. Studies on the decomposition of organic detritus derived from the turtle grass *Thalassia testudinum*. *Limnol. Oceanog.* 15: 14-20.

_____. 1972. Aspects of decomposer food chains in marine benthos. *Verh. Dtsch. Zool. Ges.* 65: 14-23.

_____. 1977. Aspects of the decomposition of seagrasses. In *Seagrass ecosystems*, eds C.P. McRoy and C. Helfferich, pp. 123-45. New York: Marcel Dekker.

Ferguson-Wood, E.J. 1959. Some east Australian seagrass communities. *Proc. Linn. Soc. NSW* 84:218-26.

Fernando, A.S., and K. Ramamoorthi. 1975. Reproductive biology of barnacles. *Bull. Dept Mar. Sci. Univ. Cochin* 71(4): 721-32.

Field, C.D. 1984. Ions in mangroves. In *Tasks for vegetation science*, ed. H.J. Teas, vol. 9, pp. 43–48. The Hague: Dr W. Junk.

Field, C.D. et al. 1983. *An investigation of natural areas, Kooragang Island, Hunter River*, ed. J. Moss. Sydney: Dept Environment and Planning.

Field, C.D., B.G. Hinwood and I. Stevenson. 1984. Structural features of salt gland of *Aegiceras*. In *Physiology and management of mangroves*, ed. H.J. Teas, pp. 37–42. The Hague: Dr W. Junk.

Flood, P.G. 1980. Tidal-flat sedimentation along the shores of Deception Bay, southeastern Queensland — a preliminary account. *Proc. Roy. Soc. Qld* 91: 77–84.

Foot, P. 1980. The biodeterioration of mangrove timber by marine fungi. Abstr. Workshop on mangrove ecosystem research in southern Queensland, Botany Dept, Univ. of Queensland.

Ford, J. 1982. Origin, evolution and speciation of birds specialised to mangroves in Australia. *Emu* 82: 12–23.

Fosberg, F.R. 1961. Vegetation-free zone on dry mangrove coasts. *US Geol. Soc. Prof. Pap. No. 424D*: 216–18.

Foster, B.A. 1970. Responses and acclimation to salinity in the adults of some balanomorph barnacles. *Phil. Trans. Roy. Soc.* (Lond.) B. 256: 377–400.

Fox, A.M. 1973. Much binding in the marsh. *Nat. Parks and Wildl. J.* 1: 86–91.

Frith, D.W., and C.B. Frith. 1978. Notes on the ecology of fiddler crab populations (Ocypodidae: Genus *Uca*) on Phuket, Surin Nua and Yao Yai Islands, Western Peninsula, Thailand. *Phuket Mar. Biol. Centre, Res. Bull.* 25: 1–13.

Fry, B., and P.L. Parker. 1979. Animal diet in Texas seagrass meadows: ^{13}C evidence for the importance of benthic plants. *Estuar. Coastal Mar. Sci.* 8: 499–509.

Fyhn, H.J. 1976. Holeuryhalinity and its mechanisms in a cirriped crustacean *Balanus improvisus*. *Comp. Biochem. Physiol.* 53A: 19–30.

Gallagher, J.L. 1979. Growth and element compositional responses of *Sporobolus virginicus* (L.) Kunth. to substrate salinity and nitrogen. *Amer. Mid. Nat.* 102: 68–75.

Galloway, R.W. 1982. Distribution and physiographic patterns of Australian mangroves. In *Mangrove ecosystems in Australia — structure, function and management*, ed. B.F. Clough, pp. 31–54. Published by AIMS with ANU Press.

Garb, S. 1961. Differential growth inhibitors produced by plants. *Bot. Rev.* 27: 422–43.

Gay, F.J., and J.A.L. Watson. 1982. The genus *Cryptotermes* in Australia (Isoptera: Kalotermitidae). *Aust. J. Zool.* (Suppl. Ser.) 88: 1–64.

Gayral, P. 1966. *Les algues des côtes françaises (Manche et Atlantique)*. Paris: Doin-Deren & Cie.

George, R.W., and D.S. Jones. 1982. A revision of the fiddler crabs of Australia (Ocypodinae: *Uca*). *Rec. West. Aust. Mus.*, (Suppl.No.) 14:1-99.

———. 1984. Notes on the crab fauna of Mangrove Bay, North West Cape. *West. Aust. Nat.* 15(8): 169-74.

Gessner, F. 1967. Untersuchungen an der Mangrove in Ost-Venezuela. *Int. Rev. ges. Hydrobiol.* 52: 769-81.

Gill, A.M. 1971. Endogenous control of growth-ring development in *Avicennia*. *For. Sci.* 17: 462-65.

———. 1975. Australia's mangrove enclaves: A coastal resource. *Proc. Ecol. Soc. Aust.* 8: 129-46.

Gill, A.M., and P.B. Tomlinson. 1969. Studies on the growth of red mangrove (*Rhizophora mangle* L.). I. Habit and general morphology. *Biotropica* 1: 1-9.

———. 1971a. Studies on the growth of red mangrove (*Rhizophora mangle* L.). II. Growth and differentiation of aerial roots. *Biotropica* 3: 63-77.

———. 1971b. Studies on the growth of red mangrove (*Rhizophora mangle* L.). III. Phenology of the shoot. *Biotropica* 3: 109-24.

———. 1975. Aerial roots: An array of forms and functions. In *The development and function of roots*, eds J.G. Torrey and D.T. Clarkson, pp. 237-60. London: Academic Press.

———. 1977. Studies on the growth of red mangrove (*Rhizophora mangle* L.). IV. The adult root system. *Biotropica* 9(3): 145-55.

Gillespie, P.A., and A.L. Mackensie. 1981. Autotrophic and heterotrophic processes on an intertidal mud-sand flat, Delaware Inlet, Nelson, New Zealand. *Bull. Mar. Sci.* 31: 648-57.

Gillham, M.E. 1965. The Fisher Island Field Station. IV. Vegetation: Additions and changes. *Proc. Roy. Soc. Tas.* 99: 71-80.

Gilmour, A.J. 1974. Impact of man on coastal systems in Victoria. In *The impact of human activities on coastal zones*. Aust. Natl. Comm. for Man and the Biosphere. Canberra: AGPS.

Golley, F.B., H.T. Odum and R.F. Wilson. 1962. The structure and metabolism of a Puerto Rican red mangrove forest in May. *Ecology* 43: 9-19.

Goodrick, G.N. 1970. A survey of wetlands of coastal New South Wales. CSIRO Div. Wildl. Res. Tech. Memo. No. 5.

Gordon, M.S., I. Boetius, D.H. Evans, R. McCarthy and L.C. Oglesby. 1969. Aspects of the terrestrial life in amphibious fishes. I. The mudskipper *Periophthalmus sobrinus*. *J. Exp. Biol.* 50: 141-49.

Gordon, M.S., K. Schmidt-Nielsen and H. Kelly. 1961. Osmotic regulation in the crab-eating frog (*Rana cancrivora*). *J. Exp. Biol.* 38(3): 659-87.

Gordon, M.S., and V.A. Tucker. 1965. Osmotic regulation in the tadpoles of the crab-eating frog (*Rana cancrivora*). *J. Exp. Biol.* 42: 437-45.

Gore, R. 1977. Wild nursery of the mangroves. *Natl. Geog.* 151: 669-89.

Gosselink, J.G., and C.J. Kirby. 1974. Decomposition of saltmarsh grass *Spartina alterniflora* Loisel. *Limnol. Oceanogr.* 19: 825-32.

Goulter, P.F.E., and W.G. Allaway. 1979. Litter fall and decomposition in a mangrove stand (*Avicennia marina* Forsk. (Vierh.)) in Middle Harbour, Sydney. *Aust. J. Mar. Freshw. Res.* 30: 541-46.

Gow, G.F. 1976. *Snakes of Australia*. Sydney: Angus & Robertson.

Graham, M., J. Grimshaw, E. Hegerl, J. McNalty and R. Timmins. 1975. Cairns wetlands — a preliminary report. *Operculum* 4 (3-4): 116-49.

Gray, I.E. 1957. A comparative study of the gill area of crabs. *Biol. Bull.* 112: 34-42.

Green, J.P., and D.F. Dunn. 1977. Osmotic and ionic balance in mangrove sipunculid *Phascolosoma arcuatum*. *Mar. Res. Indonesia* 18: 51-63.

Green, W.A. 1975a. The annual reproductive cycle of *Phascolosoma lurco* (Sipuncula). In *Proceedings of the International Symposium on the Biology of the Sipuncula and Echiura*, eds M.E. Rice and M. Todorovic, pp. 161-68. Washington, DC: Nat. Mus. Nat. History Smithson. Inst.

———. 1975b. *Phascolosoma lurco*: A semi-terrestrial Sipunculan. In *Proceedings of the International Symposium on the Biology of the Sipuncula and Echiura*, eds M.E. Rice and M. Todorovic, pp. 267-81. Washington, DC: Nat. Mus. Nat. History Smithson. Inst.

Greeson, P.E., J.R. Clark and J.E. Clark, eds. 1978. Wetland functions and values: The state of our understanding. Amer. Wat. Res. Assoc. Tech. Rep. TPS 79-2.

Griffin, D.J.G. 1968. Social and maintenance behaviour in two Australian ocypodid crabs (Crustacea: Brachyura). *J. Zool.* (Lond.) 156: 291-305.

Grigg, G.C. 1981. Plasma homeostasis and cloacal urine composition in *Crocodylus porosus* caught along a salinity gradient. *J. Comp. Physiol.* 144: 261-70.

Grigg, G.C., L.E. Taplin, P. Harlow and J. Wright. 1980. Survival and growth of hatchling *Crocodylus porosus* in saltwater

without access to fresh drinking water. *Oecologia* 47: 264–66.
Grime, J.P. 1973. Competition and diversity in herbaceous vegetation — a reply. *Nature* 244: 310–11.
———. 1977. Evidence for the existence of three primary strategies in plants and its relevance to ecological and evolutionary theory. *Amer. Natur.* 111: 1169–94.
———. 1979. *Plant strategies and vegetation processes.* Chichester: John Wiley.
Grimshaw, J.F. 1982. A checklist of spiders known from the mangrove forests and associated tidal marshes of northern and eastern Australia. *Operculum* 5(4): 158–61.
Guiler, E.R. 1951. The intertidal ecology of Pipe Clay lagoon. *Proc. Roy. Soc. Tas.* 1950: 29–41.
Guppy, H.B. 1906. *Observations of a naturalist in the Pacific between 1896 and 1899.* Vol. 2, *Plant dispersal.* London: Macmillan.
Gutteridge, Haskins and Davey. 1975. *Coastal management Queensland–New South Wales border to northern boundary of Noosa Shire.* 4 vols. Co-ordinator General's Department, Queensland.
Haberlandt, G. 1914. *Physiological plant anatomy.* London: Macmillan.
Haedrich, R.L., and C.A.S. Hall. 1976. Fishes and estuaries. *Oceanus* 19: 55–63.
Haines, E.B. 1979. Interactions between Georgia salt marshes and coastal waters. A changing paradigm. In *Ecological processes in coastal and marine systems*, ed. R.J. Livingston. Vol. 10, *Marine science*, pp. 35–47. New York and London: Plenum Press.
Haines, E.B., and R.B. Hanson. 1979. Experimental degradation of detritus made from the salt marsh plants *Spartina alterniflora* (Loisel.), *Salicornia virginicus* L. and *Juncus roemerianus* Scheele. *J. Exp. Mar. Biol. Ecol.* 40: 27–40.
Haines, E.B., and C.L. Montague. 1979. Food sources of estuarine invertebrates analysed using $^{13}C/^{12}C$ ratios. *Ecology* 60: 48–57.
Hall, J.S., and T.J. Flowers. 1973. The effect of salt on protein synthesis in the halophyte *Suaeda maritima*. *Planta* 110: 361–68.
Hall, L.S., and G.C. Richards. 1979. *Bats of eastern Australia.* Queensland Museum Booklet, No. 12.
Halle, F., R.A.A. Oldeman and P.B. Tomlinson. 1978. *Tropical trees and forests. An architectural analysis.* Berlin: Springer-Verlag.
Hamilton, L.S., and S.C. Snedaker, eds. 1984. *Handbook for mangrove area management.* Environment and Policy Institute, East-West Centre, Hawaii.

Hamlyn-Harris, R. 1933. Some ecological factors involved in the dispersal of mosquitoes in Queensland. *Bull. Ent. Res.* 24: 229-32.

Handler, S.H., and H.J. Teas. 1983. Inheritance of albinism in the red mangrove, (*Rhizophora mangle* L.). In *Tasks for vegetation science* ed. H.J. Teas, vol. 8, pp. 117-21. The Hague: Dr W. Junk.

Harbison, P. 1981. The case for the protection of mangrove swamps — geochemical considerations. *Search* 12(8): 273-76.

Harley, J.L. 1969. *The biology of mycorrhiza*. 2nd ed. London: Hill.

Harper, J.L. 1961. Approaches to the study of plant competition. In *Mechanisms in biological competition* (Symp. Soc. Exptl. Biol.) ed. F.L. Milthorpe, vol. 15, pp. 1-39. Cambridge: Cambridge Univ. Press.

Harris, V.A. 1960. On the locomotion of the mudskipper *Periophthalmus koelreuteri* (Pallas): (Gobiidae). *Proc. Zool. Soc.* (Lond.) 134: 107-35.

Harrison, P.D., and K.H. Mann. 1975. Detritus formation from eelgrass (*Zostera marina*): The relative effects of fragmentation, leaching and decay. *Limnol. Oceanogr.* 20: 924-34.

Hartog, C. Den 1970. *The sea grasses of the world*. Amsterdam: North Holland Pub. Co.

Heald, E.J. 1969. The production of organic detritus in a south Florida estuary. Doctoral dissertation, Univ. of Miami, Coral Gables, Florida.

_____. 1971. The production of organic detritus in a south Florida estuary. *Univ. Miami Sea Grant Tech. Bull. No. 6*.

Heald, E.J., and W.E. Odum. 1970. The contribution of mangrove swamps to Florida fisheries. *Proc. Gulf Carib. Fish. Inst.* 22: 730-35.

Heald, E.J., W.E. Odum and D.C. Tabb. 1974. Mangroves in the estuarine food chain. *Miami Geolog. Soc. Mem.* 2: 182-89.

Heatwole, H. 1971. Survey of mangroves of Puerto Rico. Report to the Department of Public Works, Govt. of Puerto Rico. (mimeo.)

_____. 1976. *Reptile ecology*. St Lucia: Univ. Qld Press.

_____. 1977. Voluntary submergence time and breathing rhythm in the homalopsine snake, *Cerberus rhynchops*. *Aust. Zool.* 19: 155-67.

_____. 1978. Adaptations of marine snakes. *Amer. Sci.* 66: 594-604.

_____. 1981. Role of the saccular lung in the diving of the sea krait, *Laticauda colubrina* (Serpentes: Laticaudidae). *Aust. J. Herp.* 1: 11-16.

Heatwole, H. 1985. Survey of the mangroves of Puerto Rico — a benchmark study. *Carib. J. Sci.* 21: 85-99.

Heatwole, H., and R. Seymour. 1975. Pulmonary and cutaneous oxygen uptake in sea snakes and a file snake. *Comp. Biochem. Physiol.* 51: 399-405.

_____. 1978. Cutaneous oxygen uptake in three groups of aquatic snakes. *Aust. J. Zool.* 26: 481-86.

Hegerl, E.J. 1975. The effects of flooding on Brisbane River mangroves. *Operculum* 4: 156-57.

_____. 1982. Mangrove management in Australia. In *Mangrove ecosystems in Australia — structure, function and management*, ed. B.F. Clough, pp. 275-88. Published by AIMS with ANU Press.

Hegerl, E.J., P.J.F. Davie, G.F. Claridge and A.G. Elliott. 1979. *The Kakadu National Park mangrove forests and tidal marshes.* Vol. 1, *A review of the literature and results of a field reconnaissance.* Report prepared for the Aust. Nat. Parks and Wildl. Serv. Brisbane: Australian Littoral Society.

Heinsohn, G.E., J. Wake, H. Marsh and A.V. Spain. 1977. The dugong (*Dugong dugon*) Muller in the seagrass system. *Aquaculture* 12: 235-48.

Henry, S.M. 1966. *Symbiosis.* Vol. 1, *Associations of microorganisms, plants and marine organisms.* London: Academic Press.

_____. 1967. *Symbiosis.* Vol. 2, *Association of invertebrates, birds, ruminants and other biota.* London: Academic Press.

Hering, T.F. 1965. Succession of fungi in the litter of a Lake District oakwood. *Trans. Brit. Mycol. Soc.* 48: 391-408.

Hesse, P.R. 1961a. Decomposition of organic matter in a mangrove swamp soil. *Plant and Soil* 14:249-63.

_____. 1961b. Some differences between the soils of *Rhizophora* and *Avicennia* mangrove swamps in Sierra Leone. *Plant and Soil* 14:335-46.

Hicks, D.B., and L.A. Burns. 1975. Mangrove metabolic response to alterations of natural freshwater drainage to southwestern Florida estuaries. In *Proceedings International Symposium Biology and Management of Mangroves*, eds G.E. Walsh, S.C. Snedaker and H.J. Teas, Vol. 1, pp. 238-55. Gainesville: Univ. Florida.

Hill, B.J. 1979. Aspects of the feeding strategy of the predatory crab *Scylla serrata. Mar. Biol.* 55: 209-14.

_____. 1982. *The Queensland mud crab fishery.* Qld Fisheries Inform. Ser. Fl 8201. Fisheries Research Branch, Qld Dept Primary Industries.

Hill, B.J., M.J. Williams and P. Dutton. 1982. Distribution of juvenile, subadult and adult *Scylla serrata* (Crustacea: Por-

tunidae) on tidal flats in Australia. *Mar. Biol.* 69: 117–20.
Hill, G.F. 1917. Report on some Culicidae of the Northern Territory. *Bull. Nth. Terr. Aust.* 17: 1–8.
Hodda, M., and W.L. Nicholas. 1985. Meiofauna associated with mangroves in the Hunter River Estuary and Fullerton Cove, south eastern Australia. *Aust. J. Mar. Freshw. Res.* 36: 41–50.
Hodgkin, E.P. 1974. Biological aspects of coastal zones development in Western Australia. I. General aspects. In *The impact of human activities on coastal zones*. Aust. Natl. Comm. for Man and the Biosphere. Canberra: AGPS.
––––––. 1978. *An environmental study of the Blackwood River Estuary, Western Australia, 1974–5*. A report to the Estuarine and Marine Advisory Committee of the Environmental Protection Authority, Rep. No. 1, Dept Conservation and Environment.
Hoffman, W.E., and C.J. Dawes. 1980. Photosynthetic rates and primary production by two Florida benthic red algal species from a saltmarsh and a mangrove community. *Bull. Mar. Sci.* 30: 358–64.
Hogg, R.W., and F.T. Gillan. 1984. Fatty acids, sterols and hydrocarbons in the leaves from eleven species of mangrove. *Phytochem.* 23: 93–97.
Holdich, D.M., and K. Harrison. 1980. The crustacean isopod genus *Gnathia* Leach from Queensland waters with descriptions of nine new species. *Aust. J. Mar. Freshw. Res.* 31: 215–40.
Holmes, G. 1970. Birds of the Hunter River estuary. *Hunter Nat. Hist.* 2: 13–18.
Hook, D.D., and J.R. Scholtens. 1978. Adaptation and flood tolerance of tree species. In *Plant life in anaerobic environments*, eds D.D. Hook and R.M.M. Crawford, pp. 299–332. Michigan: Ann Arbor Science.
Hosokawa, T., H. Tagawa and V.J. Chapman. 1977. Mangals of Micronesia, Taiwan, Japan, the Philippines and Oceania. In *Ecosystems of the world. I. Wet coastal ecosystems*, ed. V.J. Chapman, pp. 271–92. Amsterdam: Elsevier.
Humm, H.J. 1944. Bacterial leaf nodules. *J. NY Bot. Gdn.* 45: 193–99.
Hutchings, P.A. 1981. Polychaete recruitment onto dead coral substrates at Lizard Island, Great Barrier Reef, Australia. *Bull. Mar. Sci.* 31(2): 410–23.
––––––. 1983. The wetlands of Fullerton Cove, Hunter River, New South Wales. *Wetlands* 3(1): 12–21.
Hutchings, P.A., and C.J. Glasby. 1985. Additional nereidids (Polychaeta) from Eastern Australia, together with a

redescription of *Namanereis quadraticeps* (Gay) and the synonymising of *Ceratonereis pseudoerythraeensis* Hutchings and Turvey, with *C. aequisetis* (Augener). *Rec. Aust. Mus.* 37(2): 101-10.

Hutchings, P.A., and A. Murray. 1982. Patterns of recruitment of polychaetes to coral substrates at Lizard Island, Great Barrier Reef — an experimental approach. *Aust. J. Mar. Freshw. Res.* 33: 1029-37.

Hutchings, P.A., J. Pickard, H.F. Recher and P.B. Weate. 1977. A survey of mangroves at Brooklyn, Hawkesbury River, New South Wales. *Operculum* (Jan. '77): 105-12.

Hutchings, P.A., and H.F. Recher. 1974. The fauna of Careel Bay, with comments on the ecology of mangrove and seagrass communities. *Aust. Zool.* 18: 99-128.

―――――. 1977. The management of mangroves in an urban situation. *Mar. Res. Indonesia* 18: 1-11.

―――――. 1982. The fauna of Australian mangroves. *Proc. Linn. Soc. NSW* 106(1): 83-121.

Huxley, C.R. 1978. The ant-plants *Myrmecodia* and *Hydnophytum* (Rubiaceae) and the relationships between their morphology, ant occupants, physiology and ecology. *New Phytologist* 80: 231-68.

Hyland, S.J., B.J. Hill and C.P. Lee. 1984. Movement within and between different habitats by the portunid crab *Scylla serrata*. *Mar. Biol.* 80: 57-61.

Itai, C., A. Richmond and Y. Vaadia. 1968. The role of root cytokinins during water and salinity stress. *Israel J. Bot.* 17: 187-95.

Iyengar, M.O.T. 1965. Epidemiology of filariasis in the South Pacific. *Tech. Pap. Ser., Pacific Commission* 148: 1-183.

Janssonius, H.H. 1950. The vessels in the wood of Jarvan mangrove trees. *Blumea* 6: 464-69.

Janzen, D.H. 1974. Epiphytic myrmecophytes in Sarawak: Mutualism through the feeding of plants by ants. *Biotropica* 6: 237-59.

Jardine, F. 1925. The physiography of the Port Curtis district. *Rep. Gt. Barr. Reef. Comm.* 1: 73-100.

Jefferies, R.L. 1972. Aspects of salt marsh ecology with particular reference to inorganic plant nutrition. In *The estuarine environment*, eds R.S.K. Barnes and J. Green, pp. 61-85. London: Applied Sci. Publ.

Jeffrey, S.W. 1981. Algal pigment systems. In *Primary productivity in the sea*, no. 31, ed. Paul G. Salkowski, pp. 33-58. Brookhaven Symp. Biol. Environmental Science Research 19.

Jehne, W., and C.H. Thompson. 1981. Endomycorrhizae in plant colonization on coastal sand-dunes at Cooloola, Queensland. *Aust. J. Ecol.* 6: 221-30.

Jennings, D.H. 1968. Halophytes, succulence and sodium — a unified theory. *New Phytologist* 67: 899-911.
Jennings, J.H., and E.C.F. Bird. 1967. Regional geomorphological characteristics of some Australian estuaries. In *Estuaries*, ed. G.H. Lauff, pp. 121-28. Amer. Assoc. Adv. Sci. 83.
Jennings, J.N., and R.J. Coventry. 1973. Structure and texture of gravelly barrier island in Fitzroy Estuary, Western Australia, and the role of mangroves in shore dynamics. *Mar. Geol.* 15: 145-67.
Johnstone, I.M. 1981. Consumption of leaves by herbivores in mixed mangrove stands. *Biotropica* 13: 252-59.
———. 1983. Succession in zoned mangrove communities: Where is the climax? In *Tasks for vegetation science*, ed. H.J. Teas, pp. 131-40. The Hague: Dr W. Junk.
Jones, K. 1974. Nitrogen fixation in a salt marsh. *J. Ecol.* 62: 563-65.
Jones, S. 1967. The dugong — its present status in the seas around India with observation on its behaviour in captivity. *Int. Zool. Yearb.* 7: 215-20.
Jones, T.G.H., and J.M. Harvey. 1936. Essential oils from the Queensland flora. Part VIII. The identity of melaleucol with nerolidol. *Proc. Roy. Soc. Qld* 47: 92-93.
Jones, W.T. 1971. The field identification and distribution of mangroves in eastern Australia. *Qd. Nat.* 20(1-3): 35-51.
Joshi, A.C. 1933. A suggested explanation of the prevalence of vivipary on the seashore. *J. Ecol.* 21: 209-12.
Joshi, G.V., L.J. Bhosale, B.B. Jamale and B.A. Karadge. 1975. Photosynthetic carbon metabolism in mangroves. In *Proceedings of the International Symposium on Biology and Management of Mangroves*, eds G.E. Walsh, S.C. Snedaker and H.J. Teas, vol. 2, pp. 579-94. Gainesville: Univ. Florida.
Joshi, G.V., B.B. Jamale and L. Bhosale. 1975. Ion regulation in mangroves. In *Proceedings of the International Symposium on Biology and Management of Mangroves*, eds G.E. Walsh, S.C. Snedaker and H.J. Teas, pp. 595-607. Gainesville: Univ. Florida.
Joshi, G.V., M.D. Karekar, C.A. Jowda and L. Bhosale. 1974. Photosynthetic carbon metabolism and carboxylating enzymes in algae and mangroves under saline conditions. *Photosynthetica* 8: 51-52.
Joshi, G.V., M. Pimlaskar and L.J. Bhosale. 1972. Physiological studies in germination of mangroves. *Bot. Mar.* 15: 91-95.
Juniper, S.K. 1981. Stimulation of bacterial activity by a deposit feeder in two New Zealand intertidal inlets. *Bull. Mar. Sci.* 31: 691-701.
Kahane, I., and A. Poljakoff-Mayber. 1968. Effect of substrate

salinity on the ability for protein synthesis in pea roots. *Plant Physiol.* 43: 1115-19.

Kaplan, E.H., J.R. Walker and M.G. Kraus. 1974. Some effects of dredging on populations of macrobenthic organisms. *Fish. Bull.* 72: 445-80.

Karstedt, P., and N. Parameswaran. 1976. Anatomy and systematics of *Rhizophora* species. *Bot. Jahrb. Syst. Pflangzengesch. Pflanzengeogr.* 97(3): 317-38.

Kato, A. 1975. Brugine from *Bruguiera cyclindrica*. *Phytochem.* 14: 1458.

Kaushik, N.K., and H.B.N. Hynes. 1968. Experimental study on the role of autumn-shed leaves in aquatic environments. *J. Ecol.* 56: 229-43.

Kay, B.H., and I.D. Fanning. 1974. Brunswick Heads midge study. *Rep. Qd. Inst. Med. Res.* 29: 11-12.

Kehar, N.D., and S.S. Negi. 1953. Mangrove (*Avicennia officinalis* Linn.) leaves as cattle feed. *Science and Culture* 18: 382-83.

Kemp, E.M. 1978. Tertiary climatic evolution and vegetation history in the southeast Indian Ocean region. *Palaeogeogr. Palaeoclimatol. Palaeoecol.* 24: 169-208.

Kettle, D.S. 1977. Biology and bionomics of blood-sucking ceratopogonids. *Ann. Rev. Ent.* 22: 33-51.

Kettle, D.S., E.J. Reye and P.B. Edwards. 1979. Distribution of *Culicoides molestus* (Skuse) (Diptera: Ceratopogonidae) in man-made canals in southeastern Queensland. *Aust. J. Mar. Freshw. Res.* 30: 653-60.

Kikkawa, J., and K. Pearse. 1969. Geological distribution of land birds in Australia — a numerical analysis. *Aust. J. Zool.* 117: 821-40.

Kimball, M., and H.J. Teas. 1975. Nitrogen fixation in mangrove areas of South Florida. In *Proceedings of the International Symposium on Biology and Management of Mangroves*, eds G.E. Walsh, S.C. Snedaker and H.J. Teas, pp. 654-61. Gainesville: Univ. Florida.

King, R.J. 1981a. Mangroves and saltmarsh plants. In *Marine botany: An Australasian perspective*, eds M.N. Clayton and R.J. King, pp. 308-29. Melbourne: Longman Cheshire.

———. 1981b. The free-living *Hormosira banksii* (Turner) Decaisne associated with mangroves in temperate eastern Australia. *Bot. Mar.* 24: 569-76.

———. 1981c. The macroalgae of mangrove communities in eastern Australia. *Phycologia* 20: 107-8.

King, R.J., and M.D. Wheeler. 1985. Composition and geographic distribution of mangrove macroalgal communities in New South Wales. *Proc. Linn. Soc. NSW* 108(2):97-118.

Kirkman, H., and D.D. Reid. 1979. A study of the role of the seagrass *Posidonia australis* in the carbon budget of an estuary. *Aquat. Bot.* 7: 173-83.

Kirkpatrick, J.B., and J. Glasby. 1981. Salt marshes in Tasmania. Distribution, community composition and conservation. *Dept. Geogr. Univ. Tasm., Occ. Pap.* 8: 1-62.

Klugh, A.B. 1909. Excretion of sodium chloride by *Spartina glabra alterniflora*. *Rhodora* 11: 237-38.

Kobayashi, T., Y. Dotsu and T. Takita. 1971. Nest and nesting behaviour of the mudskipper, *Periophthalmus cantonensis* in Ariake Sound. *Bull. Fac. Fish. Nagasaki Univ.* No. 32: 27-40.

Kohlmeyer, J. 1969. Ecological notes on fungi in mangrove forests. *Trans. Brit. Mycol. Soc.* 53: 237-50.

Kolehmainen, S. 1973. Ecology of sessile and free-living organisms on mangrove roots in Jobos Bay. *Rep. Puerto Rico Nuclear Center Mayagues P.R.*

Kolehmainen, S.E., F.D. Martin and P.B. Schroeder. 1974. Thermal studies on tropical marine ecosystems in Puerto Rico. *Symposium on the Physical and Biological Effects of the Environment of Cooling Systems and Thermal Discharges at Nuclear Stations, Oslo*, 26-30 Aug. 1974.

Kratochvil, M., N.J. Hannon and L.D. Clarke. 1973. Mangrove swamp and saltmarsh communities in southeastern Australia. *Proc. Linn. Soc. NSW* 97: 262-74.

Kunze, J., and D.T. Anderson. 1979. Functional morphology of mouthparts and gastric mill in the hermit crabs *Clibanarius taeniatus* (Milne Edwards), *Clibanarius virescens* (Krauss) *Paguristes squamosus* McCulloch and *Dardanus setifer* (Milne Edwards) (Anomura: Paguridae). *Aust. J. Mar. Freshw. Res.* 30: 683-722.

Kylin, A., and R. Gee. 1970. Adenosine triphosphatase activities in leaves of the mangrove *Avicennia nitida* Jacq: Influence of sodium to potassium ratios and salt concentrations. *Pl. Physiol.* 45: 169-72.

Lanyon, J.A., I.G. Eliot and D.J. Clarke. 1982. Groundwater level variation during semidiurnal spring tidal cycles on a sandy beach. *Aust. J. Mar. Freshw. Res.* 33: 377-400.

Larkum, A.W.D. 1981. Marine primary productivity. In *Marine botany: An Australian perspective*, eds M.N. Clayton and R.J. King, pp. 369-85. Melbourne: Longman Cheshire.

Larkum, A.W.D., L.C. Collett and R.J. Williams. 1984. The standing stock, growth and shoot production of *Zostera capricorni* Aschers in Botany Bay, New South Wales, Australia. *Aquat. Bot.* 19: 307-27.

Lasserre, G., and J.L. Toffart. 1977. Echantillonnage et structure des populations ichtyologiques des mangroves de Guadeloupe en Septembre 1975. *Cybium* 2: 115-27.

Latter, P.M., and J.B. Craig. 1967. The decomposition of *Juncus squarrosus* leaves and microbial changes in the profile of *Juncus* moor. *J. Ecol.* 55: 465-82.

Law, R., and D.H. Lewis. 1983. Biotic environments and the maintenance of sex — some evidence from mutualistic symbioses. *Biol. J. Linn. Soc.* 20: 249-76.

Lear, R., and T. Turner. 1977. *Mangroves of Australia*. St Lucia: Univ. Qld Press.

Lee, B.K.H., and G.E. Baker. 1972a. An ecological study of the soil microfungi in a Hawaiian mangrove swamp. *Pac. Sci.* 26: 1-10.

―――. 1972b. Environment and the distribution of microfungi in a Hawaiian mangrove swamp. *Pac. Sci.* 26: 11-19.

―――. 1973. Fungi associated with the roots of the red mangrove, *Rhizophora mangle*. *Mycologia* 65: 894-906.

Lee, D.J., M.H. Hicks, M. Griffiths, R.C. Russell and E.N. Marks. 1982. *The Culicidae of the Australian region*. Vol. 2. *Nomenclature, synonymy, literature, distribution, biology and relation to disease*. Commonwealth Dept Health. Series Entomology Monograph No. 2. Canberra: AGPS.

―――. 1984. *The Culicidae of the Australasian region*. Vol. 3. Canberra: AGPS.

Lee, D.J., and E.J. Reye. 1955. Australasian Ceratopogonidae (Diptera, Nematocera.) Part VII. Notes on the genera *Alluaudomyia*, *Ceratopogon*, *Culicoides*, and *Lasiohelea*. *Proc. Linn. Soc. NSW* 79(5-6): 233-46.

Lee, D.J., E.J. Reye and A.L. Dyce. 1963. "Sandflies" as possible vectors of disease in domesticated animals in Australia. *Proc. Linn. Soc. NSW* 87(3): 364-76.

Leightley, L. 1980. Wood decay activities of marine fungi. *Bot. Mar.* 23: 387-95.

Lersten, N.R., and H.T. Horner. 1976. Bacterial leaf nodule symbiosis in angiosperms with emphasis on Rubiaceae and Myrsinaceae. *Bot. Rev.* 42: 145-214.

Leshem, Y., and E. Levison. 1972. Regulation mechanisms in the salt mangrove *Avicennia marina* growing on the Sinai littoral. *Oecol. Plant.* 7: 167-76.

Lewin, R.A. 1982. Symbiosis and parasitism — definitions and evaluations. BioSci. 32: 254-59.

Lewis, O.A.M., and G. Naidoo. 1970. Tidal influence on the apparent transpirational rhythms of the white mangrove. *S. Afr. J. Sci.* 66: 268-70.

Lighter, F.J. 1974. A note on a behavioural spacing mechanism of the ghost crab *Ocypode ceratophthalmus* (Pallas) (Decapoda, family Ocypodidae). *Crustaceana* 27: 312-14.

Lindall, W.N. 1974. Alterations of estuaries of South Florida: A threat to its fish resources. *Operculum* 4: 63-69.

Lindall, W.N.,and L. Trent. 1975. Housing development canal in the coastal zone of the Gulf of Mexico: Ecological consequences, regulations and recommendations. *Mar. Fish. Rev.* 37: 19-24.

Liphschitz, N., and Y. Waisel. 1974. Existence of salt gland in various genera of the Gramineae. *New Phytol.* 73: 507-13.

Little, C., and P. Stirling. 1984. Activation of a mangrove snail *Littorina scabra* (L.) (Gastropoda: Prosobranchia). *Aust. J. Mar. Freshw. Res.* 35(5): 607-10.

Littlejohn, M.J., G.F. Watson and A.I. Robertson. 1974. The ecological role of macrofauna in eelgrass communities. In *A preliminary report on the Westernport Bay Environmental Study, Victoria*, ed. M.A. Shapiro, pp. 1-6. Ministry for Conservation.

Lloyd, R.M. 1980. Reproductive biology and gametophyte morphology of New World populations of *Acrostichum aureum*. *Amer. Fern. J.* 70: 99-110.

Lockwood, A.P.M. 1967. *Aspects of the Physiology of Crustacea*. San Francisco: Freeman.

Loder, J.W., and G.B. Russell. 1969. Tumor inhibitory plants. The alkaloids of *Bruguiera sexangula* and *Bruguiera exaristata* (Rhizophoraceae). *Aust. J. Chem.* 22: 1271-75.

Loetschert, W., and F. Liemann. 1967. Die Salzspeicherung im Keimling von *Rhizophora mangle* L. wahrend der Entwicklung auf der Mutterpflanze. *Planta* 72: 142-56.

Loveridge, A. 1946. *Reptiles of the Pacific world*. New York: MacMillan.

Ludbrook, N.H. 1963. Correlations of the tertiary rocks of South Australia. *Trans. Roy. Soc. Sth Aust.* 87: 5-15.

Lugo, A.E., and M.M. Brinson. 1978. Calculations of the value of saltwater wetlands. In *Wetland functions and values: The state of our understanding*, eds P.E. Greeson, J.R. Clark and J.E. Clark, pp. 120-30. Amer. Wat. Res. Assoc. TPS 79-2.

Lugo, A.E., G. Evink, M.M. Brinson, A. Broce and S.C. Snedaker. 1975. Diurnal rates of photosynthesis, respiration and transpiration in mangrove forests in South Florida. In *Ecological studies II. Tropical ecological systems*, eds F.B. Golley and E. Medina, pp. 335-50. New York: Springer-Verlag.

Lugo, A.E., M. Sell and S.C. Snedaker. 1976. Mangrove ecosystem analysis. In *Systems analysis and simulation in ecology*, ed. B.C. Patten, pp. 113-45. New York: Academic Press.

Lugo, A.E., and S.C. Snedaker. 1973. *The role of mangrove ecosystems: Properties of a mangrove forest in south Florida*. Resource Management Systems Program. University of Florida (Gainesville) USA Report No. D1-SFEP-74-34.

———. 1974. The ecology of mangroves. *Ann. Rev. Ecol. Syst.* 5: 39–64.

———. 1975. Properties of a mangrove forest in southern Florida. In *Proceedings of the International Symposium on Biology and Management of Mangroves*, eds G.E. Walsh, S.C. Snedaker and H.J. Teas, vol. 1, pp. 170–212. Gainesville: Univ. Florida.

Lugo, A.E., and C. Zucca. 1977. The impact of low temperature stress on mangrove structure and growth. *Trop. Ecol.* 18(2): 149–61.

Lush, D.L., and H.B.N. Hynes. 1973. The formation of particles in freshwater leachates of dead leaves. *Limnol. Oceanogr.* 18(6): 968–77.

Macginitie, S.E. 1930. Natural history of the mudshrimp *Upogebia pugettensis* (Dana). *Annals Mag. Nat. Hist.* 10: 36–44.

Macnae, W. 1963. Mangrove swamps in South Africa. *J. Ecol.* 51: 1–25.

———. 1966. Mangroves in eastern and southern Australia. *Aust. J. Bot.* 14: 67–104.

———. 1967. Zonation within mangroves associated with estuaries in North Queensland. In *Estuaries*, ed. G.H. Lauff, pp. 432–41. Amer. Assoc. Adv. Sci. Publ. 83.

———. 1968. A general account of the fauna and flora of mangrove swamps and forests in the Indo-West Pacific region. *Adv. Mar. Biol.* 6: 73–270.

Macnae, W., and M. Kalk. 1962. The ecology of the mangrove swamps of Inhaca Island, Mozambique. *J. Ecol.* 50: 19–34.

Magnusson, W.E., G.C. Grigg and J.A. Taylor. 1978. An aerial survey of nesting areas of the saltwater crocodile *Crocodylus porosus* (Schneider) on the north coast of Arnhem Land, Northern Australia. *Aust. J. Wildl. Res.* 5: 401–5.

———. 1980. An aerial survey of potential nesting areas of *Crocodylus porosus* on the west coast of Cape York Peninsula. *Aust. Wildl. Res.* 7(3): 465–79.

Magnusson, W.E., G.J.W. Webb and J.A. Taylor. 1976. Two new locality records, a new habitat and a nest description for *Xeromys myoides* Thomas (Rodentia: Muridae). *Aust. Wildl. Res.* 3: 153–57.

Malaviya, C.V. 1963. On the distribution, structure and ontogeny of stone cells in *Avicennia officinalis* L. *Proc. Indian Acad. Sci.* 58: 45–50.

Malcolm, C.V. 1964. The effect of salt, temperature of seed scarification on germination of two varieties of *Arthrocnemum halocnemoides*. *J. Roy. Soc. West. Aust.* 47: 72–74.

Malcolm, W.B. 1971. Sydney rock oysters. *Aust. Nat. Hist. Mag.* 17: 46–50.

Malley, D.F. 1977. Adaptations of decapod crustaceans to life in mangrove swamps. *Mar. Res. Indonesia* 18: 63-72.
———. 1978. Degradation of mangrove leaf litter by the tropical sesarmid crab *Chiromanthes onychophorum*. *Mar. Biol.* 49: 377-86.
McConnaughey, T., and C.P. McRoy. 1979. ^{13}C label identifies eelgrass (*Zostera marina*) carbon in an Alaskan estuarine food web. *Mar. Biol.* 53: 263-69.
McRoy, C.P., and C. McMillan. 1977. Production ecology and physiology of seagrasses. In *Seagrass ecosystems*, eds C.P. McRoy and C. Helfferich, pp. 53-87. New York: Marcel Dekker.
Mangum, C., and W. Van Winkle. 1973. Responses of aquatic invertebrates to declining oxygen conditions. *Amer. Zool.* 13: 529-41.
Mantel, L.H., and L.L. Farmer. 1983. Osmotic and ionic regulation. In *Biology of the Crustacea*, ed. L.A. Mantel, vol. 5, pp. 54-161. New York: Academic Press.
Marco, H.F. 1935. Systematic anatomy of the woods of the Rhizophoraceae. *Trop. Woods* (Yale University) 44: 1-26.
Margulis, L. 1981. *Symbiosis in cell evolution*. San Francisco: Freeman.
Markley, J.L., C. McMillan and G.A. Thompson. 1982. Latitudinal differentiation in response to chilling temperatures among populations of three mangroves, *Avicennia germinans*, *Laguncularia racemosa*, and *Rhizophora mangle*, from the western tropical Atlantic and Pacific Panama. *Can. J. Bot.* 60: 2704-15.
Marks, E.N. 1953. Report on the National Mosquito Control Committee. Appendix A. *Annual Report, Health Medical Services, Queensland 1952-53*: 104-108.
———. 1966. An atlas of common Queensland mosquitoes; with a guide to common Queensland biting midges by E.J. Reye. Brisbane, Queensland (mimeo).
———. 1967. An atlas of common Queensland mosquitoes; with a guide to common Queensland biting midges by E.J. Reye. Brisbane, Queensland (mimeo). Revised edition.
Marshall, K.B., R. Stout and R. Mitchell. 1971. Selective absorption of bacteria from seawater. *Can. J. Microbiol.* 17: 1413-16.
Mazotti, F.J., and W.A. Dunson. 1984. Adaptations of *Crocodylus acutus* and *Alligator* for life in saline water. *Comp. Biochem. Physiol.* 79A: 641-46.
McCormick, W.A. 1978. The ecology of benthic macrofauna in New South Wales mangrove swamps. M.Sc. thesis, Univ. New South Wales.
McKillup, S.C., and A.J. Butler. 1979. Cessation of hole-digging

by the crab *Helograpsus haswellianus*. A resource-conserving adaptation. *Mar. Biol.* 50: 157-61.

McLeod, J. 1969. Tidal marshes of southeastern Queensland and their associated algal flora. M.Sc. thesis, Univ. Queensland.

McMillan, C. 1971. Environmental factors affecting seedling establishment of the black mangrove on the central Texas coast. *Ecology* 52: 927-30.

_____. 1974. Salt tolerance of mangroves and submerged aquatic plants. In *Ecology of halophytes*, eds R.J. Reimold and W.H. Queen, pp. 379-90. New York: Academic Press.

_____. 1975a. Adaptive differentiation to chilling in mangrove populations. In *Proceedings of the International Symposium on Biology and Management of Mangroves*, eds G.E. Walsh, S.C. Snedaker and H.J. Teas, vol. 1, pp. 62-70. Gainesville: Univ. Florida.

_____. 1975b. Interactions of soil texture with salinity tolerances of *Avicennia germinans* (L.) Lam. and *Laguncularia racemosa* (L.) Gaertn f. from North America. In *Proceedings of the International Symposium on Biology and Management of Mangroves*, eds G.E. Walsh, S.C. Snedaker and H.J. Teas, vol. 2, pp. 561-68. Gainesville: Univ. Florida.

McMillan, R.T. 1964. Studies of a recently described *Cercospora* on *Rhizophora mangle. Plant Dis. Rep.* 48: 909-11.

Mee, L.D. 1978. Coastal lagoons. In *Chemical oceanography*, eds J.P. Riley and R. Chester, vol. 7, pp. 441-90. New York: Academic Press.

Menzies, R.J., J.S. Zaneveld and R.M. Pratt. 1967. Transported turtle grass as a source of organic enrichment of abyssal sediments off North Carolina. *Deep-Sea Res.* 14: 111-12.

Mepham, R.H. 1983. Mangrove floras of the southern continents. Part 1. The geographical origin of Indo-Pacific mangrove genera and the development and present status of the Australian mangroves. *S. Afr. J. Bot.* 2(1): 1-8.

Merrill, E.D. 1945. *Plant life in the Pacific world*. London: Macmillan.

Messel, H., and S.T. Butler. 1977. *Australian animals and their environments*. Parramatta: MacArthur Press.

Messel, H. et al. 1979-82. *Surveys of tidal river systems in the Northern Territory of Australia and their crocodile population*. 19 vols. Sydney: Pergamon Press.

Meyers, S.P., P.A. Orput, J. Simms and L.L. Boral. 1965. Thalassiomycetes. VII. Observations on fungal infestations of turtle grass, *Thalassia testudinum* Koenig. *Bull. Mar. Sci.* 15: 548-64.

Middleton, M.J., J.D. Bell, J.J. Burchmore, D.A. Pollard and B.C. Pease. 1984. Structural differences in the fish communities of *Zostera capricorni* and *Posidonia australis* seagrass meadows in Botany Bay, New South Wales. *Aquat. Bot.* 18: 89–109.

Miller, B. 1972. Birds and wetlands. *Operculum* 2(3): 68–73.

Miller, I.M., I.C. Gardner and A. Scott. 1983. The development of marginal leaf nodules in *Ardisia crispa* (Thunb.) A.DC. (Myrsinaceae). *Bot. J. Linn. Soc.* 86: 237–52.

Miller, P.C. 1972. Bioclimate, leaf temperature and primary production in red mangrove canopies in south Florida. *Ecology* 53: 22–45.

——. 1975. Simulation of water relations and net photosynthesis in mangroves in southern Florida. In *Proceedings of the International Symposium on Biology and Management of Mangroves*, eds G.E. Walsh, S.C. Snedaker and H.J. Teas, vol. 2, pp. 615–31. Gainesville: Univ. Florida.

Miller, P.C., J. Hom and D.K. Poole. 1975. Water relations of three mangrove species in south Florida. *Oecol. Plant.* 10: 355–67.

Millis, N.F. 1981. Marine microbiology. In *Marine botany: An Australian perspective*, eds M.N. Clayton and R.J. King, pp. 35–60. Melbourne: Longman Cheshire.

Milne, A. 1961. Definition of competition among animals. In *Mechanisms in biological competition*, ed. F.L. Milthorpe, vol. 15, pp. 40–61. (Symp. Soc. Exptl. Biol.) Cambridge: Cambridge University Press.

Milward, N.E. 1974. Studies on the taxonomy, ecology and physiology of Queensland mudskippers. Ph.D. thesis, Univ. Queensland.

——. 1982. Mangrove-dependent biota. In *Mangrove ecosystems in Australia: structure, function and management*, ed. B.F. Clough, pp. 121–39. Published by AIMS with ANU Press.

Minocha, P.K., and K.P. Tiwari. 1981. A triterpenoidal saponin from roots of *Acanthus illicifolius*. *Phytochem.* 20: 135–37.

Mizrachi, D., R. Pannier and F. Pannier. 1980. Assessment of salt resistance mechanisms as determinant physio-ecological parameters of zonal distribution of mangrove species. I. Effect of salinity stress on nitrogen metabolism balance and protein synthesis in the mangrove species *Rhizophora mangle* and *Avicennia nitida*. *Bot. Mar.* 23: 289–96.

Mogg, A.O.D. 1963. A preliminary investigation of the significance of salinity in the zonation of species of saltmarsh and mangrove swamp associations. *S. Afr. J. Sci.* 59: 81–86.

Moore, J., and R. Gibson. 1981. The *Geonemertes* problem (Nemertea). *J. Zool.* (Lond.) 194: 175–201.

Moore, L.B. 1950. A "loose-lying" form of the brown alga *Hormosira*. *Trans. Roy. Soc. NZ* 78: 48–53.

Moore, R.T., P.C. Miller, D. Albright and L.L. Tieszen. 1972. Comparative gas exchange characteristics of three mangrove species during the winter. *Photosynthetica* 6: 387-93.

Moore, R.T., P.C. Miller, J. Ehleringer and W. Lawrence. 1973. Seasonal trends in gas exchange characteristics of three mangrove species. *Photosynthetica* 7: 387-94.

Moriarty, D.J.W. 1976. Quantitative studies on bacteria and algae in the food of mullet *Mugil cephalus* L. and the prawn *Metapenaeus bennettae* (Racek and Dall). *J. Exp. Mar. Biol. Ecol.* 22: 131-43.

―――. 1977. Quantification of carbon, nitrogen, and bacterial biomass in the food of some penaeid prawns. *Aust. J. Mar. Freshw. Res.* 28: 113-18.

―――. 1980. Measurement of bacterial biomass in sandy sediments. In *Biogeochemistry of ancient and modern environments*, ed. P.A. Trudinger, M.R. Walter and B.J. Ralph, pp. 131-38. Canberra: Aust. Acad. Science.

Moriarty, D.J.W., and M.C. Barclay. 1981. Carbon and nitrogen content of food and the assimilation efficiencies of penaeid prawns in the Gulf of Carpentaria. *Aust. J. Mar. Freshw. Res.* 32: 245-51.

Moriarty, D.J.W., and P.C. Pollard. 1981. DNA synthesis as a measure of bacterial productivity in seagrass sediments. *Mar. Ecol. Prog. Ser.* 5: 151-56.

Montgomery, J., and M.T. Price. 1979. Release of trace metals by sewage sludge and the subsequent uptake by members of a turtle grass mangrove ecosystem. *Environ. Sci. Technol.* 13(5): 546-49.

Morton, B. 1975. The diurnal rhythm and the feeding responses of the southeast Asian mangrove bivalve *Geloina proxima* Prinie 1864 (Bivalvia: Corbiculacea). *Forma et function* 3/4: 405-19.

Morton, J.F. 1965. Can the red mangrove provide food, feed and fertilizer? *Econ. Bot.* 19: 113-23.

Morton, R.M. 1984. The abundance and feeding of fishes in saltcouch marsh in Moreton Bay, Queensland. *Australian Society for Fish Biology, 11th Annual Conf., 17-20 Aug., Glenelg, South Australia*. Abstract only.

Muggeridge, P. 1979. The reproductive biology of the mangrove littorinids *Bembicium auratum* (Quoy, Gaimard) and *Littorina scabra* (Linne) (Gastropoda, Prosobranchiata) with observations on the reproductive cycles of rocky shore littorinids of NSW. Ph.D. thesis, Sydney Univ.

Mullan, D.P. 1931. On the occurrence of glandular hairs (salt glands) on the leaves of some Indian halophytes. *J. Indian Bot. Soc.* 10: 184-89.

Muller, C.H. 1966. The role of chemical inhibition (allelopathy) in vegetational composition. *Bull. Torrey Bot. Club* 93: 332-51.

Muller, J. 1964. A palynological contribution to the history of the mangrove vegetation in Borneo. In *Ancient Pacific floras*, ed. L.M. Cranwell, pp. 33-42. Honolulu: Univ. Hawaii Press.

_____. 1969. A palynological study of the genus *Sonneratia* (Sonneratiaceae). *Pollen et Spores* 11: 223-98.

Muller, J., and C. Caratini. 1977. Pollen of *Rhizophora* (Rhizophoraceae) as a guide fossil. *Pollen et Spores* 19: 361-90.

Munro, I.S.R. 1973. Fauna of the Gulf of Carpentaria. No. 1. Introduction and station lists. *Fish. Notes Qld* (N.S.) 2: 1-38.

Myers, J.G. 1935. Zonation along river courses. *J. Ecol.* 23: 356-60.

Naidoo, G. 1980. Mangrove soils of the Beachwood Area, Durban. *J.S. Afr. Bot.* 46: 293-304.

Nedwell, D.B. 1974a. Sewage treatment and discharge into tropical coastal waters. *Search* 5(5): 187-90.

_____. 1974b. Letter to the Editor. *Search* 5: 368.

Newell, B.S., and W.E. Barber. 1975. Estuaries important to Australian fisheries. *Aust. Fisheries* 34(1): 17-22.

Newell, R.C. 1965. The role of detritus in the nutrition of two marine deposit feeders: the prosobranch *Hydrobia ulvae* and the bivalve *Macoma balthica*. *Proc. Zool. Soc.* (Lond.) 144: 25-45.

Newell, S.Y. 1976. Mangrove fungi. The succession in the mycoflora of red mangrove (*Rhizophora mangle* L.) seedlings. In *Recent advances in aquatic mycology*, ed. E.B.G. Jones, pp. 51-91. London: Halstead Press.

Nichols, M.M. 1972. Effect of increasing depth on salinity in the James River estuary. In *Environmental framework of coastal plains estuaries*, ed. B.W. Nelson. Geol. Soc. Amer. Memoir No. 133.

Nishihira, M., and M. Urasaki. 1976. Production, settlement and mortality of seedlings of a mangrove, *Kandelia candel* (L.) Druce in Okinawa. Abstracts of symposia and contributed papers, *International Symposium on the Ecology and Management of Some Tropical Shallow Water Communities*, Jakarta, July 1976.

Nursall, R. 1981. Habitat and habitat changes as they affect mudskippers *Periophthalmus* spp. and *Periophthalmodon*. *Bull. Mar. Sci.* 31(3): 730-35.

Odum, E.P. 1959. Fundamentals of ecology, 2nd ed. Saunders.

_____. 1973. *A description and value assessment of South Atlantic and Gulf Coast marshes and estuaries*. Bur. Fish. and Wildl. Publ., Atlanta, Georgia.

Odum, E.P., and A. de la Cruz. 1967. Particulate organic detritus in a Georgia salt marsh estuarine ecosystem. In *Estuaries*, ed. G.H. Lauff, pp. 383-88. Amer. Assoc. Adv. Sci. Publ. 83.

Odum, W.E. 1970. Insidious alteration of the estuarine environment. *Trans. Amer. Fish. Soc.* 99: 836-47.

———. 1971. Pathways of energy flow in South Florida estuary. *Sea Grant Tech. Bull. Miami Univ.* 7: 1-162.

———. 1982. Environmental degradation and the tyranny of small decisions. *BioSci.* 32: 728-29.

Odum, W.E., and E.J. Heald. 1972. Trophic analysis of an estuarine mangrove community. *Bull. Mar. Sci.* 22: 671-38.

———. 1975a. Mangrove forests and aquatic productivity. In *Coupling of land and water systems*, ed. A.D. Hasler, pp. 129-36. Ecological Studies 10. New York: Springer-Verlag.

———. 1975b. The detritus-based food web of an estuarine mangrove community. In *Estuarine research*, vol. 1, ed. L.E. Cronin, pp. 265-86. New York: Academic Press.

Odum, W.E., C.C. McIvor and T.S. Smith. 1982. *The ecology of the mangroves of South Florida: A community profile.* US Fish and Wildlife Service, Office of Biological Services, Washington, DC. FWS/OBS — 81/24.

Ogden, J.G., and J.C. Zieman. 1977. Ecological aspects of coral reef-seagrass bed contacts in the Caribbean. *Proceedings Third International Coral Reef Symposium*, Miami, pp. 377-83.

Olexa, M.T., and T.E. Freeman. 1975. Occurrence of three unrecorded diseases on mangroves in Florida. In *Proceedings of the International Symposium on Biology and Management of Mangroves*, eds G.E. Walsh, S.C. Snedaker and H.J. Teas, vol. 2, pp. 688-92. Gainesville: Univ. Florida.

Oliver, J. 1982. The geographic and environmental aspects of mangrove communities: Climate. In *Mangrove ecosystems in Australia — structure, function and management*, ed. B.F. Clough, pp. 19-30. Published by AIMS with ANU Press.

Ong, J.E. 1982. Mangroves and aquaculture in Malaysia. *Ambio* 11: 252-57.

Onuf, C.P., J.M. Teal, and I. Valiela. 1977. Interactions of nutrients, plant growth and herbivory in a mangrove ecosystem. *Ecology* 58: 514-26.

Osborn, T.G.B., and J.G. Wood. 1923. On the zonation of the vegetation in the Port Wakefield district, with reference to the salinity of the soil. *Trans. Roy. Soc. S. Aust.* 47: 244-54.

Ostenfeld, C.H. 1916. Stray notes from tropical west Australia. *Dansk Vidensk Biol. Medd.* 2(8): 1-29.

Padmanabhan, D. 1960. The embryology of *Avicennia officinalis* L. I. Floral morphology and gametophytes. *Proc. Indian Acad. Science* 52: 131-45.
―――. 1962a. The embryology of *Avicennia officinalis* L. III. The embryo. *J. Madras Univ.* 32: 1-19.
―――. 1962b. The embryology of *Avicennia officinalis* L. IV. The seedling. *Proc. Indian Acad. Science* 56: 114-22.
Paltridge, G.W., and D. Proctor. 1977. Monthly mean solar radiation statistics for Australia. *Solar Energy* 18: 235-43.
Pannier, F. 1962. Estudio fisiologico sobre la viviparia de *Rhizophora mangle* L. *Acta Cientifica Venezolana* (Botanical Series) 13: 184-97.
Pannier, F., and R.F. Pannier. 1975. Physiology of vivipary in *Rhizophora mangle*. In *Proceedings of the International Symposium on Biology and Management of Mangroves*, eds G.E. Walsh, S.C. Snedaker and H.J. Teas, vol. 2, pp. 632-42. Gainesville: Univ. Florida.
Panshin, A.J. 1932. An anatomical study of the woods of the Philippine mangrove swamps. *Philippine J. Sci.* 48: 143-208.
Pegg, K.G., and J.L. Alcorn. 1982. *Phytophthora operculata* sp. nov., a new marine fungus. *Mycotaxon* 16: 99-102.
Pegg, K.G., and I.M. Foresberg. 1981. *Phytophthora* in Queensland mangroves. *Wetlands* 1: 2-3.
Pegg, K.G., N.C. Gillespie and L.I. Foresberg. 1980. *Phytophora* sp. associated with mangrove death in Central Coastal Queensland. *Aust. Plant Path.* 9(3): 6-7.
Penhale, P.A., and W.O. Smith. 1977. Excretion of dissolved organic carbon by eelgrass (*Zostera marina*) and its epiphytes. *Limnol. Oceanogr.* 22: 400-407.
Penhale, P.A., and G.W. Thayer. 1980. Uptake and transfer of carbon and phosphorus by eelgrass (*Zostera marina L.*) and its epiphytes. *Limnol. Oceanogr.* 22: 400-407.
Penridge, L.K. 1971. A study of the fish community of a North Queensland mangrove creek. Hons thesis. James Cook Univ. of North Queensland.
Percival, M. 1974. Floral ecology of coastal scrub in south-east Jamaica. *Biotropica* 6(2): 104-29.
Percival, M., and J.S. Womersley. 1975. *Floristics and ecology of the mangrove vegetation of Papua New Guinea*. Bot. Bull. No. 8, Dept. Forestry, Divn. Bot. Lae, Papua New Guinea.
Perry, W.J. 1946. Observations on the bionomics of the principal malaria vector in the New Hebrides, Solomon Islands. *J. Nat. Malarial Soc.* 5: 127-39.
Petit, M.G. 1922. Les Periophtalmes, Poissons fouisseurs. *Bull. Mus. Nat. Hist. Natur. 1922* No. 6: 404-408.
Pielou, E.C. 1977. *Mathematical ecology*. New York: John Wiley & Sons.

Poiner, I.R. 1980. A comparison between species diversity and community flux rates in the macrobenthos of an infaunal sand community and a seagrass community of Moreton Bay, Queensland. *Proc. Roy. Soc. Qld* 91: 21-36.

Pollard, D.A. 1973. Estuaries: Development and "Progress" versus commonsense. *Fisherman* 4: 28-32.

———. 1976. Estuaries must be protected. *Aust. Fish.* 36(6): 6-10.

———. 1981. Estuaries are valuable contributors to fisheries production. *Aust. Fish.* 40(1): 7-9.

———. 1984. A review of ecological studies on seagrass-fish communities, with particular reference to recent studies in Australia. *Aquat. Bot.* 18: 3-42.

Pomeroy, L.R. 1959. Algal productivity in salt marshes of Georgia. *Limnol. Oceanogr.* 4: 386-97.

Poole, P.J., A.E. Lugo and S.C. Snedaker. 1975. Litter production in mangrove forests of southern Florida and Puerto Rico. In *Proceedings of the International Symposium on Biology and Management of Mangroves*, eds G.E. Walsh, S.C. Snedaker and H.J. Teas, pp. 213-37. Gainesville: Univ. Florida.

Poore, G.C.B., and J.D. Kudenov. 1978. Benthos around an outfall of the Werribee sewage-treatment farm, Port Phillip Bay, Victoria. *Aust. J. Mar. Freshw. Res.* 29: 157-67.

Posford, Parvey, Sinclair and Knight. 1976. *Shoalhaven River entrance study*. Draft report to the NSW Public Works Dept, Sydney.

Post, E. 1963. Zur Verbreitung und Okologie der *Bostrychia-Caloglossa* Assoziation. *Int. Rev. ges. Hydrobiol.* 48:47-152. blue-green algae in the mangrove forests of Sinai. *Oecologia*

Potter, M.A., and B.J. Hill. 1982. Heat mortality in the Sydney Rock oyster *Saccostrea (Crassostrea) commercialis* and the effectiveness of some control methods. *Aquaculture* 29: 101-108.

Potts, M. 1979. Nitrogen-fixation (acetylene reduction) associated with communities of heterocystous and non-heterocystous blue-green algae in the mangrove forests of Sinai. *Oecologia* 39: 359-74.

Powers, L.W., and D.E. Bliss. 1983. Terrestrial adaptations. In *The biology of crustacea*, eds F.J. Vernberg and W.B. Vernberg, vol. 8, pp. 271-322. New York: Academic Press.

Pressey, R.L., and M.J. Middleton. 1982. Impacts of flood mitigation works on coastal wetlands in New South Wales. *Wetlands* 2(1): 27-44.

Primack, R.B., N.C. Duke and P.B. Tomlinson. 1981. Floral morphology in relation to pollination ecology in five Queensland coastal plants. *Austrobaileya* 1(4): 346-55.

Primack, R.B., and P.B. Tomlinson. 1978. Sugar secretions in buds of *Rhizophora* attractive to birds. *Biotropica* 10: 74-75.

Rabinowitz, D. 1975. Planting experiments in mangrove swamps of Panama. In *Proceedings of the International Symposium on Biology and Management of Mangroves*, eds G.E. Walsh, S.C. Snedaker and H.J. Teas, vol. 1, pp. 385-93. Gainesville: Univ. Florida.

_____. 1978a. Dispersal properties of mangrove propagules. *Biotropica* 10(1): 47-57.

_____. 1978b. Mortality and initial propagule size in mangrove seedlings in Panama. *J. Ecol.* 66: 45-51.

_____. 1978c. Early growth of mangrove seedlings in Panama, and an hypothesis concerning the relationships of dispersal and zonation. *J. Biogeogr.* 5: 113-33.

Rai, J.H., J.P. Tewari and K.G. Mukerji. 1969. Mycoflora of mangrove mud. *Mycopathol. Mycol. Appl.* 38: 17-31.

Rains, D.W., and E. Epstein. 1967. Preferential absorption of potassium by leaf tissue of the mangrove *Avicennia marina*: an aspect of halophytic competence in coping with salt. *Aust. J. Biol. Sci.* 20: 847-57.

Rao, A.R., and M. Sharma. 1968. The terminal sclereids and tracheids of *Bruguiera gymnorhiza* Blume and the cauline sclereids of *Ceriops roxburghiana* Arn. *Proc. Nat. Inst. Sci. India.* (Part B. Biol. Sci.) 34(6): 267-75.

Rayner, S.M. 1979. Comparisons of the salinity range tolerated by Teredinids (Mollusca: Teredinidae) under controlled conditions with that observed in an estuary in Papua New Guinea. *Aust. J. Mar. Freshw. Res.* 30: 521-33.

Rebel, T.P. 1974. *Sea turtles and the turtle industry of the West Indies, Florida and the Gulf of Mexico.* Rev. ed. Miami: Univ. Miami Press.

Redfield, J.A. 1983. Trophic relationships in mangrove communities. In *Mangrove ecosystems in Australia*, ed. B.F. Clough, pp. 259-63. Published by AIMS with ANU Press.

Redhead, T.D., and J.L. McKean. 1975. A new record of the false water rat *Xeromys myoides* (Thomas 1889) from the Northern Territory of Australia. *Aust. Mammology* 1: 347-54.

Reid, D.G. 1985. Habitat and zonation patterns of *Littoraria* species (Gastropoda, Littorinidae) in Indo-Pacific mangrove forests. *Biol. J. Linn. Soc.* (Lond.) 26(1):39-68.

_____. 1986. *The littorinid molluscs of mangrove forests in the Indo-Pacific region. The genus* Littoraria. British Museum: London.

_____. (in press). Shell colour, polymorphism of *Littoraria* species (Gastropoda: Littorinidae) from Indo-Pacific mangrove forests; and evidence for apostatic selection in *L. filiosa* (Salva). *Biol. J. Linn. Soc.* (Lond.).

Reinders-Gouwentak, C.A. 1953. Sonneratiaceae and other mangrove-swamp families, anatomical structure and water relations. *Flora Malesiana* 4:513-15.

Revelante, N., and M. Gilmartin. 1978. Characteristics of the microplankton and nanoplankton communities of an Australian coastal plain estuary. *Aust. J. Mar. Freshw. Res.* 29: 9-18.

Reye, E.J. 1969a. Mapping the habitat of *Culicoides subimmaculatus*. Univ. Queensland, Dept Entomology. Roneoed.

———. 1969b. Larval survey of *C. subimmaculatus* habitats. Univ. Queensland, Dept Entomology. Roneoed.

———. 1971. Untitled. *Ceratopogonidae, Information Exchange* 7: 3-4.

———. 1972. Pest biting midges. Some observations, December 1972. Univ. Queensland, Dept Entomology. Roneoed.

———. 1982. Midges, and the marine environment. *Operculum* 5: 152-57.

Reye, E.J., and D.J. Lee. 1961. An investigation of the possible role of biting midges (Diptera: Ceratopogonidae) in the transmission of arthropod-borne virus diseases at Townsville. *Proc. Linn. Soc. NSW* 86: 230-36.

Rhoads, D.C., and D.K. Young. 1971. Animal-sediment relations in Cape Cod Bay, Massachusetts. 11. Reworking by *Molpadia oditica* (Holothuroidea). *Mar. Biol.* 11: 255-61.

Rhodes, E.G. 1982. Depositional model for a chenier plain, Gulf of Carpentaria, Australia. *Sedimentology* 29: 201-21.

Ribelin, B.W., and A.W. Collier. 1979. Ecological considerations of detrital aggregates in the salt marsh. In *Ecological processes in coastal and marine systems*, ed. R.J. Livingstone, pp. 47-69. New York: Plenum Press.

Rice, D.L. 1979. Trace element chemistry of aging marine detritus derived from coastal macrophytes. Ph.D. thesis, Georgia Institute of Technology, Atlanta, Georgia.

Rice. D.L., and K.R. Tenore. 1981. Dynamics of carbon and nitrogen during the decomposition of detritus derived from estuarine macrophytes. *Estuar. Coastal Shelf Sci.* 13: 681-90.

Richards, P.W. 1964. *The tropical rain forest*. Cambridge: Cambridge Univ. Press.

Rickson, F.R. 1979. Absorption of animal tissue breakdown products into a plant stem — the feeding of a plant by ants. *Amer. J. Bot.* 66: 87-90.

Rintz, R.E. 1980. The peninsular Malayan species of *Dischidia* (Asclepiadaceae). *Blumea* 26: 81-126.

Robertson, A.I., and R.K. Howard. 1978. Diel trophic interactions between vertically migrating zooplankton and their fish predators in an eelgrass community. *Mar. Biol.* 48: 207-13.

Robertson, A.I., and K.H. Mann. 1980. The role of isopods and amphipods in the initial fragmentation of eelgrass detritus in Nova Scotia, Canada. *Mar. Biol.* 59: 63–69.

Robertson, M.L., A.L. Mills and J.C. Zieman. 1982. Microbial synthesis of detritus like particulates from dissolved organic carbon released by tropical seagrasses. *Mar. Ecol. Prog. Ser.* 7: 279–85.

Robinson, K.I.M., P.J. Gibbs, J.B. Barclay and J.L. May. 1983. Estuarine flora and fauna of Smiths Lake, New South Wales. *Proc. Linn. Soc. NSW* 107(1): 19–34.

Robinson, K.I.M., P.J. Gibbs, J. van der Velde and J.B. Barclay. 1983. Temporal changes in the estuarine benthic fauna of Towra Point, Botany Bay. *Wetlands* 3(1): 22–33.

Rochford, D.J. 1951. Studies in Australian estuarine hydrology. I. Introductory and comparative features. *Aust. J. Mar. Freshw. Res.* 2: 1–116.

Roessler, M.A. 1971. Environmental changes associated with a Florida power plant. *Mar. Poll. Bull.* 2(6): 87–90.

Rogers, R. 1979. The "city effect" on lichens in the Brisbane area. *Search* 8: 75–77.

Rogers, R.W., and G.N. Stevens. 1981. Lichens. In *Ecological biogeography of Australia*, ed. A. Keast, pp. 591–603. The Hague: Dr W. Junk.

Rooney, W.S., F.H. Talbot and S.S. Clark. 1978. Marine reserves: The development of policy for marine reserves. In *Environmental and urban studies Report*, No. 32, vol. 1, pp. 1–502. Centre for Environmental Studies, Macquarie Univ., Sydney.

Roth, I. 1965. Histogenese der Lentizellen am Hypocotyl von *Rhizophora mangle* L. *Ost. Bot. Zeit.* 112: 640–53.

Roughgarden, J. 1975. Evolution of marine symbiosis — a simple cost-benefit model. *Ecology* 56: 1201–8.

Roy, P.S. 1982. *Evolution of New South Wales estuary types*. Report prepared for Geological Survey of NSW, Department of Mineral Resources, GS 1982/024.

Rudov, D. 1977. Bacterial epiphytes on *Zostera muelleri* in Westernport Bay. B.Sc. (Hons) thesis, University of Melbourne.

Ruello, N.V. 1973. The influence of rainfall on the distribution and abundance of the school prawn *Metapenaeus maclayi* in the Hunter River (Australia). *Mar. Biol.* 23: 221–28.

Ryther, J.H. 1975. Mariculture: How much protein and for whom? *Oceanus* 18: 10–22.

Saenger, P. 1967. Some littoral plants of Flinders Island. *Vic. Nat.* 84: 168–71.

———. 1979. Morphological and reproductive adaptations of

Australian mangroves. Paper given at National Mangrove Workshop, AIMS.

———. 1981. Who protects Queensland's mangroves? *Bull. Aust. Litt. Soc.* 4: 5-9.

———. 1982. Morphological, anatomical and reproductive adaptations of Australian mangroves. In *Mangrove ecosystems in Australia*, ed. B.F. Clough, pp. 153-91. Published by AIMS with ANU Press.

Saenger, P., and M.S. Hopkins. 1975. Observations on the mangroves of the south-eastern Gulf of Carpentaria, Australia. In *Proceedings of the International Symposium on Biology and Management of Mangroves*, eds G.E. Walsh, S.C. Snedaker and H.J. Teas, vol. 1, pp. 126-36. Gainesville: Univ. Florida.

Saenger, P., and C.C. McIvor. 1975. Water quality and fish populations in a mangrove estuary modified by residential canal developments. In *Proceedings of the International Symposium on Biology and Management of Mangroves*, eds G.E. Walsh, S.C. Snedaker and H.J. Teas, vol. 2, pp. 753-65. Gainesville: Univ. Florida.

Saenger, P., and J. Moverley. 1985. Vegetative phenology of mangroves along the Queensland coastline. *Proc. Ecol. Soc. Aust.* 13:257-65.

Saenger, P., and J. Robson. 1977. Structural analysis of mangrove communities on the Central Queensland coastline. *Mar. Res. Indonesia* 18: 101-18.

Saenger, P., E.J. Hegerl and J.D.S. Davie. 1983. Global status of mangrove ecosystems. *Environmentalist* 3 (Suppl. No. 3): 1-88.

Saenger, P., M.M. Specht, R.L. Specht and V.J. Chapman. 1977. Mangal and coastal saltmarsh communities in Australia. In *Wet coastal ecosystems*, ed. V.J. Chapman, pp. 293-345. Amsterdam: Elsevier.

Saenger, P., W. Stephenson and J. Moverley. 1979. The subtidal fouling organisms of the Calliope River and Auckland Creek, Central Queensland. *Mem. Qd. Mus.* 19: 399-412.

———. 1980. The estuarine macrobenthos of the Calliope River and Auckland Creek, Queensland. *Mem. Qd. Mus.* 20(1): 143-61.

———. 1982. Macrobenthos of the cooling water discharge canal of the Gladstone Power Station, Queensland. *Aust. J. Mar. Freshw. Res.* 33:1083-95.

Saila, S.B., S.D. Pratt and T.T. Polgar. 1972. *Dredge spoil disposal in Rhode Island Sound*. Mar. Tech. Rep. No. 2, Univ. Rhode Island, Kingston, Rhode Island.

Sale, P.F. 1983. Temporal variability in the structure of reef fish communities. In *Proceedings Inaugural Great Barrier Reef Conference*, Townsville, Aug. 28–Sept. 1 1983, eds J.T. Baker, R.M. Carter, P.W. Sammarco and K.P. Stark, pp. 239–44. Townsville: James Cook Univ. Press.

Salmon, M. 1984. The courtship, aggression, and mating system of a "primitive" fiddler crab (*Uca vocans*: Ocypodidae). *Trans. Zool. Soc.* (Lond.) 1984: 371–450.

Salmon, M., and G.W. Hyatt. 1983. Communication. In *Biology of the Crustacea*, eds. F.J. Vernberg and W.B. Vernberg, vol. 7, pp. 1–40. New York: Academic Press.

Salmon, M., G. Hyatt, K. McCarthy and J.D. Costlow. 1978. Display specificity and reproductive isolation in the fiddler crabs *Uca panacea* and *U. pugilator*. *Z. Tierpsychol.* 48: 251–76.

Sandifer, P.A. 1975. The role of pelagic larvae in recruitment to populations of adult decapod crustaceans in the York River estuary and adjacent lower Chesapeake Bay, Virginia. *Estuar. Coastal Mar. Sci.* 3: 269–79.

Sandison, E.E., and M.B. Hill. 1966. The distribution of *Balanus pallidus stutsburi* Darwin, *Gryphaea gasar* ((Adanson) Dautzenberg) *Mercierella enigmatica* Fauvel and *Hydroides uncinata* (Phillippi) in relation to salinity in Lagos harbour and adjacent creeks. *J. Anim. Ecol.* 35(1): 235–50.

Sasekumar, A. 1974. Distribution of macrofauna on a Malayan mangrove shore. *J. Anim. Ecol.* 43: 51–69.

Sauer, J.D. 1965. Geographic reconnaissance of Western Australia seashore vegetation. *Aust. J. Bot.* 13: 39–69.

Schodde, R., I.J. Mason and H.B. Gill. 1982. The avifauna of the Australian mangroves. A brief review of composition, structure and origin. In *Structure, function and management of mangrove ecosystems in Australia*, ed. B.F. Clough, pp. 141–50. Published by AIMS with ANU Press.

Scholander, P.F. 1968. How mangrove desalinate seawater. *Physiol. Plantarum* 21: 251–61.

Scholander, P.F., E.D. Bradstreet, H.T. Hammel and E.A. Hemmingsen. 1966. Sap concentrations in halophytes and some other plants. *Plant Physiol.* 41: 529–32.

Scholander, P.F., H.T. Hammel, E.D. Bradstreet and E.A. Hemmingsen. 1965. Sap pressure in vascular plants. *Science* 148: 339–46.

Scholander, P.F., H.T. Hammel, E. Hemmingsen and W. Garey. 1962. Salt balance in mangroves. *Plant Physiol.* 37: 722–29.

Scholander, P.F., L. Van Dam and S.I. Scholander. 1955. Gas exchange in the roots of mangroves. *Amer. J. Bot.* 42: 92–98.

Schottle, E. 1932. Morphologie und Physiologie der Atmung bei wasserschlamm-und Landeslebenden Gobiiformes. *Z. Wiss Zoologie* 140: 1-114.

Segal, E., and W.D. Burbanck. 1963. Effects of salinity and temperature on osmoregulation in two latitudinally separated populations of an estuarine isopod *Cyathura polita* (Stimpson). *Physiol. Zool.* 36: 250-63.

Semeniuk, V. 1980. Mangrove zonation along an eroding coastline in King Sound, North-Western Australia. *J. Ecol.* 68: 789-812.

———. 1983. Mangrove distribution in northwestern Australia in relationship to regional and local freshwater seepage. *Vegetatio* 53: 11-31.

Semeniuk, V., K.F. Kenneally and P.G. Wilson. 1978. *Mangroves of Western Australia*. Handbook No. 12, WA Naturalist's Club, Perth.

Shanco, P., and R. Timmins. 1975. Reconnaissance of southern Bustard Bay tidal wetlands. *Operculum* 4: 149-54.

Shapiro, M.A. 1975. *Westernport Bay Environmental Study 1973-1974*. Ministry for Conservation, Victoria.

Shepherd, S.A. 1970. *Preliminary report upon degradation of seagrass beds at North Glenelg, SA*. Report South Aust. Fisheries.

Shields, L.M. 1950. Leaf xeromorphy as related to physiological and structural influences. *Bot. Rev.* 16: 399-447.

Shine, R., C.P. Ellway and E.J. Hegerl. 1973. A biological survey of Tallebudgera Creek estuary. *Operculum* 3 (5-6): 59-83.

Sidhu, S.S. 1975a. Culture and growth of some mangrove species. In *Proceedings of the International Symposium on Biology and Management of Mangroves*, eds G.E. Walsh, S.C. Snedaker and H.J. Teas, vol. 1, pp. 394-401. Gainesville: Univ. Florida.

———. 1975b. Structure of epidermis and stomatal apparatus of some mangrove species. In *Proceedings of the International Symposium on Biology and Management of Mangroves*, eds G.E. Walsh, S.C. Snedaker and H.J. Teas, pp. 569-78. Gainesville: Univ. Florida.

Simberloff, D.S., and E.O. Wilson. 1969. Experimental zoogeography of islands: The colonization of empty islands. *Ecology* 50: 278-96.

———. 1970. Experimental zoogeography of islands: A two-year record of colonization. *Ecology* 51: 934-37.

Simberloff, D.S., B.J. Brown and S. Lowrie. 1978. Isopod and insect root borers may benefit Florida mangroves. *Science* 201: 630-32.

Simmons, H.B., and F.A. Herrmann. 1972. Effects of man-made works on the hydraulics, salinity and shoaling regimes of estuaries. In *Environmental framework of coastal plains estuaries*, ed. B.W. Nelson, Geol. Soc. Amer. Mem. No. 133.

Sinclair, P. 1976. Notes on the biology of the salt marsh mosquito *Aedes vigilax* (Skuse) in south eastern Queensland. *Qld. Nat.* 21: 134-39.

Smillie, R.M. 1984. Cold and heat tolerances of mangroves and seagrass species. Unpubl. Final Report (MST Grant) No. 81/032IT.

Smillie, R.M., and S.E. Hetherington. 1983. Stress tolerance and stress-induced injury in crop plants measured by chlorophyll fluorescence in vivo. Chilling, freezing, ice cover, heat and high light. *Plant Physiol.* 72: 1043-50.

Smith, B., N. Coleman and J.E. Watson. 1975. The invertebrate fauna of Westernport Bay. *Proc. Roy. Soc. Vic.* 87: 149-55.

Smith, C.J., and R.D. Delaune. 1984. Influence of the rhizosphere of *Spartina alterniflora* Loisel. on the nitrogen loss from a Louisiana Gulf coast salt marsh. *Environ. Exptl. Bot.* 24: 91-93.

Smith, L.S. 1969. New species of and notes on Queensland plants. V. *Contrib. Qld Herbarium* No. 6: 1-25.

Smith, S.T., and C.V. Malcolm. 1959. Bringing wheatbelt saltland back into production. *J. Agric. W. Aust.* (3rd series) 8: 263-67.

Smith, W.O., and P.A. Penhale. 1980. The heterotrophic uptake of dissolved organic carbon by eelgrass (*Zostera marina* L.) and its epiphytes. *J. Exp. Mar. Biol. Ecol.* 48: 233-42.

Smith-White, A.R. 1981. Physiological differentiation in a salt-marsh grass. *Wetlands* 1: 20.

Snedaker, S.C. 1982. Mangrove species zonation: Why? In *Contribution to the ecology of halophytes*, eds D.N. Sen and K.S. Rajpurohit, vol. 2, pp. 111-25. The Hague: Dr W. Junk.

Snedaker, S.C., J.A. Jimenez and M.S. Brown. 1981. Anomalous aerial roots in *Avicennia germinans* (L.) L. in Florida and Costa Rica. *Bull. Mar. Sci.* 31: 467-70.

Snelling, B. 1959. The distribution of intertidal crabs in the Brisbane River. *Aust. J. Mar. Freshw. Res.* 10(1): 67-83.

Sokoloff, B., J.B. Redd and R. Dutcher. 1950. Nutritive value of mangrove leaves (*Rhizophora mangle* L.). *J. Fla. Acad. Sci.* 12: 191-94.

Spaargaren, D.J. 1977. On the metabolic adaptations of *Carcinus maenas* to reduced oxygen tensions in the environment. *Neth. J. Sea. Res.* 11: 325-33.

Specht, R.L. 1958. The climate, geology, soils and plant ecology of the northern portion of Arnhem Land. In *Records of the American-Australian Scientific Expedition to Arnhem Land*, eds R.L. Specht and C.P. Mountford, vol. 3, pp. 333-414. Melbourne: Melb. Univ. Press.

_____. 1970. Vegetation. In *The Australian environment*, 4th ed., ed. G.W. Leeper, pp. 44–67. Melbourne: CSIRO and Melb. Univ. Press.

_____. 1981a. Ecophysiological principles determining the biogeography of major vegetation formations in Australia. In *Ecological biogeography of Australia*, ed. A. Keast, pp. 299–332. The Hague: Dr W. Junk.

_____. 1981b. Biogeography of halophytic angiosperms (saltmarsh, mangrove and seagrass). In *Ecological biogeography of Australia*, ed. A. Keast, pp. 577–89. The Hague: Dr W. Junk.

_____. 1981c. Growth indices — their role in understanding the growth, structure and distribution of Australian vegetation. *Oecologia* 50: 347–56.

Specht, R.L., and D.G. Morgan. 1981. The balance between the foliage projective covers of overstorey and understorey strata in Australian vegetation. *Aust. J. Ecol.* 6: 193–202.

Specht, R.L., R.B. Salt and S.T. Reynolds. 1977. Vegetation in the vicinity of Weipa, North Queensland. *Proc. Roy. Soc. Qld* 88: 17–38.

Spenceley, A.P. 1976. Unvegetated saline tidal flats in North Queensland. *J. Trop. Geogr.* 42: 78–85.

_____. 1982. Sedimentation patterns in a mangal on Magnetic Island near Townsville, North Queensland, Australia. *Singapore J. Trop. Geogr.* 3(1): 100–107.

_____. 1983. Aspects of the development of mangals in the Townsville region, North Queensland, Australia. In *Tasks for vegetation science*, ed. H.J. Teas, pp. 47–56. The Hague: Dr W. Junk.

Spencer, K.A. 1977. A revision of the Australian Agromyzidae (Diptera). *West Aust. Mus.* (Special Publ.) 8: 1–255.

Spencer, R. 1956. Studies in Australian estuarine hydrology. II. The Swan River. *Aust. J. Mar. Freshw. Res.* 7: 193–253.

Stace, C.A. 1966. The use of epidermal characters in phylogenetic considerations. *New Phytol.* 65: 304–18.

Staples, D.J. 1980a. Ecology of juvenile and adolescent banana prawns *Penaeus merguiensis* (de Man) in a mangrove estuary and adjacent offshore area of the Gulf of Carpentaria. I. Immigration and settlement of post larvae. *Aust. J. Mar. Freshw. Res.* 31: 635–52.

_____. 1980b. Ecology of juvenile and adolescent banana prawns *Penaeus merguiensis* in a mangrove estuary and adjacent offshore area of the Gulf of Carpentaria. II. Emigration, population structure and growth of juveniles. *Aust. J. Mar. Freshw. Res.* 31: 653–65.

Stebbins, R.C., and M. Kalk. 1961. Observations on the natural history of the mud skipper, *Periophthalmus sobrinus*. *Copeia* (1961), 1: 18-27.

Steers, J.A. 1977. Physiography. In *Ecosystems of the world. Vol. I. Wet coastal ecosystems,* ed. V.J. Chapman, pp. 31-60. Amsterdam: Elsevier.

Steinke, T.D. 1975. Some factors affecting disperal and establishment of propagules of *Avicennia marina* (Forsk.) Vierh. In *Proceedings of the International Symposium on Biology and Management of Mangroves*, eds G.E. Walsh, S.C. Snedaker and H.J. Teas, vol. 2, pp. 402-14. Gainesville: Univ. Florida.

———. 1979. Apparent transpirational rhythms of *Avicennia marina* (Forsk.) Vierh. at Inhaca Island, Moçambique. *J. S. Afr. Bot.* 45: 133-38.

Steinke, T.D., G. Naidoo and L.M. Charles. 1983. Degradation of mangrove leaf and stem tissues in situ in Mgeni Estuary, South Africa. In *Tasks for Vegetation Science*, ed. H.J. Teas, vol. 8, pp. 141-49. The Hague: Dr W. Junk.

Stephens, A. 1973. Noosa Inlet: An unstable estuary. *Operculum* 3: 35-37.

Stephenson, W., R. Endean and I. Bennett. 1958. An ecological survey of the marine fauna of Low Isles, Qld. *Aust. J. Mar. Freshw. Res.* 9: 261-318.

Stern, W.L., and G.K. Voigt. 1959. Effect of salt concentration on growth of red mangrove in culture. *Bot. Gaz.* 121: 36-39.

Stevens, G.N. 1979. Distribution and related ecology of macrolichens on mangroves on the east Australian Coast. *Lichenologist* 11: 293-305.

———. 1981. The macrolichen flora on mangroves of Hinchinbrook Island, Queensland. *Proc. Roy. Soc. Qld* 92: 75-84.

Stevens, G.N., and R.W. Rogers. 1979. The macrolichen flora from the mangroves of Moreton Bay. *Proc. Roy. Soc. Qld* 90: 33-49.

Stocker, G.C. 1976. *Report on cyclone damage to natural vegetation in the Darwin area after cyclone Tracey, 25 December 1974*. Forestry and Timber Bureau, Leaflet No. 127.

Stoddart, D.R. 1980. Mangroves as successional stages, inner reefs of the northern Great Barrier Reef. *J. Biogeogr.* 7: 269-84.

Stoddart, D.R., G.W. Bryan and P.E. Gibbs. 1973. Inland mangroves and water chemistry, Barbuda, West Indies. *J. Nat. Hist.* 7: 33-46.

Stokey, A.G., and L.R. Atkinson. 1952. The gametophyte of *Acrostichum speciosum* Wield. *Phytomorphology* 2: 105-13.

Storch, V., and U. Welsch. 1972. Ultrastructure and histochemistry of the integument of air-breathing polychaetes from mangrove swamps of Sumatra. *Mar. Biol.* 17: 137-44.

Storr, G.M. 1973. List of Queensland birds. *West Aust. Mus.* (Special Publ.) 5: 1-177.

———. 1977. Birds of the Northern Territory. *West Aust. Mus.* (Special Publ.) 7: 1-130.
Straughan, R., ed. 1983. *The Australian Museum complete book of Australian mammals.* (The national photographic index of Australian wildlife.) Sydney: Angus & Robertson.
Strogonov, B.P., V.V. Kavanov, N.Y. Shevjakova, L.P. Lapina, R.I. Komizerko, B.A. Popov, R.K. Dostanova and L.S. Prykhod'k. 1970. *Structure and function of plant cells under salinity.* Moscow: Nauka.
Subrahmangam, C.B., and C.L. Coutlas. 1980. Studies on the animal communities in two North Florida salt marshes. Part III. Seasonal fluctuations of fish and macroinvertebrates. *Bull. Mar. Sci.* 30: 790-818.
Sundararaj, D.D. 1954. Mangrove (*Avicennia offinalis* Linn.) leaves as cattle feed. *Science and Culture* 19: 556.
Sussex, I. 1975. Growth and metabolism of the embryo and attached seedling of the viviparous mangrove *Rhizophora mangle*. *Amer. J. Bot.* 62: 948-53.
Sutherland, J.P. 1980. Dynamics of the epibenthic community on roots of the mangrove *Rhizophora mangle*, at Bahia de Buche, Venezuela. *Mar. Biol.* 58: 75-84.
Swanson, S. 1976. *Lizards of Australia.* Sydney: Angus & Robertson.
Swart, H.J. 1958. An investigation of the mycoflora in the soil of some mangrove swamps. *Acta Bot. neerl.* 7: 741-68.
———. 1963. Further investigations of the mycoflora in the soil of some mangrove soils. *Acta Bot. neerl.* 12: 98-111.
Taplin, L.E. 1984. Homeostasis of plasma electrolytes, water and sodium pools in the estuarine crocodile, *Crocodylus porosus*, from fresh, saline and hypersaline waters. *Oecologia* 63: 63-70.
Taplin, L.E., and G.D. Grigg. 1981. Salt glands in the tongue of the estuarine crocodile, *Crocodylus porosus*. *Science* 212: 1045-47.
Taplin, L.E., G.C. Grigg and L. Beard. 1985. Salt gland function in fresh water crocodiles: evidence for a marine phase in eosuchian evolution. In *Biology of Australasian frogs and reptiles*, eds G. Grigg, R. Shine and H. Ehnmann. Chipping Norton: Surrey Beatty and Sons.
Taplin, L.E., G.C. Grigg, P. Harlow, T.M. Ellis and W.A. Dunson. 1982. Lingual salt glands in *Crocodylus acutus* and *C. johnstoni* and their absence from *Allgator mississipiensis* and *Caiman crocodilus*. *J. Comp. Physiol.* 149: 43-47.
Taylor, B.W. 1959. The classification of lowland swamp communities in north-eastern Papua. *Ecology* 40: 703-11.
Taylor, J.A. 1979. The foods and feeding habits of subadult

Crocodylus porosus (Schneider) in Northern Australia. *Aust. Wildl. Res.* 6: 347-59.

Teal, J.M. 1962. Energy flow in the salt marsh ecosystem of Georgia. *Ecology* 43: 614-24.

Teal, J.M., and F.G. Carey. 1967a. The metabolism of marsh crabs under conditions of reduced oxygen pressure. *Physiol. Zool.* 40: 83-91.

―――. 1967b. Skin respiration and oxygen debt in the mud skipper *Periophthalmus sobrinus*. *Copeia* 3:677-79.

Teas, H.J. 1979. Silviculture with saline water. In *The biosaline concept*, ed. A. Hollaender, pp. 117-61. New York: Plenum.

Tenore, K.R. 1975. Detrital utilization by the polychaete *Capitella capitata*. *J. Mar. Res.* 33: 261-74.

―――. 1977. Food chain pathways in detrital feeding benthic communities: a review with new observations on sediment resuspension and detrital recycling. In *Ecology of marine benthos*, ed. B.C. Coull, pp. 37-53. Columbia: Univ. South Carolina Press.

Tenore, K.R., L. Cammen, S.E.G. Findlay and N. Phillips. 1982. Perspectives of research of detritus: Do factors controlling the availability of detritus to macroconsumers depend on its source? *J. Mar. Res.* 40: 473-90.

Tenore, K.R., and W.M. Dunstan. 1973. Comparison of feeding and biodeposition of three bivalves at different food levels. *Mar. Biol.* 21: 190-95.

Tenore, K.R., J.C. Goldman and J.P. Clarner. 1973. The food chain dynamics of the oyster, clam and mussel. *J. Exp. Mar. Biol. Ecol.* 12: 157-65.

Tenore, K.R., and U.K. Gopalan. 1974. Food chain dynamics of the polychaete *Nereis virens* cultured on animal tissue and detritus. *J. Fish. Res. Bd. Can.* 31: 1675-78.

Tenore, K.R., and R.B. Hanson. 1980. Availability of detritus of different types and ages to a polychaete macroconsumer *Capitella capitata*. *Limnol. Oceanogr.* 25: 553-58.

Tenore, K.R., and D.L. Rice. 1980. A review of trophic factors affecting secondary production of deposit feeders. In *Marine benthic dynamics*, eds K.R. Tenore and B.C. Coull, pp. 325-40. Belle W. Baruch. Library and Marine Science No. 11. Univ. S. Carolina Press.

Thayer, G.W., P.L. Parker, M.W. la Croix and B. Fry. 1978. The stable carbon isotope ratio of some components of an eelgrass *Zostera marina* bed. *Oecologia* 35: 1-12.

Theede, H. 1973. Comparative studies on the influence of oxygen deficiency and hydrogen sulphide on marine bottom invertebrates. *Neth. J. Sea. Res.* 7: 244-52.

Thom, B.G. 1967. Mangrove ecology and deltaic geomorphology, Tabasco, Mexico. *J. Ecol.* 55: 301-43.
_____. 1975. Mangrove ecology from a geomorphic viewpoint. In *Proceedings of the International Symposium on Biology and Management of Mangroves*, eds G.E. Walsh, S.C. Snedaker and H.J. Teas, vol. 2, pp. 469-81. Gainesville: Univ. Florida.
_____. 1982. Mangrove ecology — a geomorphological perspective. In *Mangrove ecosystems in Australia — structure, function and management*, ed. B.F. Clough, pp. 3-17. Published by AIMS with ANU Press.
Thom, B.G., L.D. Wright and J.M. Coleman. 1975. Mangrove ecology and deltaic-estuarine geomorphology; Cambridge Gulf-Ord River, Western Australia. *J. Ecol.* 63: 203-32.
Thomas, C., and J.E. Ong. 1984. Effect of heavy metals zinc and lead on *Rhizophora mucronata* Lam and *Avicennia alba* Bl. seedlings. In *Proceedings of the Asian Symposium Mangrove Environments — Research and Management*, eds. E. Soepadmo, A.N. Rao and D.J. Macintosh. UNESCO.
Thompson, J.N. 1981. Reversed animal-plant interactions: The evolution of insectivorous and ant-fed plants. *Biol. J. Linn. Soc.* 16: 147-55.
Thompson, R.K., and A.W. Pritchard. 1969. Respiratory adaptations of two burrowing crustaceans *Callianassa californiensis* and *Upogebia pugettensis* (Decapoda: Thalassinidae). *Biol. Bull.* 136: 274-87.
Thorhaug, A., D. Segar and M.A. Roessler. 1973. Impact of a power plant on a subtropical estuarine environment. *Mar. Poll. Bull.* 4: 166-69.
Tolken, H.R. 1967. The species of *Arthrocnemum* and *Salicornia* in Southern Africa. *Bothalia* 9: 255-307.
Tomlinson, P.B. 1971. The shoot apex and its dichotomous branching in the *Nypa* palm. *Annals Bot.* 35: 865-79.
Tomlinson, P.B., J.S. Bunt, R.B. Primack and N.C. Duke. 1978. *Lumnitzera rosea* (Combretaceae) — Its status and floral morphology. *J. Arnold Arboretum* 59(4): 342-51.
Tomlinson, P.B., R.B. Primack and J.S. Bunt. 1979. Preliminary observations on floral biology in mangrove Rhizophoraceae. *Biotropica* 11: 256-77.
Tomlinson, P.B., and D.W. Wheat. 1979. Bijugate pyllotaxis in *Rhizophora* (Rhizophoraceae). *Bot. J. Linn. Soc.* 78: 317-21.
Trench, R.K. 1979. The cell biology of plant-animal symbiosis. *Ann. Rev. Pl. Physiol.* 30: 485-531.
Treub, M. 1883. Notes sur l'embryon, le sac embryonaire et l'ovule de *Avicennia officinalis*. *Annales du Jardin Botanique de Buitenzorg* 3: 79-87.
Troughton, E. 1943. *Furred animals of Australia*. Sydney: Angus & Robertson.

Turner, R. et al. 1972. *Survey of marine borers. The family Teredinidae in Australian waters*. CSIRO Project, pp. 5–11. University of New South Wales. Project 12-045-15. 1970–1972.

Turner, R., and J. McKoy. 1979. *Bankia neztalia* n. sp. Mollusca, Bivalvia, Teredinidae from Australia and New Zealand and its relationships. *J. Roy. Soc. NZ* 9: 453–73.

Tweedie, M.W.F. 1935a. Two new species of *Squilla* from Malayan waters. *Bull. Raffles Mus.* 10: 45–52.

Tweedie, M.W.F. 1935b. Notes on the genus *Ilyoplax* Stimpson (Brachyura, Ocypodidae). *Bull. Raffles Mus.* 10: 53–61.

Udovenko, G.V., and I.F. M'Inko. 1966. The nature of the effect of potassium and chloride on nitrogen metabolism in plants. *Fiziol. Rast.* 13: 236–45.

Uhl, N.W. 1972. Inflorescence and flower structure in *Nypa fruticans* (Palmae). *Amer. J. Bot.* 59: 729–43.

Uhl, N.W., and H.E. Moore. 1977. Correlations of inflorescence, flower structure, and floral anatomy with pollination in some palms. *Biotropica* 9(3): 170–90.

Ulken, A. 1970. Phycomyceten aus der Mangrove bei Cananeia (Sao Paulo, Brasilien). *Veröff. Inst. Meeresforsch. Bremerhafen* 12: 313–19.

———. 1975. Further studies on *Phlyctochytrium mangrovis* Ulken. In *Proceedings of the International Symposium on Biology and Management of Mangroves*, eds G.E. Walsh, S.C. Snedaker and H.J. Teas, vol. 2, pp. 608–87. Gainesville: Univ. Florida.

———. 1981. The phycomycete flora of mangrove swamps in the South Pacific. *Veröff. Inst. Meeresforsch. Bremerhafen* 19: 45–59.

———. 1983. Distribution of phycomycetes in mangrove swamps with brackish waters and waters of high salinity. In *Tasks for vegetation science*, ed. H.J. Teas, vol. 8, pp. 111–16. The Hague: Dr W. Junk.

Uphof, J.C.T. 1941. Halophytes. *Bot. Rev.* 7: 1–58.

Valiela, I., and S. Vince. 1976. Green borders of the sea. *Oceanus* 19: 10–17.

Vanderplank, F.L. 1960. The bionomics and ecology of the red tree ant *Oecophylla* sp. and its relationship to the coconut bug *Pseudotheraptus wayi* Brown (Coreidae). *J. Anim. Ecol.* 29: 15–33.

Van der Valk, A.G., and P.M. Attiwill. 1984a. Acetylene reduction in an *Avicennia marina* community in Southern Australia. *Aust. J. Bot.* 32: 157–64.

———. 1984b. Decomposition of leaf and root litter of *Avicennia marina* at Westernport Bay, Victoria, Australia. *Aquat. Bot.* 18: 205–21.

Van Dijk, D.E. 1960. Locomotion and attitudes of the mudskipper *Periophthalmus*, a semi terrestrial fish. *S. Afr. J. Sci.* 56: 158–62.

Van Dyck, S., W.W. Baker and D.D. Gillette. 1979. The false water rat *Xeromys myoides* on Stradbroke Island, a new locality in southeastern Queensland. *Proc. Roy. Soc. Qld* 90: 84.

Van Hove, C. 1976. Bacterial leaf symbiosis and nitrogen fixation. In *Symbiotic nitrogen fixation in plants*, ed. P.S. Nutman, vol. 7, pp. 551–60. London: Cambridge Univ. Press.

Van Royen, P. 1956. Notes on *Tecticornia cinerea* (F. v. M.) Bailey Chenopodiaceae. *Nova Guinea* 7: 175–80.

Van Steenis, C.G.G.J. 1979. Plant geography of east Malesia. *Bot. J. Linn. Soc.* 79: 97–178.

Van Steveninck, R.F.M., W.D. Armstrong, P.D. Peters and T.A. Hall. 1976. Ultrastructural localization of ions: III. Distribution of chloride in mesophyll cells of mangrove (*Aegiceras corniculatum* Blanco). *Aust. J. Plant Physiol.* 3(3): 367–76.

Van Weel, P.B. 1970. Digestion in crustacea. In *Chemical Zoology*, eds M. Florkin and B.T. Sheer, vol. 5, pp. 97–115. New York: Academic Press.

Venkateswarlu, J. 1935. A contribution to the embryology of the Sonneratiaceae. *Proc. Indian Acad. Science* 5: 23–29.

Venkateswarlu, J., and R.S.P. Rao. 1964. The wood anatomy and taxonomic position of Sonneratiaceae. *Current Science* 33: 6–9.

Venkateswarlu, J., and P.N. Rao. 1975. A contribution to the embryology of the tribe Hippomaneae of the Euphorbiaceae. *J. Indian Bot. Soc.* 54(1/2): 98–103.

Vermeij, G.J. 1973. Morphological patterns in high intertidal gastropods: adaptive strategies and their limitations. *Mar. Biol.* 20: 319–46.

———. 1974. Molluscs in mangrove swamps: Physiognomy, diversity and regional differences. *Syst. Zool.* 22: 609–24.

Vernberg, F.J. 1983. Respiratory adaptations. In the *Biology of the Crustacea*, ed. F.J. Vernberg and W.B. Vernberg, vol. 8, pp. 1–42. New York: Academic Press.

Vernberg, W.B., and F.J. Vernberg. 1972. *Environmental physiology of marine animals*. New York: Springer-Verlag.

Verwey, J. 1930. Einiges über die Biologie ost-indischer Mangrove Krabben. *Treubia* 12(2): 169–261.

Vizioli, J. 1923. Some pyrenomycetes of Bermuda. *Mycologia* 15: 107–19.

Volz, P.A., and D.E. Jerger. 1972. A preliminary study of marine fungi from Abaco Island, the Bahamas. *Mycopathol. Mycol. Appl.* 48: 271–74.

Vu-van-Cuong, H. 1964. Flore et vegetation de la mangrove de la region de Saigon-Cap Saint Jacques, Sud Viet-Nam. Ph.D. dissertation, Univ. Paris.

Waits, E.D. 1967. Net primary production of an irregularly flooded North Carolina salt marsh. Ph.D. thesis, North Carolina State Univ.

Walker, D., ed. 1972. *Bridge and barrier — The natural and cultural history of Torres Strait*. Canberra: ANU Press.

Walls, G.L. 1942. *The vertebrate eye and its adaptative radiation*. Bloomfield Hills, Michigan: Cranbrook Institute.

Walsh, G.E. 1967. An ecological study of a Hawaiian mangrove swamp. In *Estuaries*, ed. G.H. Lauff, pp. 420-21. Amer. Assoc. Adv. Sci. Publ. No. 83.

———. 1974. Mangroves: A review. In *Ecology of halophytes*, eds R.J. Reimhold and W.H. Queen, pp. 51-174. New York: Academic Press.

Walsh, G.E.R., K.A. Ainsworth and R. Rigby. 1979. Resistance of red mangrove (*Rhizophora mangle* L.) seedlings to lead, cadmium and mercury. *Biotropica* 11: 22-27.

Walter, H., and M. Steiner. 1936. Die Okologie der Ost-Afrikanischen Mangroven. *Zeitschrift für Botanik* 30: 65-193.

Warner, G.F. 1967. The life history of the mangrove tree crab *Aratus pisoni*. *J. Zool.* 153: 321-35.

———. 1969. The occurrence and distribution of crabs in a Jamaican mangrove swamp. *J. Anim. Ecol.* 38: 379-89.

Watson, J.G. 1928. Mangrove forests of the Malay Peninsula. *Malayan Forest Records* 6: 1-275.

Weate, P. 1975. A study of the wetlands of the Myall River. *Operculum* 4(3-4): 105-13.

Webb, G.J.W., H. Messel and W. Magnusson. 1977. The nesting of *Crocodylus porosus* in Arnhem Land, Northern Australia. *Copeia* 2: 238-49.

Wells, A.G. 1982. Mangrove vegetation of northern Australia. In *Mangrove ecosystems in Australia — structure, function and management*, ed. B.F. Clough, pp. 57-78. Published by AIMS with ANU Press.

———. 1983. Distribution of mangrove species in Australia. In *Tasks for vegetation science*, ed. H.J. Teas, vol. 8, pp. 57-76. The Hague: Dr W. Junk.

Wells, F.E. 1984. Comparative distribution of macromolluscs and macrocrustaceans in a North Western Australia mangrove system. *Aust. J. Mar. Freshw. Res.* 35: 591-96.

Wells, F.E., and S.M. Slack-Smith. 1981. Zonation of molluscs in a mangrove swamp in the Kimberley, Western Australia. In *Biological survey of Mitchell Plateau and Admiralty Gulf*,

Kimberley, Western Australia, part 9: 265-74. Perth: West. Aust. Museum.

West, R.J., and A.W.D. Larkum. 1979. Leaf productivity of the seagrass *Posidonia australis* in Eastern Australian waters. *Aquat. Bot.* 7: 57-65.

West, R.J., C.A. Thorogood and R.J. Williams. 1983. Environmental stress causing mangrove "dieback" in NSW. *Aust. Fish.* (Aug. 83): 42.

West, R. J., C. A. Thorogood, T. R. Walford and R. J. Williams. 1985. *An estuarine inventory for NSW, Australia*. Fisheries Bulletin No. 2. Department of Agriculture.

Weste, G., D. Cahill and D.J. Stamps. 1982. Mangrove dieback in North Queensland, Australia. *Trans. Brit. Mycol. Soc.* 79: 165-67.

Wester, L.L. 1967. The distribution of the mangrove in South Australia. Honours thesis, Univ. Adelaide.

Westman, W.E. 1975. Ecology of canal estates. *Search* 6: 491-97.

Wetzel, R.G., and P.A. Penhale. 1979. Transport of carbon and excretion of dissolved organic carbon by leaves and roots/rhizomes in seagrasses and their epiphytes. *Aquat. Bot.* 6: 149-58.

Whitney, D.E., G.M. Woodwell and R.W. Howarth. 1975. Nitrogen fixation in Flax Pond: A Long Island salt-marsh. *Limnol. Oceanogr.* 20: 640-43.

Whittaker, R.H., G.E. Likens and H. Lieth. 1975. Scope and purpose of this volume. In *Ecological studies 14. Primary productivity of the Biosphere*, eds H. Lieth and R.H. Whittaker, pp. 3-5. New York: Springer-Verlag.

Whyte, P.J.A. 1979. Elements of social behaviour in *Periophthalmus vulgaris* Eggert. Hons. thesis, James Cook Univ.

Williams, R.B., and M.B. Murdock. 1972. Compartment analysis of the production of *Juncus roemerianus* in a North Carolina salt marsh. *Chesapeake Sci.* 13: 69-79.

Wilson, E.O., and D.S. Simberloff. 1969. Experimental zoogeography of Islands. Defaunation and monitoring techniques. *Ecology* 50(2): 267-78.

Wilson, P.G. 1980. A revision of the Australian species of Salicornieae (Chenopodiaceae). *Nuytsia* 3: 1-154.

Wolanski, E., and P. Collis. 1976. Aspects of aquatic ecology of the Hawkesbury estuary. I. Hydrodynamical processes. *Aust. J. Mar. Freshw. Res.* 27: 565-82.

Wolanski, E., and R. Gardiner. 1981. Flushing of salt from mangrove swamps. *Aust. J. Mar. Freshw. Res.* 32: 681-83.

Womersley, H.B.S., and S.J. Edmonds. 1958. A general account of the intertidal ecology of South Australian coasts. *Aust. J. Mar. Freshw. Res.* 9: 217-60.

Womersley, J.S. 1975. Management of mangrove forests: Utilisation versus conservation with special reference to the forests of the Papuan Gulf. In *Proceedings of the International Symposium on Biology and Management of Mangroves*, eds G.E. Walsh, S.C. Snedaker and H.J. Teas, vol. 2, pp. 732–41. Gainesville: Univ. Florida.

———. 1983. An introduction to the nomenclature and taxonomy of the mangrove flora in Papua New Guinea and adjacent areas. In *Tasks for vegetation science*, ed. H.J. Teas, vol. 8, pp. 87–90. The Hague: Dr W. Junk.

Wong, C.H., and J.E. Ong. 1984. Electron microscopy study on the salt glands of *Acanthus ilicifolius* L. In *Proceedings of the Asian Symposium Mangrove Environments — Research and Management*, eds. E. Soepadmo, A.N. Rao and D.J. Macintosh, pp. 172–82. UNESCO.

Wood, E.J.F. 1953. Heterotrophic bacteria in marine environments of eastern Australia. *Aust. J. Mar. Freshw. Res.* 4: 160–200.

———. 1959. Some aspects of the ecology of Lake Macquarie, NSW, with regard to an alleged depletion of fish. IV. Bacterial and fungal studies. *Aust. J. Mar. Freshw. Res.* 10: 304–15.

Woodroffe, C.D. 1982. Litter production and decomposition in the New Zealand mangrove *Avicennia marina* var. *resinifera*. *NZ. Mar. Freshw. Res.* 16: 179–88.

———. 1983. Development of mangrove forests from a geological perspective. In *Tasks for vegetation science*, ed. H.J. Teas, vol. 8, pp. 1–17. The Hague: Dr W. Junk.

———. 1985a. Studies of a mangrove basin, Tuff Crater, New Zealand. II. Comparison of volumetric and velocity area methods of estimating tidal flux. *Estuar. Coastal and Shelf Sci.* 20(4): 431–46.

———. 1985b. Studies of a mangrove basin, Tuff Crater, New Zealand. III. The flux of organic and inorganic particulate matter. *Estuar. Coastal and Shelf Sci.* 20(4): 447–62.

Woodruff, R.E. 1970. A mangrove borer *Poecilips rhizophorae* (Hopkins) (Coleoptera: Scolytidae). *Fla. Dept. Agric. Consumer Service Entomol. Circular No. 98*.

Woods, F.W. 1960. Biological antagonisms due to phytotoxic root exudates. *Bot. Rev.* 26: 546–69.

Wright, D. 1977. Pollen morpology of Australian mangroves. Honours thesis, James Cook Univ., North Queensland.

Wright, L.D. 1978. River deltas. In *Coastal sedimentary environments,* ed. R.A. Davis, pp. 5–68. New York: Springer-Verlag.

Wright, L.D., J.M. Coleman and M.W. Erickson. 1974. *Analysis of major systems and their deltas: Morphologic and process comparisons.* Coastal Studies Institute, Louisiana State University, Tech. Report 156.

Wylie, R.B. 1949. Differences in foliar organization among leaves from four locations in the crown of an isolated tree (*Acer plantanoides*). *Proc. Iowa Acad. Sci.* 56: 189-98.

Wyn Jones, R.B., and R. Storey. 1981. Betaines. In *Physiology and biochemistry of drought resistance in plants*, eds L.G. Paleg and D. Aspinall, pp. 171-204. Sydney: Academic Press.

Yânez-Arancibia, A., F.A. Linares and J.W. Day. 1980. Fish community structure and function in Terminos Lagoon, a tropical estuary in the southern Gulf of Mexico. In *Estuarine perspectives*, ed. V.S. Kennedy, pp. 465-82. New York: Academic Press.

Yarish, C., P. Edwards and S. Casey. 1979. Acclimation responses to salinity of three estuarine red algae from New Jersey. *Mar. Biol.* 51: 289-94.

Yates, R.W. 1978. Aspects of the ecology and reproductive biology of crabs in a mangrove swamp at Patonga Creek, NSW. M.Sc. thesis, Univ. Sydney.

Young, D.L. 1973. Studies of Florida Gulf Coast *Spartina alterniflora* and *Juncus roemarianus*. Salt marshes receiving thermal discharges. Presented at Thermal Ecology Symposium, Augusta, Ga.

Zanders, I.P. 1978. Ionic regulation in the mangrove crab *Goniopsis cruentata*. *Compar. Biochem. Physiol.* 60: 293-302.

Ziegler, H., and U. Luttge. 1966. Die Salzdrusen von *Limonium vulgare*. I. Die Feinstruktur. *Planta* 70: 193-206.

Zieman, J.C., G.W. Thayer, M.B. Robblee and R.T. Zieman. 1979. Production and export of seagrasses from a tropical bay. In *Ecological processes in coastal and marine systems*, ed. R.J. Livingston, pp. 21-35. New York: Plenum Press.

Zilch, A. 1959. Gastropoda von Wilhelm Weuz Teil 2. Euthyneura fortgesetz von Adolph Zilch 1959 — Handbuch der Palaozoologie Band 6. Berlin.

Zuberer, D.A., and W.S. Silver. 1978. Biological dinitrogen fixation (acetylene reduction) associated with Florida mangroves. *Appl. Environ. Microbiol.* 35: 567-75.

———. 1975. Mangrove associated nitrogen fixation. In *Proceedings of the International Symposium on Biology and Management of Mangroves*, eds G.E. Walsh, S.C. Snedaker and H.J. Teas, vol. 2, pp. 643-53. Gainesville: Univ. Florida.

Zucker, N. 1977. Neighbour dislodgement and burrow filling activities by male *Uca musica terpsichores* — a spacing mechanism. *Mar. Biol.* 41: 281-86.

———. 1978. Monthly reproductive cycles in three sympatric hood building tropical fiddler crabs (Genus *Uca*). *Biol. Bull.* 155: 410–24.

Zucker, N., and R. Denny. 1979. Interspecific communication in fiddler crabs: Preliminary report on a female rejection display directed towards courting heterospecific males. *Z. Tierpsychol.* 50: 9–17.

Index

Acacia, 8, 147
Acanthopagrus berda, 181, 287
Acanthus
   dominance, 101-2
   fruit, 40-41
   light response, 33, 54
   roots, 26, 28
   salt glands, 15, 16, 18, 102
Acanthus ilicifolius
   distribution, 84, 111
   dominance, 101
   pollination, 38
   salinity, 71
   salt glands, 15
   saponins, 95, 107
   soil water, 65
   temperature and, 51, 52
Acrochordus granulatus, 180, 207-8, 210
Acrodipsas illidgei, 176
Acrostichum
   light response, 33, 54
   salinity, 70, 72, 114, 118, 119
   salt exclusion, 17
   spores, 40, 41
Acrostichum aureum, 10
Acrostichum speciosum, 105
   distribution, 84, 111
   salinity, 71
   soil water, 65
   temperature and, 52
adaptation, 1, 98, 100-2, 107, 259, 262. See also competition; distribution, mangrove; dominance; fauna, adaptation; leaves; salinity; stress; temperature; zonation
   light response, 31-34
   physical damage, 34-36
   reproduction, 36-44
   roots, 25-31, 41, 63-64, 103-4
   salt, 14-20, 25, 102-4, 153, 260
   transpiration, 24-25
   xeromorphic features, 20-24
Aedes alternans, 172-73

Aedes vigilax, 172-73
Aegialitis, 35, 99
   light response, 33, 54
   reproduction, 37, 38, 40-41, 43, 44
   roots, 26, 28
   salt glands, 15, 16, 17
Aegialitis annulata, 43, 85
   distribution, 7, 11, 111
   leaves, 23, 48, 256-57
   molluscs and, 189
   salinity, 70, 71
   salt and, 16
   temperature and, 48, 51
Aegiceras, 35
   dominance, 101-2
   leaves, 31, 257
   light response, 31, 33, 54
   reproduction, 37, 38, 40-41, 43, 44
   roots, 28
   salinity, 66, 118, 119
   salt and, 15-16, 17, 18
   waterlogging, 118, 119
   xeromorphic features, 21, 31
Aegiceras corniculatum, 84
   distribution, 11, 111-14
   dominance, 101-2
   fauna and, 175, 176, 177
   leaves, 48, 50, 97, 256
   parasitism, 93
   reproduction, 39
   salinity, 68, 70, 72, 114
   salt glands, 15
   soil water, 65
   temperature and, 48, 50, 51, 53
aerenchyma, 30
Aizoaceae, 150, 152, 153
albinism, 39
algae, 8, 130, 197, 275, 286
   benthic, 246, 268-72, 279
   epiphytic, 139-40, 187, 245-46, 269, 271
   habitat, 136-40, 171, 279
   mats, 59, 136-37, 187

INDEX

multicellular, 271
alluvial plains, 125-27
Alpheus, 193
Alternaria, 132, 133, 134
ammensalism. *See* antagonism
ammonification, 79, 80
ammonium enrichment, 83. *See also* nutrients
amphibians, 207
Amphibolus, 41
amphipods, 184, 187, 193, 199, 201, 239, 274, 289
Amyema, 37, 38, 40. *See also* mistletoes
   congener, 147, 149
   conspicuum, 147, 149
   glabrum, 147, 149
   mackayense, 23, 147, 148-49, 176
   thalassium, 147, 149
Anabaena torulosa, 138
Anacystis marina, 138
angiosperms, 7, 271
angiosperm tissue, 234-35
Anodontostoma chacunda, 181, 287
Anopheles, 173
Anous tenuirostris, 107, 168
antagonism, 94-95, 97
ants, 104-5, 107-8, 144-46, 176-77, 178, 211. *See also* insects
Apatococcus lobatus, 140
Apis mellifera, 37, 177
aquaculture, 308
Araneidae, 178
Aratus, 218
   pisonii, 106, 231-32
Arenigobius, 181
ascidians, 183
Asclepiadaceae, 145
ascomycetes, 134, 135
Aspergillus, 133-34
Assiminiea, 202
Atriplex, 152, 153
Aureobasidium, 132, 134
Australoplax tridentata, 196, 240-41, 243
autotrophs, 131
auxin, 63
Avicennia, 94, 133
   competition, 98-99, 101-2
   fauna and, 106, 171, 186, 196, 198-99, 228
   fossils, 7, 8
   light response, 32-34, 54
   litter, 261, 270, 277, 280-83
   reproduction, 37-38, 40-44
   roots, 26, 28-29, 228
   salinity, 66, 70, 72
   salt and, 15-18, 20
   soil conditions, 64
   temperature and, 13, 47
   wood, 35-36, 156, 159, 259, 262
   xeromorphic features, 21, 24, 25
   zonation, 99, 120, 259
Avicennia alba, 304
Avicennia germinans, 20, 245. *See also Avicennia nitida*
   dominance, 124
   primary production, 259
   reproduction, 37, 42
   roots, 28, 63, 79
   salinity, 66, 72
   temperature and, 35, 51
Avicennia marina, 58, 84, 273
   distribution, 11, 13, 50-51, 111-14, 159
   dominance, 94, 101-2
   fauna and, 159, 165, 176, 177, 184, 201
   flora and, 93-94, 132, 135, 140, 142, 145, 147
   leaves, 22, 31-32, 48, 140, 256-57, 263-64, 266, 276-78
   light response, 31-32, 53-54
   litter, 249, 250, 256, 266, 276-78
   primary production, 248-50, 256, 259, 263-64
   reproduction, 39, 42, 43
   roots, 29, 32, 135, 184
   salinity, 66, 67, 70, 71, 114
   salt and, 15-20, 140
   soil water, 64-66, 70, 71
   sunflecks and, 32, 43
   temperature and, 35, 48, 50-51, 52-53
   transpiration, 24-25
   trunks, 132, 142, 184, 264
Avicennia nitida, 19-20, 28, 34, 250. *See also Avicennia germinans*
Avicennia officinalis, 3, 70-71, 111

bacteria, 8, 129-30, 131, 269, 287
   detritus and, 272, 273-74
   as food, 197, 288, 289-90
   leaves, 96-97, 130, 266, 276
   nitrogen fixation, 79-80, 96
Balanus amphitrite, 184, 187
banana prawns, 289-90
bananas, squirter's disease in, 132
bandicoots, 166
Banksia, 202
bark, 141-42, 156. *See also* trunks

barnacles, 43, 184, 187, 201, 230, 239
barriers, 3, 126, 127
Barringtonia, 72, 95
   acutangula, 85, 111
   racemosa, 85
basidiomycetes, 131, 132, 133, 135
basin mangrove forests, 123-24
Batidaceae, 150, 152-53
Batis, 153
   argillicola, 151
bats, 37, 107, 156, 166
bees, 37, 177
beetles, 201
Bembicium auratum, 188, 206, 233
Bembicium melanostomum, 184
benthic organisms, 193, 246, 268-72, 275, 276, 279, 289, 291, 307
biogeographic regions, 6, 56, 70, 84. *See also* distribution, mangrove; mangrove communities; zonation
biomass, 81-83, 246-48, 253-55, 261, 263-64, 266-67, 279, 288. *See also* primary production
bioturbation, 274, 275
birds, 8, 83, 107-8, 165
   breeding, 298, 299
   distribution, 157, 160-61, 166-70, 203
   feeding, 164, 166, 199, 207, 216, 221, 225, 284-85, 291, 298, 299
   as food, 174, 175, 179, 180
   mistletoes and, 41, 148, 150
   pollination by, 36-38, 148
biting midges, 171, 172, 173-75, 305
bivalves, 187, 192, 193, 201-2, 232, 237, 275
bole damage, 35-36
Boleophthalmus, 181
borers, 35, 43, 105, 107, 176, 201-3
Bostrychia, 136, 139-40
   tenella, 139
boundary conditions, 115-18. *See also* zonation
Bruguiera, 105
   fauna and, 177, 188, 192, 199, 202
   flora and, 133, 140, 145, 147
   leaves, 21, 106, 277, 283
   light response, 33, 54
   reproduction, 37-38, 40, 41
   roots, 26, 28, 159, 192
   salt and, 17, 18
   soil conditions, 64
   wood, 36, 133
   xeromorphic features, 21, 24
   zonation, 99, 120
Bruguiera cylindrica, 10, 38, 85, 95
Bruguiera exaristata, 85, 96
   distribution, 7, 11, 111
   dominance, 101-2
   salinity, 71
   soil water, 65
   temperature and, 51, 52
Bruguiera gymnorhiza, 84
   butterflies and, 176
   distribution, 111
   dominance, 101
   epiphytes and, 145
   reproduction, 37-39
   salinity, 70, 71
   soil water, 65
   temperature and, 51
Bruguiera parviflora, 35, 85, 165
   distribution, 11
   dominance, 101
   pollination, 38
   salinity, 71, 72
   soil water, 65
   temperature and, 51
Bruguiera sexangula, 70-72, 85, 95, 111
Bubalus bubalis, 106, 166
buffalo, 106, 166
bund walls, 307-8
burrows, 193, 232, 235, 237, 239
   air in, 177, 190
   crab, 75-76, 105, 173, 175, 192, 193, 199, 210, 214, 219-20, 222, 224-25, 227, 229, 236-37
   mud lobster, 105, 193, 222
   mudskipper, 181, 227, 230
   polychaete, 159, 192, 230
Butorides striatus, 160, 170
butterflies, 38, 104-5, 145, 148, 176

calcium, 73, 96
Callianassa, 193, 224
   australiensis, 300
Callistemon, 147
Caloglossa, 136, 139, 140
Calothrix crustacea, 138
Camptostemon, 21, 28, 40, 50, 118, 187
   schultzii, 11, 35, 71, 85, 111, 132
canal developments, 175, 305-6
canopy, 56, 58, 102, 112, 121
   damage of, 55, 107-8
   fauna and, 107-8, 155, 157, 159, 187, 205, 207
   flora and, 93, 131-32, 136, 139, 146, 187

leaf position, 34, 253, 262
light and, 32, 34, 43, 53, 252, 258, 262
primary production, 248, 253, 256
temperature and, 34, 253, 262
Canthium latifolium, 154
Capitella capitata, 274
capitellids, 192
Capparadaceae, 145
carbon, 32-33, 122, 246, 249, 252, 263, 272-73, 279, 288
DOC (dissolved organic carbon), 284
POC (particulate organic carbon), 284-85, 289, 293, 294
Carcinus maenas, 237
Carettochelys insculpta, 179
carnivores, 181, 198, 287, 289
Cassidula, 188
auris-felis, 233
mustelina, 156
Cassiopea, 229-30
Casuarina, 94-96, 147
glauca, 75, 94-96, 118, 147, 154
Catanella, 136, 139, 140
catchments, 88-92, 302, 306. *See also* drainage; rainfall
cattle, 166, 308
Cenoloba obliteralis, 176
Ceratonereis aequisetis, 232
Cerbera manghas, 154
Cerberus rhynchops, 180, 207-9, 210
Cercospora, 132
Ceriops, 8, 36
competition, 98, 99, 101-2
fauna and, 172, 177, 189, 199, 211, 228
flora and, 133, 140, 147
leaves, 31, 106
light response, 31, 33-34, 54
reproduction, 37, 38, 41, 43, 44
roots, 26, 28, 228
salt and, 17, 18, 73, 74
soil conditions, 73, 74
xeromorphic features, 21, 23-24, 31
zonation, 98, 99, 120
Ceriops decandra, 11, 52, 70-71, 85, 147
Ceriops tagal, 35, 39, 85
distribution, 11, 111-14
dominance, 101-2
fauna and, 171, 176, 201-2
flora and, 93, 132, 144, 145, 147
leaves, 31, 48, 256-57

salinity, 70-73, 114
salt and, 3, 17
soil water, 64, 65, 73
temperature and, 48, 51
xeromorphic features, 23, 31
Cerithidea obtusa, 189
Cerithiidae, 188
Chamaesipho columna, 184
Chelonodon patoca, 181, 287
chenier plains, 125, 126
Chenopodiaceae, 150, 152, 153
Chiromanthes dussumieri, 235
Chiromanthes onychophorum, 235
Chizacmaea, 187
ciliates, 274
Cladium jamaicense, 286
Cladosporium, 132, 133, 134
cladosporioides, 134
Clean Water Acts, 304
Clerodendron inerme, 65, 71, 188, 154
Clibanarius taeniatus, 221, 222
Clibanarius virescens, 221, 223
Coastal Management Investigation (Queensland), 305
coasts, 3, 6-7, 9-11, 125-28, 137, 163. *See also* coral; deltas; mangrove communities; mudflats
cocci, 129
Coenobita spinosa, 196, 197
Colluricincla megarhyncha, 160, 207
Colobopsis, 176
Columbella duclosiana, 189
competition, 97-104, 109-10, 112, 119.
See also adaptation; dominance
fauna, associated, 157, 169, 205-6
flora, associated, 109, 119, 140
resources, 296
composite alluvial plains and barriers, 126-28
Conocarpus, 36
Convolvulaceae, 152
copepods, 182
Copidita nigronotata, 176
Coptotettis masticatus, 175
coral, 3, 61, 77, 154
coasts, 125, 126, 128
reefs, 8, 157, 162, 187, 289, 303
crabs
adaptations, 213-16, 219-24, 236-37
burrows, 75-76, 105, 173, 175, 192, 193, 199, 210, 214, 219-20, 222, 224-25, 227, 229, 236-37

damage by, 43, 44, 106
distribution, 163, 185, 188, 192, 193-95, 196, 199-201, 216, 218, 219, 239
feeding, 193, 197, 199, 219-22, 224, 229, 235-36, 272
as food, 165, 179, 180, 206, 207, 291
habitat, 159, 184, 196, 197, 213-15, 230
migration, 199, 229, 230-32, 289, 291-92
reproduction, 199, 225-27, 230-32, 291-92
respiration, 193, 197, 213-15, 236-37
salinity and, 188, 193, 197, 232, 238-44
Craticus quoyi, 170
spaldingi, 207
Crematogaster laeviceps, 176
crocodiles, 179, 208, 209, 291
Crocodylus acutus, 208, 209
Crocodylus johnstoni, 179, 209
Crocodylus porosus, 179, 208, 209
crustaceans, 193, 244. *See also* amphipods; barnacles; crabs; prawns; shrimps
distribution, 164, 204, 295
feeding, 287
as food, 209
habitat, 183, 186, 199, 201, 204, 224
osmoregulation, 238-39
respiration, 236-37
Cryptotermes domesticus, 171
Cryptotermes primus, 171-72
Cryptotermes secundus, 171
cryptovivipary, 41, 103
Ctenogobius criniger, 181, 287
Culex sitiens, 172, 173
Culicoides, 173-75
Curvularia, 133
Cylindrocarpon didymum, 133
Cymbidium, 145
Cynometra, 28, 37, 40-41, 50, 101, 118-19
iripa, 10, 41, 65, 70-71, 85, 96, 101, 111
cytokinin, 97

Dardanus setifer, 221, 224
Dasyhea, 173
decapods, 164, 231, 232, 235-36, 239
decomposers, 43, 93, 129, 133, 135, 201, 273. *See also* bacteria; degradation; detritivores; fungi; litter; nutrients; primary production
degradation, 133, 245-46, 272, 274, 276, 293. *See also* decomposers; detritus; litter; primary production
deltas, 3, 77, 80, 108, 125-28
Dendrobium, 145
Dendrophthoe glabrescens, 147, 149
deposit feeders, 196, 198, 275, 276
Derris, 95
trifoliata, 96, 145
desalinated water, 20-24, 92. *See also* salinity
detritivores, 272, 273, 274, 287
detritus, 94, 162, 224, 235, 285. *See also* litter; primary production
crabs and, 219, 221
cycle, 292-93
export, 81, 279, 285-86, 289
food chain, 106, 245, 259, 261, 271-76, 287
insects and, 165, 171
nutrients, 76, 81, 259, 261, 276
deuteromycetes, 134, 135
diatoms, 197, 268, 274
Dicaeum hirundinaceum, 41, 148
Didymosphaeria rhizophorae, 133
Diospyros, 262
ferrea, 85, 154
littorea, 154
Diptera, 175-76
Dischidia, 145
dispersal. *See* propagules, dispersal
Distichlis, 151-52
distribution, mangrove, 3-13, 50, 69-70, 84-88, 98-99, 102, 158, 167-70, 309. *See also* biogeographic regions; elevation; fauna; flora, associated; latitude; mangrove communities; rainfall; salinity; temperature; tides; zonation
Australia
Botany Bay, 181, 182, 246, 271
Cambridge Gulf, 70
Cape York Peninsula, 5, 83, 84, 111
Gladstone, 44, 50, 51, 56, 59, 62, 69, 74, 93, 115-16. *See also* Port Curtis
Great Barrier Reef, 9, 61, 128, 182

Gulf of Carpentaria, 5, 56, 59, 69
Hinchinbrook Island, 76, 83, 128, 249, 250, 260, 266, 286
Moreton Bay, 135
Port Curtis, 43, 51, 58-61, 94, 112, 117-18. *See also* Gladstone
Princess Charlotte Bay, 69, 98, 99, 135
Proserpine, 43, 44, 50, 65, 100-101. *See also* Repulse Bay
Repulse Bay, 43, 51, 101, 117-18, 128. *See also* Proserpine
Sydney, 61-63, 94-95, 113, 192, 196, 248-50, 266, 270, 276
Westernport Bay, 13, 80, 127, 203, 248, 250, 259, 261-64, 270, 273
world
Africa, 5, 13, 50, 61, 65, 135, 182. *See also* South Africa
Asia, 8, 308
Bangladesh, 309
Benin, 298
Brazil, 8, 179, 309
Burma, 309
Cameron, 309
Colombia, 309
Costa Rica, 124
Ecuador, 124, 298, 309
Fiji, 304
Florida, 39, 51-52, 66, 112, 133-34, 203, 262. *See also* United States of America
classification, 122-24
fauna, 106, 180
leaves, 106, 252, 256, 273
nutrients, 76, 79-83
primary production, 245, 247, 250, 256, 263, 270, 286
Gabon, 309
Gambia, 298
Hawaii, 80, 135
India, 21, 32-33, 293, 309
Indonesia, 293, 298, 309
Jamaica, 216, 218
Japan, 29
Malagasy, 309
Malaysia, 64, 156, 193, 233, 235, 247, 262, 309
Mexico, 110, 124, 309
New Zealand, 3, 50, 138, 250, 266, 270, 276, 280-82
Nigeria, 234, 309
Pakistan, 309
Panama, 42, 114, 124, 247, 309
Papua New Guinea, 7, 8, 70, 92, 128, 154, 167, 169, 180, 273, 309
Philippines, 7, 247, 298, 309
Puerto Rico, 51, 124, 247, 250-52, 253, 255, 258, 262, 263, 297
Senegal, 309
Singapore, 130, 193
South Africa, 37, 39, 42, 61, 193. *See also* Africa
South America, 5, 8, 113
Thailand, 247, 250, 253-54, 256, 261-62, 265, 276
United States of America, 34, 72, 97, 309. *See also* Florida
Venezuela, 233, 243, 309
Vietnam, 309
disturbance, 100, 104, 121, 147, 270. *See also* adaptation; stress
DOC (dissolved organic carbon), 284
Dolichandrone spathacea, 85
Dolomedes, 178
dominance, 100-2, 108, 113, 119, 124, 259. *See also* adaptation; competition
drainage, 46-47, 199, 308
basins, 86-92
patterns of, 159, 162, 297, 302, 306
soil, 60-61, 63-65, 75, 77, 78, 105
terrestrial, 81, 124
Drechslera, 133
dredging, 305, 306-7
Drosophila, 146
drosophilid flies, 37
drowned bedrock coasts, 126, 128
dugongs, 285
dwarf (scrub) mangrove forests, 123-24

Ecdlytolophora, 203
echinoderms, 183
echiuroids, 192
ectomycorrhizae, 96
eelgrass, 268, 285
Electra, 187
elevation, 10, 12, 46-47, 60, 78, 83, 112
Ellobiidae, 188, 232

Ellobium, 156, 202, 232
Elminius modestus, 184
Emoia atrocostata, 208
encrusting fauna, 184, 187, 192, 201, 205, 233-34. *See also* barnacles; oysters; serpulids
endomycorrhizae, 96
Enhalus, 42
Enteromorpha clathrata, 139
Eopsaltria pulverulenta, 160, 170, 207
epifauna, 162, 187-90, 191, 199, 201, 205
epiphytes, 8, 274, 285
    algae, 139-40, 187, 245-46, 269, 271
    mutualism and, 95, 144-46
Eriachne pallescens, 151
Eriophora, 178
erosion, 77, 83, 100, 110, 114, 259, 299, 305-7
Erythrodiplax berenice, 210-11
Eucalyptus, 8, 147
Eunicidae, 214, 217
evaporation, 46-47, 55-57, 59, 66-68, 76, 88, 159
evapotranspiration, 55, 56, 58
Excoecaria, 36, 133
    dominance, 101
    light response, 33, 54
    reproduction, 37, 38, 40, 41, 44
    roots, 26, 28
    salt and, 17, 18, 118
Excoecaria agallocha, 35, 84, 145
    distribution, 111-14
    dominance, 101-2
    fauna and, 107, 176
    flora and, 144, 147
    leaves, 48, 256-57
    salinity, 70-72
    soil water, 65
    temperature and, 48, 51

Fabaceae, 145
facultative mangrove-dwellers, 206
fauna, 163-64, 246, 260, 266. *See also* mangrove communities; zonation, fauna and
    adaptation, 181, 205-11, 244
        behavioural, 221-30
        morphological, 211-21
        physiological, 234-44
        reproductive, 225-28, 230-34
    epifauna, 162, 187-90, 191, 199, 201, 205

freshwater, 155, 178-79, 182, 210, 240. *See also* fresh water, fauna and
infauna, 162, 188, 190-99, 201, 275, 276. *See also* crabs
macrofauna, 182, 274, 275
marine, 43, 155-56, 159, 162-65, 179-206, 210, 211-44
meiofauna, 182, 272, 275, 289
microfauna, 182, 274
opportunists, 175, 178, 273
specialists, mangrove, 163, 183-84, 204
terrestrial, 155-57, 159, 160-61, 165-78, 182, 201, 205, 206-11, 240, 244
feral mammals, 166
Ficopomatus uschakovi, 187
filter feeders, 184, 196, 198, 230, 275
fish, 180-82, 292-93, 295, 307. *See also* gobies; mudskippers
    adaptation, 211-13
    feeding, 164, 181-82, 184, 225, 279, 284, 287-91, 298
    as food, 179, 291
Fish and Oyster Act, 300
Fisheries Acts, 300
Fisheries Habitat Reserves, 300-301
fishing industry, 293, 295-300
Flagellaria indica, 154
flagellates, 274
flies, 37
flooding, 59, 114, 286. *See also* drainage; fresh water; rainfall; rivers; tides, inundation; water
    fauna and, 162-63, 166, 238, 239, 244
    land development and, 305, 306, 308
    salinity and, 66, 76, 162, 238, 239
flora, associated, 8, 92-97, 109, 115-19, 129-55, 187, 268, 276, 279, 284-85, 308. *See also* algae; bacteria; epiphytes; fungi; lichens; mistletoes; salt marshes; seagrasses
flowering, 36-37, 39, 256. *See also* pollination; reproduction, mangrove
flying foxes, 166, 170, 175
foliage projective cover (FPC), 119, 121-22
food chain. *See also* detritus; primary production

detrital, 106, 245, 259, 261, 271-76, 287
  metals and, 304. *See also* waste discharge
Forcipomyia, 173
Fordonia leucobalia, 180, 208
foreshore development, 305-6
fossils, 7-8, 13, 37
fouling species, 43, 187, 233-34
FPC (foliage projective cover), 119, 121-22
Frankenia, 153
Frankia, 96
fresh water, 46-47, 83, 125, 135, 268, 291. *See also* flooding; rainfall; rivers; water
  catchments, 88-92, 302
  distribution, mangrove, 70, 84-88, 110
  fauna and, 172, 178-79, 193, 205, 230, 239-40, 242, 244, 260. *See also* freshwater fauna
  inundation, 59, 193
  run-off, 88-91, 162, 230, 266
  run-on, 65, 260, 261, 266
  soil salinity and, 59, 66, 260
freshwater fauna, 155, 178-79, 182, 210, 240. *See also* fresh water, fauna and
fringe mangrove forests, 122-24
fringing species, 118, 154
frogs, 207
frosts, 34-35, 50-51, 100, 162
fruit, 39-41, 176. *See also* propagules; reproduction, mangrove
fruiting times, 38-39, 256, 264
fungi, 34, 39, 79, 96, 130-36, 140, 201, 272
  leaves, 131-33, 275, 287
  litter, 93, 130, 133, 135, 266, 273, 277
  parasitism, 93-94, 131-32, 135
Fusarium, 133

Galeolaria caespitosa, 184
gas exchange, 21, 26, 30, 32, 48, 249-52, 259. *See also* photosynthesis; respiration
gastropods, 182, 184, 192, 199, 206, 219, 232
  feeding, 140, 272
  as food, 221
  habitat, 187, 193, 229
  microgastropods, 202
Geolycosa, 178

geomorphological classification, 2, 124-28. *See also* mangrove communities; zonation
Gerygone levigaster, 161, 170, 207
Gerygone magnirostris, 161, 170
Gerygone tenebrosa, 161, 170, 207
Gloeocapsa alpicola, 138
glycinebetaine, 16, 19
Gnathia, 201
gobies, 181, 211, 212, 287. *See also* fish
Golfingia vulgaris, 234
Goniopsis cruentata, 243
Gracilaria, 138
Grapsidae, 106, 187, 188, 239-40. *See also* crabs
grasshoppers, 175
grazing, 81, 106-7, 140, 235, 269, 274, 279, 284, 285, 287, 308. *See also* herbivores
gypsy moth, 107

Hadrachaeta aspeta, 204
Halcyon chloris, 160, 170
Halosarcia, 152, 153
haustorium, 146, 148
Heloecius cordiformis, 196, 200, 227, 243-44
Helograpsus haswellianus, 196, 225
herbivores, 100, 106-7, 159, 205-6, 235, 275, 279, 284, 287. *See also* grazing
herders, 227. *See also* crabs, reproduction
Heritiera, 36
  dominance, 101
  leaves, 21, 106
  reproduction, 40, 41
  roots, 26, 28
  salinity, 118, 119
Heritiera littoralis, 84, 145
  distribution, 10, 114
  dominance, 101
  salinity, 70, 71, 72, 114
  soil water, 65
  temperature and, 51
  xeromorphic features, 22
heterotrophs, 131
Hibiscus, 7, 40, 118, 119
  tiliaceus, 21, 52, 72, 111, 154
Hormosira, 268
  banksii, 138
Hoya, 145
  australis, 145
human influence, 100, 253, 261, 293, 296-310

humidity, 51-52 56, 58, 66, 142, 144, 258
hummock forests, 123-24
Hydnophytum, 145
    formicarium, 145, 146
Hydromys, 166
Hypochrysops
    apelles, 176
    apollo, 145, 176
    digglesii, 148
    epicurus, 176
    narcissus, 148, 176
hysteresis effect, 50

Ilyoplax, 214
    delsmani, 224, 232
Incisitermes barretti, 172
infauna, 162, 188, 190-99, 201, 275, 276. *See also* crabs; fauna
insects, 165, 167. *See also* ants
    damage by, 39, 43, 106, 107-8, 201, 203
    feeding, 106, 287
    as food, 179-80, 182
    habitat, 156-57, 171-77, 210-11
    pollination by, 36-38
insolation, 33, 46, 47, 53-56, 66, 81, 257-59, 263. *See also* light; photosynthesis
intertidal landforms, 77, 236. *See also* mangrove communities
Intsia bijuga, 154
inundation, 56
    freshwater, 59, 193
    tidal, 26, 46, 59-61, 65-69, 75-76, 109, 110-12, 128, 137, 159, 184, 187-88, 191, 205, 211, 232-33, 308
invertebrates, 8, 163, 165, 171-78, 182-204, 287, 305. *See also* fauna
ionic balance, 41
Iravadia, 202
Iridomyrmex cordatus, 145, 146
isopods, 184, 199, 201, 239, 274

Juncus, 94, 157, 174, 279, 286
    maritimus, 95
    roemerianus, 271, 279

Kandelia, 40
    candel, 29, 39

lagoons, 126, 127
Laguncularia, 15, 34, 36, 40-41, 42-43
    racemosa, 72, 79, 97, 124, 250

Laminaria, 265
Lasaea australis, 187
Lates calcarifer, 290-91
Laticauda colubrina, 208, 210
latitude, 260, 308. *See also* distribution; mangrove; zonation
    fauna, 160-62, 163, 167, 168, 184, 187, 203
    flora, 141-42, 143, 150-51, 308
    frost, 35, 50-51, 162
    light, 54-55
    primary production, 253, 259, 270
    temperature, 35, 52-53
leaf scorch, 35
leaves, 103-4, 276-77. *See also* canopy; litter; photosynthesis; primary production; respiration; transpiration
    damage, 35, 36
    fauna and, 199, 235, 287. *See also* grazing; herbivores
    flora and, 96-97, 130-33, 135, 140, 266, 275, 287
    growth, 47-48, 259. *See also* production
    hairs, 21, 22, 25, 56, 140
    light response, 31-32, 33-34, 252, 262
    photosynthesis, 21, 32-34, 48, 249, 251, 252, 258, 262
    production, 25, 39, 247-67, 273
    respiration, 249, 251, 262-63
    salinity and, 114, 252
    scales, 21, 22, 25, 56
    stomata, 21, 24-25, 48-49, 58, 63
    succulence, 21, 23-24
    temperature and, 21, 25, 46-52, 58, 250-53, 258, 262
    transpiration, 21, 25, 24-25, 48, 56, 58, 258, 260
    xeromorphic features, 20-25, 31
Leguminosae, 96
lenticels, 29
Lepidonotus, 156
Lepidoptera, 107, 175, 176
Leptosphaeria australiensis, 133
Liasis, 170
Lichenostomus versicolor, 161, 170
lichens, 140-44
light. *See also* insolation; leaves; photosynthesis; sun; temperature
    fauna and, 183, 184, 187, 234-35
    flora and, 139-40, 142, 144, 187
    latitude 54-55

photosynthesis, 32-34, 53, 54, 252, 257, 262
primary production, 252, 253, 255, 258, 263
response, 25, 31-34, 43, 53-54
shortage, 44, 100
sunflecks, 32, 43
lignin, 133, 275
Limonium, 153
limpets, 201
litter, 236. *See also* detritus; leaves; primary production
export, 266-67, 285-86, 289
fall, 252, 256, 259, 265, 276, 278, 280-81
fungi and, 93, 130, 133, 135, 266, 273, 277
production, 249, 250, 252, 261, 264-65, 266, 270, 271, 276
Littoraria, 199
Littorina, 156
scabra, 156, 189, 199, 232-33
Littorinidae, 187, 232-33
lizards, 165, 180
lobsters, 105, 193, 222
Lophognathus temporalis, 179
Loranthaceae, 146
Lulworthia, 133-34
grandispora, 133, 134
Lumnitzera
dominance, 101
leaves, 21, 31
light response, 33-34, 54
reproduction, 37, 38, 40, 41, 43, 44
roots, 28
salt and, 18
Lumnitzera littorea, 36, 85, 147
leaves, 97
reproduction, 38
salinity, 70, 71
Lumnitzera racemosa, 35, 85
distribution, 111-13
dominance, 101-2
flora and, 145, 147
leaves, 23, 48, 97, 256-77
reproduction, 38, 43
salinity, 66, 70, 71, 118
soil water, 65, 66
temperature and, 48, 51
Lymantia dispar, 107
Lyrodus, 202
Lysiana, 40, 147, 148, 149

Macrocyttara expressa, 176
macrofauna, 182, 274, 275

Macrophthalmus, 193, 200, 214
crassipes, 240, 241, 243
depressus, 199, 220
setosus, 200, 240, 241, 243
Maireana, 152, 153
malic acid accumulation, 30
malic dehydrogenase, 18, 33
mammals, 165-66, 179, 206. *See also* terrestrial fauna
mangrove communities, 1-5, 45, 163, 268, 308. *See also* adaptation; biogeographic regions; distribution, mangrove; primary production; zonation
classification, 2, 108, 119, 121-28
definition, 1
interactions
plant-animal, 104-8. *See also* fauna
plant-physico-chemical, 45, 92
plant-plant, 92, 104. *See also* flora, associated
management, 296-310
opportunism, 100, 102
marine fauna, 43, 155-56, 159, 162-65, 179-206, 210, 211-44
Martesia striata, 202
Mastotermes darwiniensis, 172
mathematical models, 81-82, 262-63
Maytenus emarginata, 154
meiofauna, 182, 272, 275, 289
Melaleuca, 107, 111, 118, 154, 157
acacioides, 111, 154
quinquenervia, 111, 154
viridiflora, 144, 154
Melanagromyza avicenniae, 175-76
Meliphaga fasciogularis, 170
Meliphaga gracilis, 38
Melithreptus albogularis, 148
Metapenaeus bennettae, 288
Metaplax, 214
meteorological features, 6-7. *See also* mangrove communities
methanogenesis, 269
Microcoleus lyngbyaceus, 137
microfauna, 182, 274
micro-organisms, 95-97, 269, 272, 274-76, 285
Mictyris, 159, 194-95
longicarpus, 200, 229, 240, 242-43
platycheles, 175, 193
mineralization, 261
mistletoes, 41, 92, 146-50, 176
mitochondria, 16
Modiolus auriculatus, 187

molluscs, 162, 232-33
  adaptation, 216, 218
  distribution, 164, 182-83, 202-4, 295
  feeding, 193, 196, 198, 287
  habitat, 186-90, 193, 196, 198, 202
Monocanthus chinensis, 284
Monostroma crepidinum, 137
Morelia spilotes, 170
Morula, 187
mosquitoes, 172-73, 182
moths, 37, 38, 107, 176
mud, 3, 80, 125, 128
  fauna and, 105, 159, 172, 175, 177, 178, 187-88, 196-97, 202, 205-6, 211, 222, 230, 234-36, 287
  flora and, 129-30, 135-36, 138, 233, 279, 288
mud crabs, 105, 193. *See also* crabs; mudflats; *Scylla serrata*
mudflats, 116, 130, 137, 173, 180, 186, 188-92, 198, 204
  crabs and, 159, 192, 196, 221, 225, 227, 229, 291
mud lobsters, 105, 193, 222
mudskippers, 179, 227-29. *See also* fish
  adaptations, 181, 211-13, 222
  reproduction, 228, 230
Mugil cephalus, 288, 290
Mugilogobius, 181
mushrooms, 136
mutualism, 95-97, 104-5, 144-46
mycorrhizae, 96
Myiagra alecto, 161
  melvillensis, 170
Myiagra ruficollis, 160, 170
Myoporum, 37, 38, 154
Myrmecodia, 145-46
  beccarii, 144-45
myrmecophytes, 144-45
Myron richardsonii, 180, 208
Myrsinaceae, 97
Mytilus edulis planulatus, 184
Myzomela erythrocephala, 161, 207

Nacaduba kurava, 176
Napier grass, 265
Nasutitermes graveolus, 172
Nausitora sauli, 202
Nectarinia jugularis, 38, 148
nematodes, 182, 272, 287
nemerteans, 192, 201-2
Neotermes insularis, 171

nereidids, 187, 192, 214, 217, 230, 232
Nereis virens, 276
Nerita, 203
  atramentosa, 203
  lineata, 189, 199
Neritidae, 187
Nigrospora, 132
nitrification, 79, 80
nitrogen, 18-19, 83, 138, 273, 275-77, 283. *See also* nutrients
  fixation, 79-80, 96-97, 261
Nostoc, 140
NPP (net primary production). *See* primary production
nutrients, 63, 96, 298, 304. *See also* detritus; nitrogen; phosphorus; roots
  availability, 2, 79, 81
  fauna and, 107, 144, 146, 234-35, 275, 276
  flora and, 96, 142, 144, 146
  input, 78, 81-83, 107
  parasitism, 41, 92
  primary production, 81-83, 107, 260-61, 265-66
  recycling, 77, 82, 122
  shortage of, 100, 124, 162
  tides and, 80-81, 259-61
nutritional parasitism, 41. *See also* parasitism
Nypa, 92
  fossils, 7, 8
  reproduction, 37, 40, 41
  roots, 26, 28
Nypa fruticans, 10, 72, 85

obligatory mangrove-dwellers, 160-61, 206
ocean currents, 2, 5, 13
Ochetostoma australiense, 192
Ocypode, 213, 215, 225, 238
  ceratophthalma, 196, 197
  convexa, 196, 197
ocypodids, 196, 224
Oecophylla smaragdina, 107, 156, 177
Ogyris amaryllis, 148
  hewitsoni, 148, 176
omnivores, 199, 235, 272, 287-88
Omobranchus, 181
Onchidium, 188
Oncis, 199
Ophicardelus, 188
opportunism, 100, 102
  fauna, 175, 178, 273

Orchidaceae, 145
Osbornia
   distribution, 7
   dominance, 101
   light response, 33-34, 54
   reproduction, 37, 38, 40, 41
   roots, 28, 29
   salt and, 17, 18
Osbornia octodonta, 84, 147
   distribution, 7, 11, 111
   dominance, 101-2
   leaves, 23, 48, 50, 256-57
   salinity, 71
   soil water, 65
   temperature and, 48, 50, 51
osmoregulation, 237-44
osmotic pressure gradient, 25, 260. *See also* adaptation, salt; transpiration
overwash mangrove forests, 122-23
oysters, 192, 234-35
   Fish and Oyster Act, 300
   production, 293, 295, 296, 298
   zonation, 184, 187, 206, 230

Pachycephala lanioides, 160, 170, 207
Pachycephala melanura, 160, 170
   dahli, 170
Pachycephala simplex, 170
Pachygrapsus, 218
Paguristes squamosus, 221, 223
Palaeobruguiera, 8
Pantinonemertes winsori, 201
Paracleistostoma mcneilli, 240, 241, 243
Paragrapsus gaimardi, 200
Paragrapsus laevis, 196, 199, 230
parasitism, 92, 175-76
   fungi, 93-94, 131-32, 135
   mistletoes, 92, 146-47, 148, 150
   nutrients and, 41, 92
Parmotrema, 142-44
Pelliciera, 40-41, 42-43
Pemphis, 118
   acidula, 85, 154
Penaeus merguiensis, 289-90
Penicillium, 132, 133, 134, 135
Periophthalmodon, 181, 228-29
Periophthalmus, 181, 211-13, 228-29, 231
Pestalotia, 131-32, 134
Phalacrocorax melanoleucos, 168
Phalacrocorax varius, 167
Phascolosoma agassizii, 234
Phascolosoma arcuatum, 192, 230, 234, 244

Phascolosoma lurco, 156
Pheidole myrmecodiae, 177
Phialophora, 133
Phlyctochytrium mangrovis, 135
Phoma, 133
Phormidium angustissimum, 138
phosphate, 80, 83, 285
phosphorus, 79, 80, 83-84, 96, 261. *See also* nutrients
photosynthesis, 18, 47, 81-82, 100, 262-63, 271, 307. *See also* gas exchange; primary production
   flora and, 79, 137-38
   FPC, 119, 121
   leaves, 21, 32-34, 48, 249, 251, 252, 258, 262
   light and, 32-34, 53, 54, 252, 257
   salinity, 110-12
   temperature and, 48, 52, 258, 259
   tides, 259-60
phycobiont, 140
phycomycetes, 133, 135
Phyllosticta, 132
physical damage. *See also* stress
   frost, 34-35, 50, 100
   water, 35-36, 44. *See also* waterlogging
   waves, 35, 41, 43-44, 307. *See also* tides
   wind, 35-36, 100, 122, 264
physiographic classification, 122-24. *See also* mangrove communities
Phytophthora, 58, 93-94, 133 135
   vesicula, 133, 134
plankton, 172, 184, 272
   phytoplankton, 246, 265, 268-69, 271-73
   zooplankton, 289
plant cover, 46, 56, 66
plasmodesmata, 16
Pleistocene, 8, 157, 169
Plumbaginaceae, 152
pneumatophores, 248. *See also* roots
   adaptations, 26-29
   fauna and, 159, 184, 187, 199, 201, 229
   flora and, 136, 139, 171, 187, 245
   respiration, 60, 249
Poaceae, 150, 152
POC (particulate organic carbon), 284-85, 286, 289, 293, 294. *See also* primary production
Poecilips rhizophorae, 107
pollination, 36-38, 55, 104, 148, 177. *See also* flowering; reproduction, mangrove
pollen, 7-8, 37, 156

polychaetes, 183, 184, 214, 236, 273, 300. *See also* worms
  feeding, 272, 274-75, 276, 287
  as food, 175
  habitat, 159, 187, 192, 201, 204, 230, 232
Polypodiaceae, 145
Polypodium, 145
Polyporaceae, 132, 133
Portunus pelagicus, 196, 197
Posidonia, 42, 276, 279, 284
  australis, 246, 271, 284
possums, 150, 156, 166
Potamides obtusa, 199
Potamididae, 188, 218
prawns, 193, 286, 292
  feeding, 164, 184, 273, 288, 289-91
  as food, 179, 181-82
  production, 293, 295, 298, 299
Premna acuminata, 154
primary production, 51-52, 100, 122, 268-73, 292-93, 294. *See also* detritus; leaves; nutrients; photosynthesis; POC; respiration
  biomass, 81-83, 246-48, 253-55, 261, 263-64, 266-67, 279, 288
  budgets, 262-66, 279, 284
  definition of, 245-46
  fate of, 266-67, 273-76, 279, 285-86, 289-90
  measurement of, 246-53
  nutrients, 81-83, 107, 260-61, 265-66
  seasonal variation, 256-59, 285
  wood, 262, 264, 265
primary succession, 109
Primulaceae, 152
Procalyptis parooptera, 176
propagules, 176
  damage, 43-44, 107
  dispersal, 5, 41-42, 43, 55, 98, 112
  establishment, 1, 42-44, 109, 112
  production, 38-39, 254, 264
protozoans, 182, 272
Pseudendoclonium submarinum, 140
Pteropus alecto, 107, 166
Pteropus conspicillatus, 107, 166
Pteropus poliocephalus, 166
Pucciniella, 151-52

Quoyia, 188

radiation. *See* insolation
rainfall, 46, 47, 64, 75, 83, 121, 159, 253. *See also* catchments; drainage; flooding; storms; watertable
  catchments, 88-92
  distribution and, 10, 12, 13, 56-57, 83-85, 87, 144
  nutrients, 81, 83, 261
  primary production, 262, 266, 285
  salinity, 66-68, 76, 162, 237
rainforest, 158, 160-61, 169
Ramalina, 142-44
Rana cancrivora, 207
rats, 165-66, 206
Rattus rattus, 206
reclamation, 163, 292, 306-8
recombination, 245-46
redox potential, 78, 83, 261
Relicina, 142, 143
reproduction
  fauna, 225-28, 230-34
  mangrove, 36-44, 103-4
reptiles, 165, 170-71, 179-80, 207-10
respiration, 18, 25, 47, 81, 252. *See also* gas exchange; primary production
  anaerobic, 30, 219, 236-37. *See also* waterlogging
  fauna, 190, 193, 197, 210, 212-14, 216, 219, 222, 230, 236-37
  leaves, 34, 249, 251, 262-63
  roots, 30, 45, 60, 249
  salinity, 110, 260
  zonation, 259
Rhipidura fuliginosa, 167
Rhipidura phasiana, 161, 207
Rhipidura rufifrons, 167
  dryas, 161
Rhizobium, 96
Rhizoclonium, 137
rhizosphere, 95-96
Rhizophora, 8, 36, 105, 120, 262-63
  distribution, 7, 99
  dominance, 101-2
  fauna and, 106, 156, 159, 171-72, 184, 186-89, 196, 198, 202, 211
  flora and, 133, 140, 147
  light response, 33, 34, 54
  reproduction, 37, 38, 40-44, 109
  roots, 26, 28, 29, 109, 159, 184, 228
  salt and, 3, 15, 17, 18
  soil conditions, 61, 64
  xeromorphic features, 21, 24
Rhizophora apiculata, 11, 36, 39, 52, 70-72, 84, 111-14, 250, 253-54

Rhizophoraceae, 7, 24, 29, 107-8
Rhizophora lamarckii, 84
Rhizophora mangle, 34, 304
   dominance, 122, 124, 245
   flora and, 131-34, 273
   leaves, 48, 49, 51-52, 106, 114
      252, 259, 266, 273
   nutrients and, 79, 83
   primary production, 51-52,
      248-50, 259, 266
   reproduction, 39, 43
   roots, 28, 79, 201, 203
   salinity, 66, 114, 252
   salt and, 17, 19, 20
   temperature and, 48, 49, 51-52, 252
Rhizophora mucronata, 10, 70, 72, 85, 114, 304
Rhizophora stylosa, 35, 84
   distribution, 111-14
   dominance, 101-2
   fauna and, 165, 175-77
   leaves, 48, 256-57
   light response, 34, 52
   propagules, 39, 43, 112
   roots, 177
   salinity, 68, 70-72
   salt and, 17
   soil conditions, 64-65
   temperature and, 34, 48, 51, 52
ribulose diphosphate carboxylase, 18
riverine mangrove forests, 123-24
rivers, 88, 90-92, 124, 306-7. *See also* drainage; flooding; fresh water
   mangrove communities and, 70, 85-87, 108, 110-15, 124-28. *See also* zonation, upriver
   roots, 109. 228. *See also* pneumatophores; soil; waterlogging
      adaptation, 25-31, 41, 63-64, 103-4
      branching, 201, 203
      damage, 35, 36, 307
      fauna and, 159, 173, 184, 188, 192-93, 201, 206, 233
      flora and, 79-80, 93, 96, 131-33, 135-36, 139
      nutrients, 63, 259
      primary production, 248, 252-54, 262-64, 266
      respiration, 30, 45, 60, 249
      temperature and, 17, 28, 45, 250-52
      tides and, 26, 29, 122, 259
Rubiaceae, 97, 145

Saccostrea commercialis, 184, 187, 234, 235
Saccostrea echinata, 187
Salinator, 188
   burmana, 189
salinity, 55, 88, 98, 102-4, 153, 252, 302, 307-8. *See also* desalinated water; salt
   distribution and, 70-71, 84-87, 94, 110, 112-14, 159, 162
   fauna and, 159, 162, 174, 181, 184, 187-88, 193, 197, 205, 207-12, 230, 232, 233-34, 237-44
   flora and, 135, 140. *See also* salt marshes
   propagules and, 42, 44
   river systems and, 70, 112-15, 124
   soil, 46, 47, 56, 61, 66-74, 76, 78, 83, 110-19, 144, 260-61
   structural formation and, 121
   tides and, 66-69, 72, 76, 117
Saliticidae, 178
salt, 25, 34, 41, 76, 102-4, 205, 260, 308. *See also* adaptation; desalinated water; osmoregulation; salinity; xeromorphic features
   accumulation, 14, 18-20
   distribution and, 2, 3, 110
   exclusion, 14, 17, 30, 45
   secretion, 14-17, 45, 72
   tolerance, 125, 127-28, 151, 153. *See also* salt marshes
saltcouch marshes, 182
salt flats, 56, 59-60, 69-70, 72-74, 116-20, 136-37, 154, 268, 308
salt flushing, 77, 259. *See also* tides, flushing
salt marshes, 6, 62, 64, 192, 246, 252, 268, 305, 308-10. *See also* flora, associated
   distribution, 94, 150-52, 153-54
   fauna and, 172, 181, 295
   flora and, 130, 135, 136, 138, 270
   primary production, 270-72, 274, 279, 298
   salinity, 69, 94, 117-19, 153
   watertable and, 75
   zonation, 116, 120
salt-sieving, 73
Samolus, 153
sand dredging, 306-7
sand flats, 130
saprophytes, 131-32, 134, 135

INDEX

Sarcocornia, 94, 152, 174, 175
   quinqueflora, 94, 152
sawgrass flats, 286
Scartelaos, 181
Schizochtytrium, 133
Schizothrix, 137
Sclerostegia, 152
Scopimera, 194, 195
Scrophulariaceae, 152
scrub (dwarf) mangrove forests, 123-24
Scylla serrata, 105, 179, 193, 196-97, 221-22, 224, 229-31, 291-92
Scyphiphora, 24, 28, 33, 40, 50, 54
   hydrophyllacea, 11, 71, 85, 97
seagrasses, 41, 42, 157, 265, 266, 268, 269, 304-5
   fauna and, 159, 162, 181-82, 229, 273, 279, 284-85, 289, 295
   flora and, 130, 135
   food chain, 271-74, 276
   sediment, 192, 288
sea-level, 55, 58-60, 77, 110, 127, 128
sea snakes, 179
seaweed, 271, 274
secondary succession, 109-10
sediment, 28-29, 77-78, 80-81, 110, 115, 125, 285, 298-300. *See also* infauna; soil
   dispersion, 29, 55, 60, 127, 259
   dredging, 306-7
   fauna and, 105, 159, 162, 166, 177, 184, 192-95, 218-20, 230, 274-75, 288
   flora and, 272, 288
   seagrasses, 192, 288
   terrigenous sedimentary coasts, 125-28
   tides, 55, 60, 127, 259
seedlings, 5, 43, 159, 199, 304. *See also* propagules
   development, 39-41, 95, 103, 104, 112
   fungi and, 131, 133-34
   light response, 31-32, 33
   mortality, 39, 51, 112
   primary production, 251, 255
   salinity, 72
   salt and, 18, 19
   sunflecks and, 32, 43
serpulids, 43, 184, 187, 230
Sesarma, 43, 156, 185, 196, 197, 200, 218, 232, 238
   darwinensis, 193
   erythrodactyla, 196, 199, 200, 230
sesarmids, 179, 199, 231, 235-36

Sesuvium, 153, 192
   portulacastrum, 151, 152
shade tolerance. *See* light, response
shallow shores, 2, 5, 272. *See also* coasts; zonation, shoreline
shrimps, 193, 272
sipunculans, 192, 230, 234, 237
snails, 233
snakes, 170, 179, 180
soil. *See also* erosion; mud; nutrients; sediment; substrate
   clay, 60-61, 72-73, 75-76, 80, 125, 129, 194-95
   drainage, 46-47, 60-61, 63-65, 75, 77, 78, 105
   flora and, 96, 131, 133-36, 138
   nature of, 46, 47, 60-61, 77-84, 286
   salinity, 46, 47, 56, 61, 66-74, 76, 78, 83, 110-19, 144, 260-61
   water and, 25-26, 30, 46-47, 56, 60-78, 259. *See also* waterlogging; watertable
Sonneratia, 35-36, 120, 156
   distribution, 99
   fauna and, 177, 187, 189
   fossils, 7, 8
   light response, 33, 54
   reproduction, 37, 40, 41
   roots, 26, 28
   salt and, 15, 17, 18
   temperature and, 50
   xeromorphic features, 21, 24
Sonneratia alba, 11, 39, 70-72, 85, 111, 114, 166
Sonneratia caseolaris, 23, 70-72, 85
Spartina, 153, 265, 279
specialists, mangrove, 163, 183-84, 204
Sphaeroma, 239
   terebrans, 201, 203
spiders, 177-78
spionids, 183, 187, 192, 214, 217, 230, 232
spirorbids, 232
sponges, 183
Sporobolus, 94, 151-52, 153, 192
   virginicus, 118, 153, 174
Squilla choprai, 224
squirter's disease, 132
Stilobezzia, 173
stomata, 21, 24-25, 48-49, 58, 63
storms, 35, 58, 77, 162, 266, 285. *See also* rainfall

stranding, 42, 43, 114. *See also* propagules
stress, 28, 36, 78, 100, 102, 104. *See also* adaptation; physical damage; salinity; water discharge
   thermal, 100, 250-52, 305
   water, 25, 28, 58, 60, 63-64, 93, 100. *See also* water, damage by
structural classification, 119, 121-22, 124. *See also* mangrove communities
Strychnos lucida, 154
Suaeda, 94, 152, 192
   arbusculoides, 151
   australis, 152, 174
substrate, 56, 60, 77, 124, 206. *See also* mud; sediment; soil
   build-up, 109
   change in, 36, 78
   crabs and, 219, 225. *See also* crabs, burrows
   roots and, 25-26, 29
   stranding on, 43
succession, plant, 109-10
succulence, 21, 23-24. *See also* xeromorphic features
   saltmarsh, 153
sugarcane, 265, 298, 308
sulphur, 105, 130, 131, 269
sun, 172, 269. *See also* light; temperature
   sunflecks, 32, 43
surface raspers, 196, 198
Syringodium, 273, 285

Taenioides, 181
tannins, 31, 34, 95, 107, 165, 205, 236, 293, 296
Tatea rufilabris, 187
Tecticornia australasica, 150-51, 152
Teloschistes, 142, 143
   flavicans, 143, 144
temperature, 42, 45-53, 80, 88. *See also* frosts; light
   distribution and, 2-5, 10, 13, 35, 47, 50-53
   fauna and, 172, 184, 205, 219, 230, 234, 235
   leaves, 21, 25, 46-52, 58, 250-53, 258, 262
   light and, 33, 34
   primary production and, 253, 258, 259, 263
   roots and, 17, 28, 45, 250-52
   stress, 100, 250-52, 305

terebellids, 204, 230
Terebralia palustris, 188
Terebralia sulcata, 188, 189
teredinids, 201, 202, 237-38
Teredo, 202, 203
Terminalia subacropta, 154
termites, 105, 171-72, 201
terrestrial fauna, 155-57, 159, 160-61, 165-78, 182, 201, 205, 206-11, 240, 244. *See also* fauna
terrigenous sedimentary coasts, 125-28
Tetragnatha, 178
tettigoniids, 106, 175
Thalamita crenata, 196, 197
Thalassia, 273, 274, 285, 289
   testudinum, 265
Thalassina, 159, 185, 222
   anomala, 105, 193
Thalassodendron, 41, 42
Thelephoraceae, 132
Therdiidae, 178
Thespesia, 95, 111, 154
Thomisidae 178
Thraustochytrium, 133, 134
Threskiornis molucca, 107
tidal plains, 126-27, 173. *See also* coasts
tides, 47, 268, 302. *See also* inundation; water
   distribution and, 2, 3, 5-7, 87, 110-19
   fauna and, 157, 159, 187-88, 191-92, 199, 206, 211, 232-33, 236
   flora and, 59, 133, 136-40
   flushing, 122, 259-61, 266, 285-86, 306
   inundation, 26, 46, 59-61, 65-69, 75-76, 109, 110-12, 128, 137, 159, 184, 187-88, 191, 205, 211, 232-33, 308
   levels, 26, 76, 112, 115, 117, 136, 140, 191
   mangrove communities and, 122, 124-28, 268
   nutrients and, 80-81, 259-61
   propagule dispersal, 41, 42-43, 114
   range, 2, 3, 5, 127, 128, 286
   roots and, 26, 29, 122, 259
   salinity and, 66-69, 72, 76, 117
   sediment dispersion, 55, 60, 127, 259
tortoises, 179

toxins, 94-95, 259, 266, 307. *See also* waste discharge
transpiration, 47, 63-64. *See also* evapotranspiration; xeromorphic features
   leaves, 21, 24-25, 34, 48, 56, 58, 258, 260
   salinity, 110, 260-61
   temperature, 48, 258
   wind, 56, 58
   zonation, 259
Trebouxia, 140
Trentepohlia, 140
   odorata, 140
   rigidula, 140
Trichoderma, 133
   viride, 134
Trichomya hirsuta, 187
Triglochin, 94
   procera, 118
trunks. *See also* bark; wood
   fauna and, 156, 159, 172, 184, 187, 188, 189, 199, 201
   flora and, 131-32, 136, 139-40, 142, 187
   primary production, 252-54, 266
tubeworms, 43. *See also* polychaetes
Tubulanus polymorphus, 192
Turbellarians, 192
turtles, 179, 285

Uca, 197, 213-15, 219-20, 242. *See also* crabs
   distribution, 185, 193-96, 199, 200, 219
   reproduction, 225-27, 230, 232
   respiration, 236-37
Upogebia, 156, 224
Valanga irregularis, 175
varanid lizards, 165
Varanus indicus, 180, 208
Varanus semiremex, 180
Vaucheria, 136
vertebrates, 163. *See also* fauna
   freshwater, 178-79
   marine, 179-82
   terrestrial, 165-71
vivipary, 39, 40, 41, 103, 109

wallabies, 166, 174
waste discharge, 301, 304-5, 308. *See also* toxins
water, 41-42, 100, 142, 302. *See also* catchments; desalinated water; drainage; fresh water; tides; waves
   damage by, 35-36, 44. *See also* stress, water
   fauna and, 159, 183, 222, 224, 230
   loss, 33-34, 63-64, 159, 260. *See also* transpiration
   soil and, 25-26, 30, 46-47, 56, 60-78, 259. *See also* waterlogging; watertable
   temperature, 3, 10, 13
water dragon, 179
waterlogging, 30, 63, 65, 77, 94, 98, 103-4, 110, 115-19. *See also* respiration, anaerobic; roots
   bunding and, 308
   salt marshes, 153-54
   structural formation and, 121
watertable, 46, 47, 73, 75-77
waves, 2, 55, 125, 127-28. *See also* tides; water
   damage by, 35, 41, 43-44, 307
wind, 41, 46, 47, 66. *See also* evaporation; evapotranspiration
   damage by, 35-36, 100, 122, 264
   pollination by, 36-38
   sea-level and, 55, 58-60
   transpiration, 56, 58
windthrow, 35
wood, 7, 95, 259, 293, 296. *See also* bark; borers; trunks
   anatomy, 24, 36
   fauna and, 35, 105, 159, 171, 176, 201-3, 273
   flora and, 132, 133, 273
   primary production, 262, 264, 265
worms, 201, 214, 236, 244, 300. *See also* nereidids; polychaetes; spionids
   feeding, 192, 272
   as food, 175
   habitat, 187, 192, 230, 232

Xenostrobis securis, 187
Xerochloa barbata, 151
xeromorphic features, 20-25, 31. *See also* adaptation
Xeromys myoides, 165, 206
Xylocarpus, 24, 36, 102, 147
   distribution, 99
   light response, 33, 54
   reproduction, 37, 38, 40-42
   roots, 26, 28
   salt and, 18
Xylocarpus australasicus, 35, 84, 147
   distribution, 110-11

dominance, 101
leaves, 48, 256-57
roots, 28
salinity, 70, 71
soil water, 65
temperature and, 48, 51
Xylocarpus granatum, 84
distribution, 111, 113
dominance, 101
flora and, 132, 145
salinity, 70, 71, 118, 119
soil water, 65
temperature and, 50, 51

yabby, 300

Zalerion varium, 133, 134
zonation, 5, 108, 259. *See also* biogeographic regions; distribution, mangrove; latitude; mangrove communities; salinity; temperature
boundary conditions, 115-18
fauna and, 159-62, 188-91, 196, 199, 203, 218. *See also* latitude, fauna
shoreline, 98-99, 108-10, 115, 118-19, 120
upriver, 108, 110-15, 118-19, 120. *See also* rivers, mangrove communities and
zooplankton, 289
Zostera, 192
 capricorni, 246, 271
 marina, 265, 285
Zygosporium, 132, 134